CLASSICAL FIELD THEORY

Davison E. Soper
Institute of Theoretical Science
University of Oregon

DOVER PUBLICATIONS, INC.
Mineola, New York

Copyright

Bibliographical Note

This Dover edition, first published in 2008, is an unabridged, corrected republication of the work originally published in 1976 by John Wiley & Sons, Inc., New York. The author has provided a new Preface for the Dover edition.

Library of Congress Cataloging-in-Publication Data

Soper, Davison E., 1943–
Classical field theory / Davison E. Soper.—Dover ed.
p. cm.
Originally published: New York : Wiley, 1976.
Includes index.
ISBN-13: 978-0-486-46260-8
ISBN-10: 0-486-46260-9
1. Field theory (Physics) 2. Last action. I. Title.

QC174.45.S65 2008
530.14—dc22

2007033642

Manufactured in the United States of America
Dover Publications, Inc., 31 East 2nd Street, Mineola, N.Y. 11501

TO

DOROTHY

Preface to the Dover Edition

This book covers the mechanics of fluids and elastic solids and related subjects from the point of view of field theory, in which one starts with a Lagrangian density and derives equations of motion from the Lagrangian density using the principle of stationary action. This is the point of view of quantum field theory, but it works just as well for classical fields as for quantum fields. In the classical case, this same framework is commonly adopted to describe the electromagnetic field and the gravitational field (that is, the metric tensor of general relativity). The purpose of this book is to use this framework to describe the mechanics of continuous media, which are more commonly described using just the equations of motion rather than the principle of stationary action. I am pleased that Dover Publications has reprinted the book so that it can be easily available.

The description used is, for the most part, based on the special theory of relativity, so that the equations are covariant under Lorentz transformations. This makes it easy to couple the fluids or solids to the electromagnetic field. It is then simple to specialize to slowly moving matter so as to obtain the non-relativistic approximation. To couple the fluids and solids to gravity as well as the electromagnetic field, we use a formulation that is covariant under the general coordinate transformations of general relativity.

Most of the book is concerned with mechanics without friction, or, more generally, without entropy generation. In the last section, the equations are modified so as to include dissipative processes in the limit that the dissipation is small and all the functions involved are slowly varying. The formulation is Lorentz invariant, but no attention is paid to the requirement that the equations of motion not allow disturbances to travel faster than the speed of light. For this reason, there is a limitation on the domain of applicability of the resulting equations.

Davison E. Soper
Eugene, Oregon,
November, 2007

Preface

This is a book about continuum mechanics, electrodynamics and the mechanics of electrically polarized media, and gravity. It differs from other books on these subjects in that the theories are formulated by means of the principle of least action.

The chief advantage of this formulation is that it is simple and easy, since the physical input into the theory consists of choosing only one scalar function of the fields involved—the Lagrangian density. Thus my chief motivation for writing this book was to make the physical content of the classical subjects discussed available to students of physics in a concentrated, easy to digest form.

My second goal has been to explain the general structure of field theories based on the least action formulation. I hope that this discussion of general principles in a purely classical setting will be useful to students of quantum field theory, since quantum field theory is almost without exception based on the principle of least action.

It is also my hope that my colleagues in quantum field theory will enjoy seeing their everyday methods applied to such subjects as the motion of fluids.

Although I have borne the applications to quantum field theory in mind, I have tried to remain true to the subject of *classical* field theory. I have therefore avoided interesting diversions that do not bear on real classical phenomena and have omitted such subjects as spin $\frac{1}{2}$ fields, non-Abelian gauge fields, and Poisson brackets.

Physicists often regard Hamilton's principle of least action as the starting point for particle mechanics. However, modern workers in continuum mechanics almost always prefer Newton's formulation of mechanics, in which one writes the equations of motion directly instead of writing an action and deriving the equations of motion. Their preference is not without good reason, for they seek to describe both mechanics and thermodynamics—that is, both entropy conserving phenomena, which generally can be described by an action principle, and entropy increasing phenomena, which generally cannot. In this book, I have followed the usual physicists' inclination to sweep "friction" under the rug and discuss conservative forces only—at least to begin with. In the final chapter, dissipation is resurrected and added onto some simple theories. When dissipation is included, the Lagrangian plays a less central role, but it is a substantial computational aid for managing the nondissipative terms in the field equations.

This book began as a graduate level course on classical field theory that I taught at Princeton University in the fall of 1972. I have tried to write it so that it can be used either as a textbook for such a course or as a supplementary text for a course on classical mechanics or classical electrodynamics. There are therefore a few problems at the end of each chapter. They ask the reader to do a calculation to derive a result used in the text or to extend or amplify some idea in the text. A course in classical field theory should, in my opinion, also include some practice in solving the partial differential equations that are derived, but I have not attempted to provide material on the solution of these equations. Such material is readily available in a number of books [such as Jackson's text, *Classical Electrodynamics*, 2nd edition (Wiley, 1975)].

I would like to thank Dr. G. Farrar for suggesting to me the optimum choice of fields for a relativistic theory of continuum mechanics. I am also indebted to E. Witten, Professor C. Callan, Professor A. C. Eringen, Professor E. Hamilton, and Professor A. Wightman for very helpful conversations. Finally, I would like to thank Professor D. Towne for arousing my initial interest in some of the subjects discussed in this book.

DAVISON E. SOPER

Princeton, New Jersey
September 1975

Contents

CHAPTER 1

Fields and Transformation Laws

The field theories to be discussed in this book are, for the most part, *relativistic* field theories. We therefore begin with a brief review of the special theory of relativity. The reader is presumed to have had some previous exposure to relativity, so we will present a more abstract development of the subject than would be appropriate for a first introduction.

1.1 ROTATIONAL COVARIANCE

The reader is no doubt familiar with vectors in three dimensional Euclidean space—the space in which we live. The prototype of all vectors is the coordinate difference $\mathbf{\Delta}_{PQ}$ between a point P and another point Q, which may be thought of as an arrow whose tail lies at P and whose head lies at Q. A vector is usually represented by its components Δ_1, Δ_2, Δ_3 in a Cartesian coordinate system. However, we must be careful when we use the components Δ_k of a vector in a formula expressing a physical law, since the components Δ_k of $\mathbf{\Delta}$ in one coordinate system will differ from the components $\bar{\Delta}_k$ of the same vector $\mathbf{\Delta}$ in a rotated coordinate system. We must ensure that an equation involving components of vectors is of such a form that its truth in one coordinate system implies its truth in any other coordinate system. For instance, the equation $a_k = 7b_k$ between two vectors $\mathbf{a}$ and $\mathbf{b}$ is a candidate for a physical law, while the relation $a_1 = 7b_1$, $a_2 = 7b_2$, $a_3 = 6b_3$ is not. This requirement of rotational covariance is not a matter of mathematical consistency, but rather a matter of physics. It reflects the fact that physicists whose laboratories are rotated with respect to one another always observe the same physical laws.

Fortunately, it is easy to use components of vectors in equations while still satisfying the requirement of rotational covariance in an automatic way. To see how the method works we begin by writing the transformation law for the components Δ_k of a coordinate difference vector $\mathbf{\Delta}$. If Δ_k are the components of $\boldsymbol{\Delta}$ in one coordinate system and $\bar{\Delta}_k$ are the components of $\boldsymbol{\Delta}$ in a rotated coordinate system, then the relation between $\bar{\Delta}_k$ and Δ_k has the form

$$\bar{\Delta}_k = \sum_{l=1}^{3} \mathcal{R}_{kl} \Delta_l, \tag{1.1.1}$$

where the matrix $\mathcal{R}_{kl}$ is an "orthogonal matrix":

$$\sum_{k=1}^{3} \mathcal{R}_{ki} \mathcal{R}_{kj} = \delta_{ij}, \tag{1.1.2}$$

$$\det \mathcal{R} = +1. \tag{1.1.3}$$

Here δ_{ij} is the Kroneker delta: $\delta_{11} = \delta_{22} = \delta_{33} = 1$, $\delta_{12} = \delta_{13} = \delta_{23} = \delta_{21} = \delta_{31} = \delta_{32} = 0$. The essence of the orthogonality condition (1.1.2) is that the distance between two points as given by the Pythagorean formula

$$[\text{distance}(P,Q)]^2 = \sum_{k=1}^{3} (\Delta_k)^2,$$

is the same no matter what orthogonal coordinate system is used to measure the Δ_k:

$$\sum_{k=1}^{3} (\bar{\Delta}_k)^2 = \sum_{ijk=1}^{3} \Delta_i (\mathcal{R}_{ki} \mathcal{R}_{kj}) \Delta_j$$

$$= \sum_{ij=1}^{3} \Delta_i \delta_{ij} \Delta_j$$

$$= \sum_{j=1}^{3} (\Delta_j)^2.$$

The orthogonality condition (1.1.2) implies that the determinant of the matrix $\mathcal{R}$ is either $+1$ or -1. The additional requirement $\det \mathcal{R} = +1$ eliminates those $\mathcal{R}$ that express the transformation between a right handed coordinate system and a left handed coordinate system.

The same method shows that the dot product $\mathbf{A} \cdot \mathbf{B} = \Sigma_j A_j B_j$ between two vectors is invariant under rotations of the coordinate systems. We write out the proof using a simplified notation in which all summation signs are omitted; any latin index i, j, k, etc. that appears twice in a product is assumed to be summed from 1 to 3:

$$\bar{A}_k \bar{B}_k = A_i \mathfrak{R}_{ki} \mathfrak{R}_{kj} B_j = A_i \delta_{ij} B_j = A_i B_i.$$

Tensors

In the development that follows we need to use objects called tensors, which are a simple generalization of vectors. As with vectors, it is not the tensor itself that usually appears in equations, but rather its components in a particular coordinate system. For instance, a "fourth rank" tensor **T** has $(3)^4$ components in a particular coordinate system, which are labeled by four indices, each of which takes the values 1, 2, 3: T_{ijkl}. The components of **T** in two coordinate systems are related by the simple rule

$$\bar{T}_{ijkl} = \mathfrak{R}_{ii'} \mathfrak{R}_{jj'} \mathfrak{R}_{kk'} \mathfrak{R}_{ll'} T_{i'j'k'l'}. \tag{1.1.4}$$

A first rank tensor is the same as a vector and can be represented physically as an arrow. Unfortunately, there is no intuitively appealing physical representation for a higher rank tensor. One is forced to the mathematically respectable but physically effete definition of a tensor **T** as a rule for assigning components T_{ijkl} to each coordinate system such that the components obey the transformation law (1.1.4). Despite the difficulty of visualization, a tensor **T**, like a vector, is to be thought of as a real physical object, whereas the components T_{ijkl} measured in a particular coordinate system depend on which coordinate system was used.

One can easily give examples of tensors. If **A** and **B** are vectors, then the equation

$$T_{ij} = A_i B_j$$

defines a second rank tensor, since

$$\begin{aligned}\bar{T}_{ij} = \bar{A}_i \bar{B}_j &= (\mathfrak{R}_{ii'} A_{i'})(\mathfrak{R}_{jj'} B_{j'})\\ &= \mathfrak{R}_{ii'} \mathfrak{R}_{jj'} T_{i'j'}.\end{aligned}$$

Indeed, higher rank tensors can always be formed from lower rank tensors in this way. If **A** is a tensor of rank N and **B** is a tensor of rank M, then the equation $T_{i\ldots kl\ldots n} = A_{i\ldots k} B_{l\ldots n}$ defines a tensor **T** of rank $N+M$. New tensors can also be formed from old tensors of the same rank by multiplying by real numbers and adding: if **A** and **B** are tensors of rank N then $T_{i\ldots k} = \alpha A_{i\ldots k} + \beta B_{i\ldots k}$ defines a tensor $\mathbf{T} \equiv \alpha \mathbf{A} + \beta \mathbf{B}$.

Fortunately it is also possible to form lower rank tensors from higher rank tensors, so that we will not be caught in an inevitable population explosion of tensor indices. If $\mathbf{T}$ is a tensor of rank 3, say, then

$$A_i = T_{ijj}$$

defines a tensor of rank 1, $\mathbf{A}$. The proof is easy, and has an obvious generalization to any rank:

$$\begin{aligned}\bar{A}_i = \bar{T}_{ijj} &= \mathcal{R}_{ik}(\mathcal{R}_{jl}\,\mathcal{R}_{jn})T_{kln}\\ &= \mathcal{R}_{ik}\,\delta_{ln}T_{kln} = \mathcal{R}_{ik}\,T_{kll}\\ &= \mathcal{R}_{ik}A_k.\end{aligned}$$

This method for forming new tensors is called contraction of indices.

An important special case is the formation of a scalar (that is, an ordinary real number) by contracting the indices of a second rank tensor. For example, from two vectors $\mathbf{A}$ and $\mathbf{B}$ we can form the tensor with components A_iB_j, then the scalar $\mathbf{A}\cdot\mathbf{B} = A_iB_i$.

Invariant Tensors

There are two special tensors with the remarkable property that their components do not vary from one coordinate system to another. The first is the "unit tensor" $\boldsymbol{\delta}$, whose components in any coordinate system are δ_{ij}, the Kroneker delta. To see that this definition obeys the transformation law (1.1.4) write

$$\mathcal{R}_{ik}\,\mathcal{R}_{jl}\,\delta_{kl} = \mathcal{R}_{ik}\,\mathcal{R}_{jk} = \delta_{ij}.$$

The other invariant tensor is called the completely antisymmetric tensor $\boldsymbol{\epsilon}$ and has components

$$\begin{aligned}\epsilon_{123} = \epsilon_{231} = \epsilon_{312} &= +1,\\ \epsilon_{321} = \epsilon_{213} = \epsilon_{132} &= -1.\end{aligned} \tag{1.1.5}$$

All other components are zero. This rule can be summarized by saying that $\epsilon_{123} = +1$ and ϵ_{ijk} is antisymmetric under interchange of any two indices (e.g., $\epsilon_{ijk} = -\epsilon_{ikj}$). To see that this rule obeys the transformation law (1.1.4) we note that the same object ϵ_{ijk} appears in the definition of the determinant of a matrix:

$$\det \mathcal{R} = \mathcal{R}_{1i}\,\mathcal{R}_{2j}\,\mathcal{R}_{3k}\,\epsilon_{ijk}.$$

Thus, using $\det \mathcal{R} = 1$,

$$\mathcal{R}_{li}\,\mathcal{R}_{mj}\,\mathcal{R}_{nk}\,\epsilon_{ijk} = (\det \mathcal{R})\epsilon_{lmn}$$
$$= (+1)\epsilon_{lmn}.$$

The invariant tensor ϵ appears in the vector cross product. For example, the definition of angular momentum, $\mathbf{L} = \mathbf{R} \times \mathbf{P}$, is written in the present notation as $L_i = \epsilon_{ijk} R_j P_k$.

The method for ensuring that equations expressing physical laws are covariant under rotations should now be clear. One uses scalars, vectors, and tensors to represent physical entities and writes physical laws of the form "(scalar) = (scalar)", "(vector) = (vector)", or, in general, "(N^{th} rank tensor) = (N^{th} rank tensor)" using the component notation.

1.2. LORENTZ COVARIANCE

Two observers whose reference frames are rotated with respect to each other will discover the same physical laws. We have seen how to write equations expressing physical laws in a way that make this rotational invariance manifest. According to the principle of relativity, two observers whose reference frames are moving with constant velocity with respect to one another must also discover the same physical laws. We will now investigate how to write equations that make this "Lorentz" invariance manifest.

Observers and Inertial Reference Frames

The special theory of relativity begins by focusing attention on the space coordinates x, y, z and the time coordinate t that an observer in gravity free space assigns to an event (for instance, the collision of two billiard balls, or, in the canonical example, the exploding of a firecracker). We imagine that the observer lays out a grid of meter sticks at right angles to one another and that at the intersections of the meter sticks he places confederates with clocks. The coordinates (t, x, y, z) of an event are to be determined by noting which confederate was next to the event and asking him for his coordinates (x, y, z) and the time t shown on his clock when the event occurred. We imagine that the coordinate system is an "inertial" coordinate system. That is, a body placed at rest in front of any of the confederates and subjected to no forces will remain at rest.

We will shortly make use of a remarkable physical law: the speed of light is a certain constant $c \cong 3 \times 10^8$ m/sec, independent of any characteristics of the emitting or detecting devices. As a matter of convenience, we will assume that our observers choose their time and distance scales in such a way that this constant is $c=1$. (For instance, 10^{-9} sec is a convenient unit of length, being about equal to 1 ft).

To express this light propagation law in a compact form, let us rename the coordinates of an event x^μ, $\mu=0,1,2,3$, with $x^0=t$, $x^1=x, x^2=y, x^3=z$. Suppose two events labeled by x_1^μ and x_2^μ could be connected by a light signal. That is, a pulse of light emitted at x_1, y_1, z_1 at time t_1 would reach x_2, y_2, z_2 at time t_2. Then the fact that the speed of light is 1 implies that $(t_2-t_1)^2=(x_2-x_1)^2+(y_2-y_1)^2+(z_2-z_1)^2$. In a more compact notation this is

$$(x_2^\mu - x_1^\mu)\, g_{\mu\nu} (x_2^\nu - x_1^\nu) = 0. \tag{1.2.1}$$

Here we have adopted the "Einstein summation convention" that repeated Greek indices are to be summed over the values $0,1,2,3$. The matrix $g_{\mu\nu}$, called the "metric tensor," is

$$\begin{aligned} g_{00} &= -1, \qquad g_{11}=g_{22}=g_{33}=+1, \\ g_{\mu\nu} &= 0 \quad \text{if} \quad \mu\neq\nu. \end{aligned} \tag{1.2.2}$$

Lorentz Transformations

Consider now two inertial coordinate systems, $\bar{\mathcal{O}}$ and $\mathcal{O}$. The coordinates $(\bar t, \bar x, \bar y, \bar z)$ and (t,x,y,z) assigned by the two sets of observers to the same event are related by a set of functions F^μ: $\bar x^\mu = F^\mu(x^0,x^1,x^2,x^3)$. We wish to determine the form of the functions $F^\mu(x)$ by imposing the requirement that both observers find that the speed of light is 1:

$$[F^\mu(x_2)-F^\mu(x_1)]\, g_{\mu\nu} [F^\nu(x_2)-F^\nu(x_1)]=0$$

if and only if

$$(x_2^\mu - x_1^\mu)\, g_{\mu\nu} (x_2^\nu - x_1^\nu) = 0. \tag{1.2.3}$$

The requirement (1.2.3) is very restrictive. Let us make the additional reasonable technical assumptions that the functions $F^\mu(x)$ are twice continuously differentiable for all x^μ and that the x^μ are related to the $\bar x^\mu$ by a twice continuously differentiable inverse mapping $x^\mu = \bar F^\mu(\bar x)$. One can show* that the only such functions F^μ which satisfy (1.2.3) have the simple form

$$\bar x^\mu = F^\mu(x) = \lambda \Lambda^\mu{}_\nu x^\nu + a^\mu,$$

*The proof is a bit involved if one does not assume at the start that the transformation is linear. See Problems 4 and 5.

where a^μ are any four numbers, λ is a positive number, and $\Lambda^\mu{}_\nu$ is a matrix with the property

$$\Lambda^\mu{}_\alpha \Lambda^\nu{}_\beta g_{\mu\nu} = g_{\alpha\beta}. \qquad (1.2.4)$$

The matrix Λ is called a Lorentz transformation matrix; we will discuss the structure of Lorentz transformations in some detail presently. The numbers a^μ apparently tell the coordinates of the origin $x^\mu = 0$ of the $\mathcal{O}$ coordinate system as seen in the $\bar{\mathcal{O}}$ system. The positive constant λ is a scale factor that arises because nothing we have said so far requires both observers to use the same scale of length.

Let us now require that all observers choose the same scale of length—for instance, by standardizing their meter sticks against the wavelength of light emitted in a certain atomic transition. Then we can argue that the scale factor λ must always be 1.

For this purpose, consider the class of observers who agree on a single event to be the origin of their coordinate systems. Any two such observers are related by $\bar{x}^\mu = \lambda \Lambda^\mu{}_\nu x^\nu$. Let "me" be one of the observers. There cannot be two observers with coordinates related to mine by the same matrix $\Lambda^\mu{}_\nu$ but different factors λ, say $\bar{x}^\mu = \lambda_1 \Lambda^\mu{}_\nu x^\nu$ and $\bar{\bar{x}}^\mu = \lambda_2 \Lambda^\mu{}_\nu x^\nu$, for these two observers would be related to each other by $\bar{\bar{x}}^\mu = (\lambda_1/\lambda_2)\bar{x}^\mu$ and would manifestly not agree on the standard of length. Thus the parameters of the coordinate transformations that relate my system to other physically allowed coordinate systems must be related by an equation $\lambda = \lambda(\Lambda)$. Furthermore, any other observer must obtain the same function $\lambda(\Lambda)$, since there is nothing special about "me" as an observer.

We must now ask whether $\lambda(\Lambda)$ can have a nontrivial dependence on the Lorentz transformation Λ. By comparing $\bar{\bar{x}} = \lambda(\Lambda_1)\Lambda_1 \bar{x}$, $\bar{x} = \lambda(\Lambda_2)\Lambda_2 x$ with $\bar{\bar{x}} = \lambda(\Lambda_1\Lambda_2)\Lambda_1\Lambda_2 x$ we find that the function $\lambda(\Lambda)$ obeys a multiplication law

$$\lambda(\Lambda_1)\lambda(\Lambda_2) = \lambda(\Lambda_1\Lambda_2).$$

But it is a simple mathematical fact (see Problem 6) that the only function which satisfies this law is

$$\lambda(\Lambda) = 1 \qquad \text{for} \quad \text{all } \Lambda.$$

Thus the coordinate transformations that keep $c = 1$, preserve the length scale, and leave the origin of coordinates unchanged are the "Lorentz transformations":

$$\bar{x}^\mu = \Lambda^\mu{}_\nu x^\nu. \qquad (1.2.5)$$

If we allow the origin of coordinates to be changed we have a "Poincaré transformation":

$$\bar{x}^{\mu} = \Lambda^{\mu}{}_{\nu} x^{\nu} + a^{\mu}. \tag{1.2.6}$$

We now turn our attention to the structure of the matrices Λ.

The Structure of Lorentz Transformations

Since Lorentz invariance is an important tool in this book, it pays us to investigate what kind of matrices Λ have the property (1.2.4). This property is very similar to the defining property of a rotation matrix:

$$\mathfrak{R}_{ik}\,\mathfrak{R}_{jl}\,\delta_{ij} = \delta_{kl}. \tag{1.2.7}$$

In fact, it is clear from comparing (1.2.7) and (1.2.4) that rotations are a subset of all Lorentz transformations. That is, if $\mathfrak{R}$ is a 3×3 rotation matrix, and the 4×4 matrix $\Lambda(\mathfrak{R})$ has the form

$$\Lambda(\mathfrak{R})^{\mu}{}_{\nu} = \begin{bmatrix} 1 & 0 & 0 & 0 \\ 0 & \mathfrak{R}_{11} & \mathfrak{R}_{12} & \mathfrak{R}_{13} \\ 0 & \mathfrak{R}_{21} & \mathfrak{R}_{22} & \mathfrak{R}_{23} \\ 0 & \mathfrak{R}_{31} & \mathfrak{R}_{32} & \mathfrak{R}_{33} \end{bmatrix},$$

so

$$\bar{x}^{0} = x^{0}, \tag{1.2.8}$$

$$\bar{x}^{j} = \sum_{k=1}^{3} \mathfrak{R}_{jk}\, x^{k},$$

then Λ is a Lorentz transformation. Another important special example of Lorentz transformations, called a "boost" in the z-direction, is

$$\Lambda(\omega)^{\mu}{}_{\nu} = \begin{bmatrix} \cosh\omega & 0 & 0 & \sinh\omega \\ 0 & 1 & 0 & 0 \\ 0 & 0 & 1 & 0 \\ \sinh\omega & 0 & 0 & \cosh\omega \end{bmatrix} \tag{1.2.9}$$

where the "angle" ω can be any real number. Imagine two coordinate systems related by a boost $\Lambda(\omega)$. The path through space-time of the first observer, who sits at the origin of his space coordinates and watches his clock run, is given by $(x,y,z)=(0,0,0)$, $t=$anything. This path is given in terms of the second coordinate system by

$$\begin{cases} \bar{x} = \bar{y} = 0, \\ \bar{t} = \cosh \omega t, \\ \bar{z} = \sinh \omega t, \\ t = \text{anything}, \end{cases}$$

or $\bar{x} = \bar{y} = 0$, $\bar{z} = v\bar{t}$ where $v = \tanh \omega$. Thus the first coordinate system is moving with a constant velocity $v = \tanh \omega$ in the z-direction with respect to the second.

One might expect on the basis of physical intuition that the most general Lorentz transformation can be built from three successive transformations: first a rotation of the axes of the $\mathcal{O}$ system so that the new z-axis is aligned with the direction of motion, then a boost in the (new) z-direction, then another rotation. Thus any Lorentz transformation matrix $\mathbf{\Lambda}$ should have the form

$$\Lambda^{\mu}{}_{\nu} = \Lambda(\mathcal{R}')^{\mu}{}_{\alpha}\Lambda(\omega)^{\alpha}{}_{\beta}\Lambda(\mathcal{R})^{\beta}{}_{\nu} \tag{1.2.10}$$

where $\mathbf{\Lambda}(\mathcal{R})$ and $\mathbf{\Lambda}(\mathcal{R}')$ are rotations as in (1.2.8) and $\mathbf{\Lambda}(\omega)$ is a boost as in (1.2.9).

An explicit construction (see Problem 8) shows that (1.2.10) is almost the most general transformation satisfying (1.2.4). The exception arises because we have not considered the transformations

$$\Lambda_P{}^{\mu}{}_{\nu} = \begin{pmatrix} 1 & 0 & 0 & 0 \\ 0 & -1 & 0 & 0 \\ 0 & 0 & -1 & 0 \\ 0 & 0 & 0 & -1 \end{pmatrix}$$

and

$$\Lambda_T{}^{\mu}{}_{\nu} = \begin{pmatrix} -1 & 0 & 0 & 0 \\ 0 & 1 & 0 & 0 \\ 0 & 0 & 1 & 0 \\ 0 & 0 & 0 & 1 \end{pmatrix} \tag{1.2.11}$$

The transformation $\mathbf{\Lambda}_P$ relates the coordinates in a right handed coordinate system to those in a left handed system, and is called the parity transformation. The transformation $\mathbf{\Lambda}_T$ relates a system in which clocks run forward to a system in which clocks run backward, and is called the time reversal transformation. The most general transformation satisfying (1.2.4) is either one of the type (1.2.10) or else one of this type followed by $\mathbf{\Lambda}_P$, $\mathbf{\Lambda}_T$ or $\mathbf{\Lambda}_P\mathbf{\Lambda}_T$.

We will often refer to the transformations (1.2.10) made up from rotations and boosts as "Lorentz transformations." A more precise (but somewhat stuffy) name is "proper orthochronous Lorentz transformations," to distinguish them from transformations containing $\mathbf{\Lambda}_P$ or $\mathbf{\Lambda}_T$ or both. The proper orthochronous transformations can be distinguished by the requirements

$$\det \mathbf{\Lambda} = +1,$$

$$\Lambda^0{}_0 \geqslant 1 \tag{1.2.12}$$

in addition to the requirements that $\Lambda^\mu{}_\alpha \Lambda^\nu{}_\beta g_{\mu\nu} = g_{\alpha\beta}$. (See Problem 7.)

We will demand that physical laws be invariant under (proper orthochronous) Lorentz transformation of the coordinates. The reason that we do not at the outset require invariance under $\mathbf{\Lambda}_P$ and $\mathbf{\Lambda}_T$ as well is that the real world does not have this extra invariance in the so called "weak interactions" of nuclear physics. Nevertheless, the theories we discuss in all but the last chapter of this book turn out to be invariant under $\mathbf{\Lambda}_P$ and $\mathbf{\Lambda}_T$. (In the last chapter we introduce irreversible processes like friction, and thus lose invariance under $\mathbf{\Lambda}_T$.)

Four-Vectors

We can write Lorentz covariant equations by generalizing the familiar methods discussed in the preceding section for writing rotationally covariant equations. We begin by defining a four-vector $\mathbf{V}$ to be a rule which assigns to each inertial reference frame four numbers V^0, V^1, V^2, V^3, called the components of $\mathbf{V}$ in that reference frame; the components V^μ in one reference frame are to be related to the components $\bar{V}^\mu$ in another frame by the transformation law

$$\bar{V}^\mu = \Lambda^\mu{}_\nu V^\nu, \tag{1.2.13}$$

where $\mathbf{\Lambda}$ is the Lorentz transformation matrix relating the two reference frames: $\bar{x}^\mu = \Lambda^\mu{}_\nu x^\nu + a^\mu$. Thus the archetype four-vector is the vector whose components are the coordinate differences $x^\mu - y^\mu$ between two events.

The fundamental property (1.2.4) of Lorentz transformations implies that the product

$$\mathbf{U} \cdot \mathbf{V} = U^\mu g_{\mu\nu} V^\nu$$

of two four-vectors is a scalar (i.e., independent of reference frame). It is often convenient to write such products as $\mathbf{U} \cdot \mathbf{V} = U^\mu V_\mu$ where

$$V_\mu = g_{\mu\nu} V^\nu. \tag{1.2.14}$$

The components V_0, V_1, V_2, V_3 of $\mathbf{V}$ defined in this way are often called the covariant components of $\mathbf{V}$; the original components V^0, V^1, V^2, V^3 are called the contravariant components. The transformation law for the covariant components of a vector is

$$\bar{V}_\mu = g_{\mu\nu}\bar{V}^\nu = g_{\mu\nu}\Lambda^\nu{}_\alpha V^\alpha$$

$$= (\Lambda^{-1})^\beta{}_\mu g_{\beta\alpha} V^\alpha$$

or

$$\bar{V}_\mu = (\Lambda^{-1})^\nu{}_\mu V_\nu. \tag{1.2.15}$$

The components V^μ can be recovered from the components V_μ by multiplying by the inverse matrix to $g_{\mu\nu}$, which is called $g^{\mu\nu}$ and has the same numerical values:

$$g^{00} = -1, \qquad g^{11} = g^{22} = g^{33} = +1,$$

$$g^{ij} = 0 \text{ otherwise.} \tag{1.2.16}$$

Thus

$$V^\mu = g^{\mu\nu} V_\nu. \tag{1.2.17}$$

The notation is arranged so that when two indices are summed over, one is always an upper index and the other is a lower index.

Let us mention here two important four-vectors associated with a moving point particle, its four-velocity and its momentum. The motion of a particle is conveniently described by giving a parametric equation for its path in space time $x^\mu = x^\mu(\sigma)$. The tangent vector $dx^\mu/d\sigma$ to the path is clearly a four-vector. Since the length of this vector has no significance, it is convenient to define a normalized tangent vector u^μ with length $[-u^\mu u_\mu]^{1/2} = 1$

$$u^\mu(\sigma) = \left[-\frac{dx^\mu}{d\sigma}\frac{dx_\mu}{d\sigma} \right]^{-1/2} \frac{dx^\mu}{d\sigma}. \tag{1.2.18}$$

The vector u^μ is called the four-velocity of the particle. The ordinary three-velocity of the particle is

$$v_k = \frac{dx^k}{dt} = \frac{dx^k}{dx^0} = \frac{u^k}{u^0}. \tag{1.2.19}$$

Thus

$$u^0 = (1 - \mathbf{v}^2)^{-1/2},$$

$$u^k = (1 - \mathbf{v}^2)^{-1/2} v_k. \tag{1.2.20}$$

If we multiply u^μ by the mass of the particle we obtain a vector $P^\mu = mu^\mu$ called the four-momentum of the partcle. P^0 is identified as the energy of the particle and P^1, P^2, P^3 are the components of its three-momentum. There are two main reasons for this indentification. First, if the velocity of the particle is small compared to the speed of light we recover the nonrelativistic expressions for momentum and energy, except for a constant term $E_0 = mc^2$ in the energy:

$$P^k = \frac{mv^k}{\sqrt{1 - \mathbf{v}^2}} \cong mv^k,$$

$$E = \frac{m}{\sqrt{1 - \mathbf{v}^2}} \cong m + \tfrac{1}{2} m\mathbf{v}^2.$$

Second, it is found from experiment that when fast moving particles collide the total four-momentum $P^\mu = P_1^\mu + P_2^\mu + \ldots$ of the system is conserved.

Tensors

Given the development so far, the definition of four-tensors will come as no surprise. A third rank tensor $\mathbf{T}$, for instance, has components $T^{\alpha\beta\gamma}$ that transform according to

$$\overline{T}^{\alpha\beta\gamma} = \Lambda^\alpha{}_\mu \Lambda^\beta{}_\nu \Lambda^\gamma{}_\rho T^{\mu\nu\rho}. \tag{1.2.21}$$

Any of the components of $\mathbf{T}$ can be lowered using the metric tensor, as in $T^\alpha{}_\beta{}^\gamma = g_{\beta\delta} T^{\alpha\delta\gamma}$. The components $T^\alpha{}_\beta{}^\gamma$ transform according to

$$\overline{T}^\alpha{}_\beta{}^\gamma = \Lambda^\alpha{}_\mu (\Lambda^{-1})^\nu{}_\beta \Lambda^\gamma{}_\rho T^\mu{}_\nu{}^\rho. \tag{1.2.22}$$

We can form higher rank tensors by multiplying lower rank tensors together, as in $T^{\alpha\beta} = A^\alpha B^\beta$, just as we did with three-tensors. We can also form lower rank tensors from higher rank tensors by multiplying by $g_{\mu\nu}$ and summing, as in $S^\alpha = T^{\alpha\mu\nu} g_{\mu\nu}$. Such equations are usually written in the more compact form $S^\alpha = T^{\alpha\mu}{}_\mu$ or $S^\alpha = T^\alpha{}_\mu{}^\mu$.

Up to this point, we have been careful to distinguish between a tensor $\mathbf{T}$ and its components $T^{\mu\nu}, T^\mu{}_\nu, T_{\mu\nu}$ or $T_\mu{}^\nu$ in some coordinate system. In the sequel, however, we will economize on words by writing "the tensor

$T^{\mu\nu}=A^{\mu}B^{\nu}$" instead of "the tensor **T** whose components in each coordinate system are related to the components of **A** and **B** by $T^{\mu\nu}=A^{\mu}B^{\nu}$". The reward in simplicity of language is considerable.

Invariant Tensors

There are two important tensors that are invariant under Lorentz transformations (i.e., they have the same components in all coordinate systems). The first is the metric tensor $g_{\mu\nu}$. Indeed, the equation

$$(\Lambda^{-1})^{\mu}{}_{\alpha}(\Lambda^{-1})^{\nu}{}_{\beta}g_{\mu\nu}=g_{\alpha\beta}$$

is the defining property of a Lorentz transformation.

The second invariant tensor is the completely antisymmetric tensor $\epsilon^{\mu\nu\rho\sigma}$, which is defined by

$$\epsilon^{0123}=+1, \tag{1.2.23}$$

$\epsilon^{\mu\nu\rho\sigma}$ is antisymmetric under interchange of any pair of indices.

Using the definition of the determinant of a matrix we find that

$$\Lambda^{\mu}{}_{\alpha}\Lambda^{\nu}{}_{\beta}\Lambda^{\rho}{}_{\gamma}\Lambda^{\sigma}{}_{\delta}\epsilon^{\alpha\beta\gamma\delta}=(\det\Lambda)\epsilon^{\mu\nu\rho\sigma}. \tag{1.2.24}$$

The determinant of a Lorentz transformation matrix Λ is $+1$ provided that Λ does not include a parity of time reversal transformation. Thus the components of ϵ are unchanged under proper orthochronous Lorentz transformations. One must be careful about the sign of ϵ: clearly the covariant components $\epsilon_{\mu\nu\rho\sigma}$ of ϵ also form a completely antisymmetric array of numbers, but $\epsilon_{0123}=-1$.

There are no unexpected new invariant tensors waiting to surprise us. Every invariant tensor can be formed from sums of products of $g_{\mu\nu}$ and $\epsilon^{\mu\nu\rho\sigma}$. For example, the most general fourth rank invariant tensor is

$$\begin{aligned}T^{\mu\nu\rho\sigma}&=\alpha g^{\mu\nu}g^{\rho\sigma}+\beta g^{\mu\rho}g^{\nu\sigma}\\&\quad+\gamma g^{\mu\sigma}g^{\nu\rho}+\delta\epsilon^{\mu\nu\rho\sigma},\end{aligned}$$

where $\alpha,\beta,\gamma,\delta$ are arbitrary scalars. (See Problems 9 and 10.)

1.3 SCALAR, VECTOR, AND TENSOR FIELDS

A function that associates a scalar ϕ with each point x of space-time is called a scalar field. An observer $\mathcal{O}$ can assign the scalar ϕ to the coordinates x^μ of x in his reference frame, giving a function $\phi(x^0,x^1,x^2,x^3)$. Another observer $\bar{\mathcal{O}}$ will see things slightly differently. He will assign the same value of ϕ to his coordinates, $\bar{x}^\mu = \Lambda^\mu{}_\nu x^\nu$, of this point x, giving a function $\bar{\phi}(\bar{x}^0,\bar{x}^1,\bar{x}^2,\bar{x}^3)$. Apparently the two functions are related by $\bar{\phi}(\Lambda^0{}_\nu x^\nu, \Lambda^1{}_\nu x^\nu, \Lambda^2{}_\nu x^\nu, \Lambda^3{}_\nu x^\nu) = \phi(x^0,x^1,x^2,x^3)$. In a more economical notation, this transformation law for a scalar field is written

$$\bar{\phi}(\Lambda^\mu{}_\nu x^\nu) = \phi(x^\mu). \tag{1.3.1}$$

Similarly, a function that associates a tensor **T** with each event x is called a tensor field. Consider, for example, a second rank tensor field. An observer $\mathcal{O}$ will assign to the coordinates x^μ of x the components $T^{\mu\nu}$ of **T** as seen in his reference frame, giving functions $T^{\mu\nu}(x^\sigma)$. Another observer $\bar{\mathcal{O}}$ will assign to his coordinates for x, $\bar{x}^\mu = \Lambda^\mu{}_\nu x^\nu$, the components $\bar{T}^{\mu\nu} = \Lambda^\mu{}_\alpha \Lambda^\nu{}_\beta T^{\alpha\beta}$ of **T** as seen in his reference frame. Thus the functions representing a tensor field in these two reference frames are related by the transformation law

$$\bar{T}^{\mu\nu}(\Lambda^\sigma{}_\rho x^\rho) = \Lambda^\mu{}_\alpha \Lambda^\nu{}_\beta T^{\alpha\beta}(x^\sigma). \tag{1.3.2}$$

One can also, of course, represent **T** by its covariant components, giving the transformation law

$$\bar{T}_{\mu\nu}(\Lambda^\sigma{}_\rho x^\rho) = (\Lambda^{-1})^\alpha{}_\mu (\Lambda^{-1})^\beta{}_\nu T_{\alpha\beta}(x^\sigma). \tag{1.3.3}$$

The simplest example of a tensor field (with rank 1) is the gradient of a scalar field:

$$T_\mu(x^\sigma) = \frac{\partial}{\partial x^\mu}\phi(x^\sigma).$$

To verify that this formula defines a tensor field one must check the transformation law (1.3.3) using the chain rule for partial differentiation:

$$\begin{aligned}
\bar{T}_\mu(\bar{x}^\sigma) &= \frac{\partial}{\partial \bar{x}^\mu}\bar{\phi}(\bar{x}^\sigma) \\
&= \frac{\partial x^\nu}{\partial \bar{x}^\mu}\frac{\partial}{\partial x^\nu}\phi(x^\sigma) \\
&= (\Lambda^{-1})^\nu{}_\mu T_\nu(x^\sigma).
\end{aligned}$$

Apparently this result can be generalized. If $T^{\alpha\beta\cdots\gamma}$ is a tensor field, then $S^{\alpha\beta\cdots\gamma}_{\mu\nu\cdots\lambda} = (\partial/\partial x^\mu)(\partial/\partial x^\nu)\cdots(\partial/\partial x^\lambda)T^{\alpha\beta\cdots\gamma}$ is a tensor field. The gradient operator is often abbreviated $\partial/\partial x^\mu = \partial_\mu$, so that the field $S^{\alpha\beta\cdots\gamma}_{\mu\nu\cdots\lambda}$ in the example above is written $\partial_\mu \partial_\nu \cdots \partial_\lambda T^{\alpha\beta\cdots\gamma}$.

PROBLEMS

1. Show that $\epsilon_{ijk}\epsilon_{ilm} = \delta_{jl}\delta_{km} - \delta_{jm}\delta_{kl}$.
2. Use the result of Problem 1 to show that $\mathbf{A} \times (\mathbf{B} \times \mathbf{C}) = (\mathbf{A} \cdot \mathbf{C})\mathbf{B} - (\mathbf{A} \cdot \mathbf{B})\mathbf{C}$.
3. Prove that any rotationally invariant second rank tensor must equal $\lambda\delta_{ij}$ for some number λ.
4. Suppose that two coordinate systems x^μ and $\bar{x}^\mu$ are related by a *linear* transformation $\bar{x}^\mu = A^\mu{}_\nu x^\nu$ and that $\bar{x}^\mu g_{\mu\nu} \bar{x}^\nu = 0$ if and only if $x^\mu g_{\mu\nu} x^\nu = 0$. Show that $A^\mu{}_\nu = \lambda\Lambda^\mu{}_\nu$ where λ is a constant and $\Lambda^\mu{}_\nu$ is a Lorentz transformation matrix: $\Lambda^\mu{}_\alpha \Lambda^\nu{}_\beta g_{\mu\nu} = g_{\alpha\beta}$.
5. Suppose that two coordinate systems x^μ and $\bar{x}^\mu$ are related by a transformation $\bar{x}^\mu = F^\mu(x)$ which is twice continuously differentiable and invertable, at least in some open region containing $x^\mu = 0$. Suppose also that the transformation preserves light cones in the sense of (1.2.3). Show that the transformation has the form

$$\bar{x}^\mu = \lambda\Lambda^\mu{}_\nu \frac{x^\nu + x^2 b^\nu}{1 + 2\mathbf{b} \cdot \mathbf{x} + \mathbf{b}^2\mathbf{x}^2} + a^\mu$$

where λ is a constant, $\Lambda^\mu{}_\nu$ is a Lorentz transformation matrix and a^μ and b^μ are fixed vectors. (If $\lambda = 1$, $\Lambda^\mu{}_\nu = g^\mu{}_\nu$, and $a^\mu = 0$ this transformation is called a "special conformal transformation"; if only $a^\mu = 0$ it is called simply a "conformal transformation.")

Argue that the transformation relating two inertial coordinate systems must have $b^\mu = 0$ because otherwise (*a*) the transformation becomes singular along the surface $1 + 2\mathbf{b} \cdot \mathbf{x} + \mathbf{b}^2\mathbf{x}^2 = 0$ and (*b*) particles which move with no acceleration in the x^μ system appear to be accelerated in the $\bar{x}^\mu$ system.
6. Prove that the only one dimensional representation of the proper orthochronous Lorentz group is the identity representation. That is, if $\lambda(\Lambda)$ is a function of proper orthochronous Lorentz transformations $\Lambda^\mu{}_\nu$ such that $\lambda(\Lambda_1)\lambda(\Lambda_2) = \lambda(\Lambda_1\Lambda_2)$, then $\lambda(\Lambda) \equiv 1$. Show also that the only one dimensional representations of the full Lorentz group are formed by combining $\lambda(\Lambda) = 1$ for a proper orthochronous Λ with $\lambda(\Lambda_P) = +1$ or -1 and $\lambda(\Lambda_T) = +1$ or -1.
7. Show that every Lorentz transformation matrix Λ satisfies $|\det\Lambda| = 1$ and $|\Lambda^0{}_0| \geqslant 1$.
8. Show that every Lorentz transformation matrix $\Lambda^\mu{}_\nu$ can be written in the form $\Lambda = \Lambda_d \Lambda_{R'} \Lambda_\omega \Lambda_R$ where Λ_R and $\Lambda_{R'}$ are rotations, Λ_ω is a boost in the z-direction, and Λ_d is one of $\mathbf{1}$, Λ_P, Λ_T, $\Lambda_P\Lambda_T$.
9. Show that a fourth rank three-tensor T^{ijkl} that is invariant under

rotations must have the form $T^{ijkl} = \alpha\delta_{ij}\delta_{kl} + \beta\delta_{ik}\delta_{jl} + \gamma\delta_{il}\delta_{jk}$. [Hint: Using the rules for combining angular momenta in quantum mechanics, show that the tensor product of four $\mathbf{J}=1$ irreducible representations of the rotation group contains the $\mathbf{J}=0$ representation exactly three times.]

10. Suppose that $T^{\mu\nu\rho\sigma}$ is a tensor that is invariant under proper orthochronous Lorentz transformations. Show that $\mathbf{T}$ has the form $T^{\mu\nu\rho\sigma} = \alpha g^{\mu\nu}g^{\rho\sigma} + \beta g^{\mu\rho}g^{\nu\sigma} + \gamma g^{\mu\sigma}g^{\nu\rho} + \delta\epsilon^{\mu\nu\rho\sigma}$.

CHAPTER 2

The Principle of Stationary Action

In the chapters that follow we formulate various field theories using the principle of stationary action. Most readers will have seen this principle as it applies to the mechanics of a finite number of point particles or, more generally, to systems with a finite number of degrees of freedom. In this chapter we outline very briefly how the method works in this familiar case and mention some of its advantages and disadvantages as compared to the "$F=ma$" formulation of mechanics. Then we use a simple example to show how a formulation of a field theory via the principle of stationary action can arise as a limiting case from a similar formulation for a system with a finite number of degrees of freedom.

2.1 LAGRANGIAN MECHANICS OF POINT PARTICLES

Consider a system consisting of N point particles that lie on a line. Call the positions of the particles $x_1, x_2, \ldots, x_N$ and the masses $m_1, m_2, \ldots, m_N$. Then the kinetic energy of the particles is

$$T(\dot{x}_1, \dot{x}_2, \ldots, \dot{x}_N) = \sum_{i=1}^{N} \tfrac{1}{2} m_i \dot{x}_i^2. \tag{2.1.1}$$

Assuming that the motion of the particles conserves energy, the forces on the particles can be derived from a potential energy function $V(x_1, x_2, \ldots, x_N)$:

$$F_i = -\frac{\partial V}{\partial x_i}. \tag{2.1.2}$$

The form of V will depend on the physical situation at hand, and is left arbitrary here. The equations of motion for this system are $F_i = m_i \ddot{x}_i$, or

$$\frac{d}{dt}\frac{\partial T}{\partial \dot{x}_i} = -\frac{\partial V}{\partial x_i}. \tag{2.1.3}$$

The equations of motion (2.1.3) are equivalent to a variational principle called the principle of stationary action or Hamilton's principle. To state Hamilton's principle, one defines a function L, called the Lagrangian, which depends on the coordinates x_i and the velocities $\dot{x}_i$:

$$L(\dot{x}_i, x_i) \equiv T(\dot{x}_i) - V(x_i). \tag{2.1.4}$$

As the system moves from an initial configuration $x_i = a_i$ at time 0 to a final configuration $x_i = b_i$ at time T, it will follow a certain path $x_i = X_i(t)$, as determined by the equations of motion. However, there are many other paths $x_i(t)$ which the system might follow to get from a_i to b_i. To each such path we associate a quantity called its action,

$$A = \int_0^T dt L(\dot{x}_i(t), x_i(t)). \tag{2.1.5}$$

Why is the physical path $X_i(t)$ different from all other paths? It is because the physical path is a stationary point of the action. That is, a small variation $\delta x_i(t) = \epsilon \xi_i(t)$ of the path away from the physical path produces a change δA of the action which is zero to first order in ϵ.

To show that the equations of motion (2.1.3) are equivalent to the principle of stationary action just enunciated, consider a suitably differentiable variation of the path, $\delta x_i(t) = \epsilon \xi_i(t)$. Since both the physical path $X_i(t)$ and the varied path $x_i(t) = X_i(t) + \epsilon \xi_i(t)$ must go from a_i to b_i, we must require that $\xi_i(0) = \xi_i(T) = 0$. By differentiating under the integral, we calculate the variation of the action:

$$\delta A = \epsilon \int_0^T dt \sum_{j=1}^N \left[\dot{\xi}_j(t) \frac{\partial}{\partial \dot{x}_j} L(\dot{X}_i(t), X_i(t)) + \xi_j(t) \frac{\partial}{\partial x_j} L(\dot{X}_i(t), X_i(t)) \right].$$

Since $\xi_i(t)$ vanishes at the endpoints, we can integrate by parts in the first term to obtain

$$\delta A = \epsilon \int_0^T dt \sum_{j=1}^N \xi_j(t) \left[-\frac{d}{dt}\frac{\partial L}{\partial \dot{x}_j} + \frac{\partial L}{\partial x_j} \right]. \tag{2.1.6}$$

It is clear from this expression that δA will be zero for every variation $\xi_j(t)$ if, and only if, the quantities in brackets are all identically zero:

$$-\frac{d}{dt}\frac{\partial}{\partial \dot{x}_j} L(\dot{X}_i(t), X_i(t)) + \frac{\partial}{\partial x_j} L(\dot{X}_i(t), X_i(t)) = 0. \tag{2.1.7}$$

The derivation is now complete, for this is precisely the equation of motion (2.1.3).

We have seen that the two equations, $F = m\ddot{x}$ and $\delta A = 0$, are equivalent. Is there any reason to believe that one formulation of mechanics is either more convenient or more fundamental than the other? There is no definitive answer to this question, but it is worthwhile to mention a few points that may help explain the wide use of the principle of stationary action in advanced classical mechanics.

One advantage of the variational formulation is that the properties of the system are compactly summarized in one function, the Lagrangian $L(\mathbf{x}, \dot{\mathbf{x}})$. This gives the action principle a certain elegance, and it also leads to important computational advantages. For instance, it is easy to make approximations in the Lagrangian before going to the equations of motion. It is also easy in a given problem to make a change of coordinates from x_i to any different set $y_i = y_i(x_1, \ldots, x_N; t)$. It is often a much more tedious task to change coordinates in each equation of motion. This ease of making coordinate changes makes the variational formulation ideal for problems involving constraints (for example, a roller coaster that is constrained to remain on the track); one uses coordinates y_i in which the constraint condition $C(\mathbf{x}) = 0$ becomes $y_N = 0$. (See Problems 3 and 4.)

Two more fundamental properties separate the variational and $F = m\ddot{x}$ approaches to mechanics, First, as we shall see later, there is a direct connection between invariances of the Lagrangian and constants of the motion. For instance, if the Lagrangian is invariant under rotations then angular momentum is conserved. Second, there is a close relation between the Lagrangian formulation of classical particle mechanics and quantum mechanics. (There is likewise a close relation between the Lagrangian formulation of classical field theory and quantum field theory. This relation is explored in books on quantum field theory, so we will not pursue it here.)

Although the principle of stationary action is a powerful tool whenever it can be used, it does not apply to systems subject to frictional forces.* To discuss such systems one must return to an $F = ma$ approach. Likewise we will have to go beyond Hamilton's principle in Chapter 13 when we discuss field theories that include dissipative processes like viscosity, heat flow, and the flow of electric current through a resistor.

*This point is further discussed in Section 9.7.

2.2 THE CONTINUUM LIMIT FOR MASS POINTS ON SPRINGS

Consider the following simple system with N degrees of freedom: N mass points are constrained to lie on a line; each mass is connected to the next one by a spring; all of the masses and springs are identical. (See Figure 2.1.) The Lagrangian for this system is

$$L=\sum_{i=1}^{N}\tfrac{1}{2}M(\dot{x}_i)^2-\sum_{i=2}^{N}\tfrac{1}{2}K(x_i-x_{i-1}-a)^2, \tag{2.2.1}$$

where x_i = coordinate of ith mass point,
M = mass of each mass point,
K = spring constant of each spring,
a = unstretched length of each spring.

We may imagine that this is a model for a crystal in a one dimensional world. If we are solid state physicists in this world, we will want to use the full Lagrangian (2.2.1) to extract all the information we can about the behavior of the crystal. The model is so simple that it is, in fact, possible to solve the equations of motion exactly without too much trouble. The equations of motion derived from the Lagrangian (2.2.1) are

$$\begin{aligned} M\ddot{x}_i &= K(x_{i+1}-2x_i+x_{i-1}) \qquad i=2,\ldots,N-1,\\ M\ddot{x}_1 &= K(x_2-x_1-a),\\ M\ddot{x}_N &= K(x_{N-1}-x_N+a). \end{aligned} \tag{2.2.2}$$

The reader will be able to verify that the general solution of this set of equations is

$$\begin{aligned} x_j(t) &= a\left[j-\tfrac{1}{2}\right]+x_0+vt\\ &\quad+\sum_{m=1}^{N-1}\alpha_m\cos\left(k_m a\left[j-\tfrac{1}{2}\right]\right)\cos(\omega_m t+\phi_m), \end{aligned} \tag{2.2.3}$$

where x_0, v, α_1, $\phi_1,\ldots,\alpha_m$, $\phi_m,\ldots$ are constants chosen to fit the initial

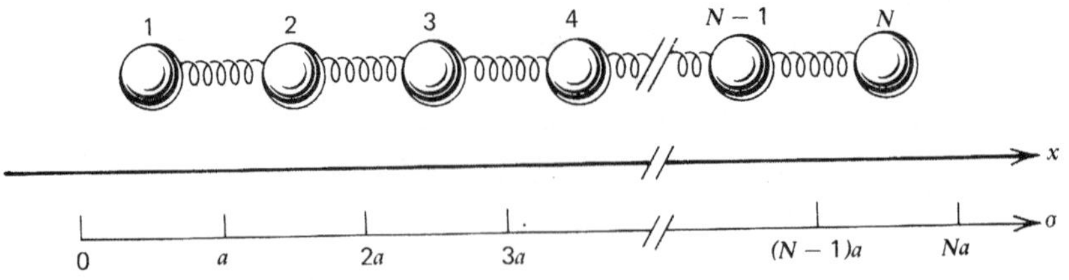

Figure 2.1 Mass points on springs.

conditions, and the wave numbers k_m and frequencies ω_m are

$$k_m = \frac{\pi m}{Na},$$

$$\omega_m^2 = 2\frac{K}{M}\left[1-\cos(k_m a)\right]. \tag{2.2.4}$$

This prediction of our model for the crystal could be compared with experiment by measuring the speed of sound, ω/k, in the crystal for sound waves with various frequencies, and comparing with the theoretical curve plotted in Figure 2.2.

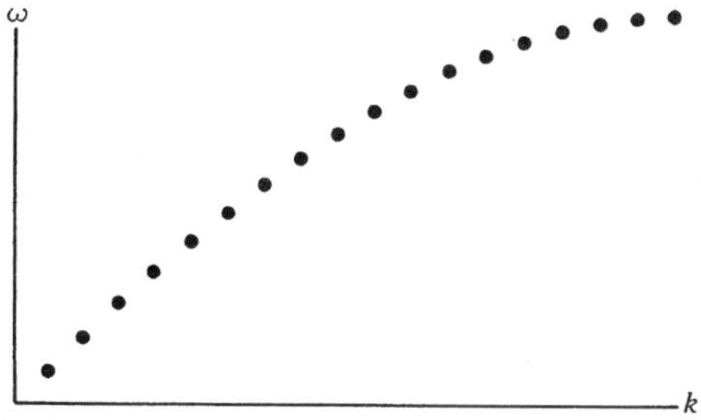

Figure 2.2 Dispersion curve for mass points on springs.

An enormous amount of information about the model crystal is contained in Figure 2.2. But there is a wide class of physical problems for which most of this information is irrelevant. In our simple example, we might want to know how much the length of the crystal changes if the ends are subjected to a certain force, or we might ask for the lowest vibration frequency of a crystal of length L. The common feature of these problems is that the distances characteristic of the phenomena being investigated are much larger than the interatomic spacing a. It is desirable to have at hand a macroscopic model that can account for this "long wavelength" behavior of the material. The macroscopic model will contain certain parameters that we will (in this case) derive from the microscopic model, but it will contain only those parameters that are relevant to macroscopic problems.

To obtain a macroscopic model we replace the discrete index i by a continuous index σ which varies over the range $0<\sigma<Na$. The variables $x_i(t)$ are replaced by a function $x(\sigma,t)$ which is defined at the points $\sigma=(j-\frac{1}{2})a$ by $x((j-\frac{1}{2})a,t)=x_j(t)$ and varies smoothly in between. The sum $\Sigma_i \frac{1}{2}M(\dot{x}_i)^2$ representing the kinetic energy can be approximated by $\int d\sigma \frac{1}{2}(M/a)(\partial x/\partial t)^2$. Similarly, the potential energy, $\Sigma_i \frac{1}{2}K(x_i - x_{i-1} - a)^2$ can be approximated by $\int d\sigma \frac{1}{2}(Ka)[(\partial x/\partial \sigma)-1]^2$. This gives us a field

theoretic Lagrangian

$$L=\int_0^{Na} d\sigma\left\{\tfrac{1}{2}\rho M\left(\frac{\partial x}{\partial t}\right)^2-\tfrac{1}{2}B\left(\frac{\partial x}{\partial \sigma}-1\right)^2\right\}, \tag{2.2.5}$$

where $\rho\equiv 1/a$, is the "particle density" and $B=Ka$ is the "compression modulus" of the crystal.

The equations of motion in this macroscopic model are to be obtained from the principle of stationary action, with the understanding that the "path" to be varied is now the function $x(\sigma,t)$ instead of the $x_i(t)$. We form the action

$$\begin{aligned}A&=\int_0^T dt\,L(t)\\&=\int_0^T dt\int_0^{Na} d\sigma\left\{\tfrac{1}{2}\rho M\left(\frac{\partial x}{\partial t}\right)^2-\tfrac{1}{2}B\left(\frac{\partial x}{\partial \sigma}-1\right)^2\right\}.\end{aligned} \tag{2.2.6}$$

Then we consider a possible physical path $x(\sigma,t)$ and a varied path $x(\sigma,t)+\delta x(\sigma,t)$, which goes through the same initial and final points. (That is, $\delta x(\sigma,0)=\delta x(\sigma,T)=0$.) We demand that the variation of the action be zero for all such variations of the path:

$$0=\delta A=\int_0^T dt\int_0^{Na} d\sigma\left\{\rho M\left(\frac{\partial x}{\partial t}\right)\frac{\partial \delta x}{\partial t}-B\left(\frac{\partial x}{\partial \sigma}-1\right)\frac{\partial \delta x}{\partial \sigma}\right\}.$$

This expression becomes useful if it is integrated by parts. The surface term from the integration by parts in the t-integral vanishes because the variation δx vanishes at $t=0$ and $t=T$. The surface term in the σ-integral does not vanish, and is physically significant:

$$\begin{aligned}0=&-\int_0^T dt\int_0^{Na} d\sigma\left\{\rho M\frac{\partial^2 x}{\partial t^2}-B\frac{\partial^2 x}{\partial \sigma^2}\right\}\delta x\\&-\int_0^T dt\left[B\left(\frac{\partial x}{\partial \sigma}-1\right)\delta x\right]_{\sigma=0}^{\sigma=Na}.\end{aligned} \tag{2.2.7}$$

Since (2.2.7) holds for all variations $\delta x(\sigma,t)$, the coefficient of $\delta x(\sigma,t)$ must vanish. Thus we get both the differential equation for $x(\sigma,t)$ and the proper boundary conditions at $\sigma=0$ and $\sigma=Na$:

$$\begin{aligned}&\rho M\frac{\partial^2 x}{\partial t^2}-B\frac{\partial^2 x}{\partial \sigma^2}=0 \qquad 0<\sigma<Na,\\&\frac{\partial x}{\partial \sigma}-1=0 \qquad \sigma=0 \text{ and } \sigma=Na.\end{aligned} \tag{2.2.8}$$

It is instructive to compare the solutions of the "macroscopic" differential equation (2.2.8) with the corresponding "microscopic" result (2.2.3).

MACROSCOPIC:

$$x(\sigma,t)=\sigma+x_0+vt + \sum_{m=1}^{\infty} \alpha_m \cos(k_m\sigma)\cos(\hat{\omega}_m t+\phi_m), \qquad (2.2.9)$$

where

$$k_m=\frac{\pi m}{Na}, \quad \hat{\omega}_m{}^2=\frac{B}{\rho M}k_m^2=\frac{Ka^2}{M}k_m^2.$$

MICROSCOPIC, $a[j-\frac{1}{2}]=\sigma$:

$$x_j(t)=\sigma+x_0+vt + \sum_{m=1}^{N-1} \alpha_m \cos(k_m\sigma)\cos(\omega_m t+\phi_m), \qquad (2.2.10)$$

where $k_m=\pi m/Na$ and

$$\omega_m^2=2\frac{K}{M}[1-\cos(k_m a)]\sim\frac{Ka^2}{M}k_m^2+0([k_m a]^4).$$

Apparently the two models give completely different results for the frequencies of vibrations whose wavelengths are comparable to the "interatomic spacing," a, of the microscopic model. Indeed, there are no normal modes of vibration in the microscopic model with wavelengths less than $2a$, although there is no such cutoff in the macroscopic model. Nevertheless, the comparison of the two results shows that the two models are indistinguishable in their description of long wavelength phenomena.

In the remainder of this book we are concerned with macroscopic phenomena, hence with continuum models similar to (2.2.6). We do not investigate the important question of how much such models can be derived from a more fundamental microscopic description, although we often notice that the measured values of the macroscopic parameters give some hint of their relation to microscopic physics. Of course a microscopic model of real materials using quantum statistical mechanics is much more complicated (and interesting) than the simple model used in this chapter.

Making x the Independent Variable

In the Lagrangian (2.2.6) the state of the system at a given time t is specified by giving the function $x(\sigma,t)$. The independent variable σ is a label that moves with the material; the function $x(\sigma, t)$ tells the laboratory coordinate x of the particle of material whose label is σ. Clearly, we could just as well specify the state of the system at time t by giving the inverse function $\sigma(x, t)$, which tells the label σ of the particle which is at the position x at time t.

What are the equations of motion for $\sigma(x, t)$? They are the differential equations that result from Hamilton's principle, $\delta A = 0$. The most straightforward way to obtain these equations is to express the action A directly as a function of $\sigma(x, t)$:

$$A = \int_0^T dt \int_{-\infty}^{\infty} dx \frac{\partial \sigma}{\partial x} \Theta(0, \sigma(x,t), Na)$$

$$\times \left\{ \tfrac{1}{2} \rho M \left(\frac{\partial \sigma / \partial t}{\partial \sigma / \partial x} \right)^2 - \tfrac{1}{2} B \left(\frac{1}{\partial \sigma / \partial x} - 1 \right)^2 \right\},$$

where the function Θ (α, σ, β) is defined to be 1 for $\alpha < \sigma < \beta$, zero otherwise.

We do not bother to write out the resulting equations of motion, since it is clear that the problem at hand has not benefited from being reformulated in this fashion. However, when we deal with the mechanics of materials in three dimensional space, we will write the action in an analogous formulation. That is, we will use fixed orthogonal coordinates x, y, z as the independent variables instead of coordinates that move and deform with the material. In this way we will avoid dealing with moving curved coordinate systems.

PROBLEMS

1. A particle of mass m is pulled toward the origin of coordinates by a spring with spring constant k: $m\ddot{\mathbf{x}} = -k\mathbf{x}$. What is the Lagrangian for this system? What are the equations of motion in spherical polar coordinates r, Θ, ϕ?
2. Consider a particle of mass m moving in the gravitational potential $V(r) = -k/r$ of the earth. Let $\boldsymbol{\Omega} = (0, 0, \omega)$ be the angular velocity vector of the earth's rotation. Show that the Lagrangian describing the motion of this particle relative to a reference frame fixed to the rotating earth is $L = \frac{1}{2} m (\dot{\mathbf{x}} + \boldsymbol{\Omega} \times \mathbf{x})^2 - V(|\mathbf{x}|)$. What are the equations of motion in terms of the coordinates x_i?

3. (Constrained systems.) Consider the variational equation

$$0=\delta A_c=\delta\int dt\left[\tfrac{1}{2}m\dot{\mathbf{x}}^2-V(\mathbf{x})-\lambda C(\mathbf{x})\right],$$

where x_1, x_2, x_3 are the coordinates of a particle, $C(x)$ is a fixed function, and $\lambda(t)$ is a "coordinate" on the same footing as x_i and is free to vary.

(a) Show that the variational equation $\delta A_c=0$ describes a particle which moves on the surface $C(x)=0$ subject to the force $-\nabla V$ and to the additional force $\mathbf{F}_c=-\lambda(t)\nabla C$, which is normal to the surface and of sufficient strength to keep the particle on the surface.

(b) Let the motion $\mathbf{x}=\mathbf{X}(t)$, $\lambda=\Lambda(t)$ satisfy the variational equation $\delta A_c=0$. Show that $x=x(t)$ also makes $\delta\int dt[\frac{1}{2}m\dot{\mathbf{x}}^2-V(\mathbf{x})]=0$ for every variation $\delta\mathbf{x}(t)$ such that the varied path $\dot{\mathbf{X}}(t)+\delta\mathbf{x}(t)$ satisfies $C(\mathbf{X}+\delta\mathbf{x})=0$.

4. Carry through the analysis of Problem 3 for the case of a pendulum bob, described by spherical coordinates r, θ, ϕ, which moves in a potential $V=-mgr\cos\theta$ and is subject to the constraint $r-c=0$.

5. Prove the converse to Problem 3b. That is, if $\mathbf{x}=\mathbf{X}(t)$ satisfies $C(\mathbf{x})=0$ and $\delta\int dt[\frac{1}{2}m\dot{\mathbf{x}}^2-V(\mathbf{x})]=0$ when $C(\mathbf{X}+\delta\mathbf{x})=0$, then there exists $\Lambda(t)$ such that $\mathbf{x}=\mathbf{X}(t)$, $\lambda=\Lambda(t)$ satisfies $\delta\int dt[\frac{1}{2}m\dot{\mathbf{x}}^2-V(\mathbf{x})-\lambda C(\mathbf{x})]=0$ for all variations $\delta\mathbf{x}$, $\delta\lambda$.

CHAPTER 3

Some General Features of Classical Field Theory

Here we begin the discussion of classical field theory proper, with a general introduction to the formulation of field theories, using the principle of stationary action.

We also show how the invariance of the action under displacements of $\mathbf{x}$ and t results in the conservation of momentum and energy. This result is a special case of a remarkable theorem of E. Noether, which says that each invariance of the action results in the existence of a conserved quantity. We discuss Noether's theorem more thoroughly in Chapter 9.

These general features are illustrated with the simplest of examples, the theory of a single scalar field which obeys the wave equation. In the chapters that follow we discuss less trivial examples.

3.1 THE FORMULATION OF A CLASSICAL FIELD THEORY

The Lagrangian

Let us suppose that we want to formulate a field theory involving certain fields we have in mind—for instance, a tensor field $F_{\mu\nu}(t,\mathbf{x})$ and two scalar fields $\rho(t,\mathbf{x})$ and $\omega(t,\mathbf{x})$. For the sake of having a uniform notation in the present discussion, we give each of the components of the fields a new name, $\phi_1(x), \phi_2(x), \ldots,$ in any convenient order. In our example we might choose $\phi_1 = \omega, \phi_2 = \rho$, $\phi_3 = F_{01}$, $\phi_4 = F_{01}$, etc.

To specify a theory involving the field components $\phi_1, \ldots, \phi_n$, we simply write down a function $\mathcal{L}$ which depends on the fields and their derivatives: $\mathcal{L} = \mathcal{L}(\phi_K, \partial_\mu \phi_K, \partial_\mu \partial_\nu \phi_K, \ldots)$. The action to be made stationary is the space-time integral of $\mathcal{L}$,

$$A = \int d^4x\, \mathcal{L}(\phi_K(x),\ \partial_\mu \phi_K(x),\ \partial_\mu \partial_\nu \phi_K(x), \ldots). \tag{3.1.1}$$

The function $\mathcal{L}$ is properly called the Lagrangian density of the theory, since $\int d^3\mathbf{x}\,\mathcal{L} = L$ is apparently the Lagrangian. Nevertheless, we usually refer to $\mathcal{L}$ simply as the Lagrangian.

An example of a field theory is the theory of a single scalar field $\phi(x)$ with the Lagrangian

$$\mathcal{L} = -\tfrac{1}{2}(\partial_\mu\phi)(\partial^\mu\phi) - \frac{m^2}{2}\phi^2. \tag{3.1.2}$$

(In quantum field theory, ϕ is known as a free scalar field with mass m, but no real classical physical system with this Lagrangian is known to the author.) We use this example as an illustration throughout the chapter.

What sort of Lagrangian should we choose? One very useful criterion is that the Lagrangian should be a Lorentz scalar, assuming that we want a Lorentz invariant theory. The Lagrangian (3.1.2) clearly passes this test. The Lagrangian (3.1.2) also has the commendable property of being simple. As we will see, the Lagrangian for electrodynamics is also simple; but Lagrangians for the mechanics of materials can be rather complicated. We will sometimes proceed by writing down the most general scalar function of the fields and their first derivatives, then looking at special cases that are interesting because of their simplicity. Of course, the ultimate test of a good Lagrangian is that the theory derived from it describe a real physical system.

It is by no means necessary to the general formalism of classical field theory that the Lagrangian be a Lorentz scalar. One can apply the methods and results of this chapter just as well to a Lorentz non-invariant Lagrangian such as $\mathcal{L} = \frac{1}{2}\{(\partial_0\phi)^2 - v^2 \sum_{j=1}^{3} (\partial_j\phi)^2 - m^2\phi^2\}$. In fact, we will use the results of this chapter when we discuss nonrelativistic systems in Chapter 7. Nevertheless, we continue here to call the time coordinate x^0 and to use the summation convention in order to have a compact notation.

Equations of Motion

Suppose we have a Lagrangian that depends on the fields ϕ_K and their first derivatives $\partial_\mu\phi_K$. (The extension to higher derivatives is simple but not commonly used in physics.) We can derive equations of motion by imposing the requirement that the variation δA of the action be zero when we make a small variation $\delta\phi_K(x)$ of the fields. The only requirements to be satisfied by the variations $\delta\phi$ are that they be differentiable and vanish outside some bounded region of space-time, so as to allow an integration by parts. We calculate δA:

$$\delta A = \int d^4x \sum_{K=1}^{n} \left\{ \frac{\partial \mathcal{L}}{\partial \phi_K} \delta\phi_K + \frac{\partial \mathcal{L}}{\partial(\partial_\mu \phi_K)} \partial_\mu \delta\phi_K \right\}$$

$$= \int d^4x \sum_{K=1}^{n} \delta\phi_K \left\{ \frac{\partial \mathcal{L}}{\partial \phi_K} - \partial_\mu \frac{\partial \mathcal{L}}{\partial(\partial_\mu \phi_K)} \right\}. \qquad (3.1.3)$$

In order that δA vanish for all admissible variations $\delta\phi_K(x)$, the quantity in braces { } must vanish for all x:

$$\frac{\partial \mathcal{L}}{\partial \phi_K} - \partial_\mu \frac{\partial \mathcal{L}}{\partial(\partial_\mu \phi_K)} = 0 \qquad K = 1, \ldots, n. \qquad (3.1.4)$$

These are the equations of motion, and are usually called the Euler-Lagrange equations.

In our example of a "free scalar field" we have

$$\mathcal{L} = -\tfrac{1}{2}(\partial_\mu \phi)(\partial^\mu \phi) - \frac{m^2}{2}\phi^2$$

so

$$\frac{\partial \mathcal{L}}{\partial(\partial_\mu \phi)} = -\partial^\mu \phi, \qquad \frac{\partial \mathcal{L}}{\partial \phi} = -m^2\phi.$$

Thus the Euler-Lagrange equation is

$$\partial_\mu \partial^\mu \phi(x) - m^2\phi(x) = 0. \qquad (3.1.5)$$

External Sources

It is often useful to be able to calculate the response of the fields to a given external disturbance. For example, one may want to know the electric and magnetic fields produced by a given electric current distribution. Such problems can be formulated by incorporating the external disturbance into the Lagrangian. In our scalar field example, one simple type of external source can be introduced by writing

$$\mathcal{L}(\phi, \partial_\mu \phi, x) = -\tfrac{1}{2}(\partial_\mu \phi)(\partial^\mu \phi) - \frac{m^2}{2}\phi^2 + \phi S(x), \qquad (3.1.6)$$

where $S(x)$ is a given function of $\mathbf{x}$ and t. We require that $\delta A = 0$ when we make a small variation of $\delta\phi$ of the field, (but not of $S(x)$!). This gives the equation of motion

$$\partial_\mu \partial^\mu \phi(x) - m^2\phi(x) = S(x). \tag{3.1.7}$$

When we have introduced external sources that depend on x^μ, the Lagrangian will depend directly on x^μ, as well as on the fields and their derivatives. In what follows we allow for this possibility and write $\mathcal{L} = \mathcal{L}(\phi_K, \partial_\mu \phi_K, x)$. If $\mathcal{L}$ does depend directly on x^μ, the equations of motion still retain the form

$$\frac{\partial \mathcal{L}(\phi_K, \partial_\mu \phi_K, x)}{\partial \phi_K} - \partial_\mu \frac{\partial \mathcal{L}(\phi_K, \partial_\mu \phi_K, x)}{\partial(\partial_\mu \phi_K)} = 0$$

with the same derivation as before.

3.2 CONSERVED CURRENTS

Let us take a closer look at the model introduced in (3.1.6), the scalar field with an external source $S(x)$. Consider the quantities $\mathcal{E}^\mu(x)$ defined by

$$\begin{aligned} \mathcal{E}^0(x) &= \tfrac{1}{2}\left[(\partial_0\phi)^2 + (\partial_j\phi)(\partial_j\phi) + m^2\phi^2\right] + \phi S, \\ \mathcal{E}^j(x) &= -(\partial_0\phi)(\partial_j\phi) \qquad j = 1,2,3. \end{aligned} \tag{3.2.1}$$

(Recall that, by convention, the repeated Latin indices in $(\partial_j\phi)(\partial_j\phi)$ are to be summed over $j = 1,2,3$.) As we will see shortly, it is appropriate to call $\mathcal{E}^0$ the energy density in the model.

If we perform the exercise of computing $\partial_\mu \mathcal{E}^\mu = \partial_0 \mathcal{E}^0 + \partial_j \mathcal{E}^j$ we find

$$\partial_\mu \mathcal{E}^\mu = (\partial_0\phi)\left[(-\partial_\mu\partial^\mu + m^2)\phi + S\right] + \phi(\partial_0 S). \tag{3.2.2}$$

When ϕ obeys its equation of motion (3.1.7) the first term vanishes, leaving $\partial_\mu \mathcal{E}^\mu = \phi(\partial_0 S)$. Suppose now that the external source $S(x)$ is independent of the time x^0. Then

$$\partial_\mu \mathcal{E}^\mu = 0. \tag{3.2.3}$$

The implication of this is that the "energy" defined by

$$E(t) = \int d\mathbf{x}\, \mathcal{E}^0(t, \mathbf{x}) \tag{3.2.4}$$

is conserved. Indeed, use of $\partial_\mu \mathcal{E}^\mu = 0$ and Gauss' theorem gives

$$\frac{d}{dt}E(t) = \int d\mathbf{x}\,\partial_0 \mathcal{E}^0 = -\int d\mathbf{x}\,\partial_j \mathcal{E}^j = 0, \tag{3.2.5}$$

provided only that $\phi(\mathbf{x}, t)$ falls off rapidly as $|\mathbf{x}|$ becomes large.

To recapitulate, when the external source $S(x)$ is constant in time, then the energy E defined by (3.2.1) and (3.2.4) is a constant of the motion.

Since the "energy" is the integral over space of $\mathcal{E}^0(x)$, it is reasonable to interpret $\mathcal{E}^0$ as the density of energy. The quantities $\mathcal{E}^j$ can then be interpreted as the energy current; that is $\boldsymbol{\mathcal{E}} \cdot d\mathbf{A}$ is the amount of energy crossing an infinitesimal surface area $d\mathbf{A}$ per unit time. If the equation $\partial_\mu \mathcal{E}^\mu = 0$ is integrated over a small volume V bounded by a surface ∂V, it reads

$$\frac{d}{dt}\int_V \mathcal{E}^0 d\mathbf{x} = -\int_{\partial V} \boldsymbol{\mathcal{E}} \cdot d\mathbf{A}. \tag{3.2.6}$$

Thus, $\partial_\mu \mathcal{E}^\mu = 0$ is a local statement of energy conservation: the rate of change of the amount of energy contained in the volume V equals the net rate of energy flow into the volume through its surface ∂V.

Apparently this is an example of a general phenomenon. Whenever we can construct out of the basic fields of a theory a set of fields $J^0(x)$, $J^1(x)$, $J^2(x)$, $J^3(x)$ which obey $\partial_\mu J^\mu = 0$, then the quantity $Q(t) = \int d\mathbf{x} J^0(x)$ is a constant of the motion, and J^0 can be interpreted as the local density of "Q." Another example is Q = electric charge, J^0 = charge density, $\mathbf{J}$ = electric current.

Transformation Laws

The discussion above applies in one particular reference frame, and applies equally if the quantities J^μ are the components of a vector field, or some particular components of a tensor field (as in $J^\mu = T^{\mu 3}$), or have no particular transformation properties at all. However, the transformation law of J^μ determines the transformation law for the associated conserved quantity Q, as we will now see.

Suppose, to begin with, that the quantities $J^\mu(x)$ are the components of a vector field, and that $\partial_\mu J^\mu = 0$. We will consider the "charge" Q calculated in a particular coordinate system

$$Q = \int d\mathbf{x} J^0(0, \mathbf{x}). \tag{3.2.7}$$

Let us analyze this simple integral in what may at first seem a perverse fashion. The integral is a surface integral over the surface $x^0=0$ in space-time. We have used the coordinates x^1, x^2, x^3 to parameterize the surface, but any other set a^1, a^2, a^3 would do just as well:

$$Q=\int d^3a\frac{\partial(\mathbf{x})}{\partial(\mathbf{a})}J^0\big(x^\mu(a^1,a^2,a^3)\big).$$

The Jacobean $\partial(\mathbf{x})/\partial(\mathbf{a})$ can be written as

$$\begin{aligned}\frac{\partial(\mathbf{x})}{\partial(\mathbf{a})}&=\det\left[\frac{\partial x^i}{\partial a^j}\right]\\ &=\epsilon_{ijk}\frac{\partial x^i}{\partial a^1}\frac{\partial x^j}{\partial a^2}\frac{\partial x^k}{\partial a^3}\\ &=-\epsilon_{0\nu\alpha\beta}\frac{\partial x^\nu}{\partial a^1}\frac{\partial x^\alpha}{\partial a^2}\frac{\partial x^\beta}{\partial a^3}.\end{aligned}$$

Now the integral takes on a covariant looking form,

$$Q=\int dS_\mu J^\mu \tag{3.2.8}$$

where

$$dS_\mu=-\epsilon_{\mu\nu\alpha\beta}\frac{\partial x^\nu}{\partial a^1}\frac{\partial x^\alpha}{\partial a^2}\frac{\partial x^\beta}{\partial a^3}da^1da^2da^3. \tag{3.2.9}$$

It should be apparent that one can define a surface integral like (3.2.8) for any three dimensional surface $\mathcal{S}$ given parametrically by equations of the form $x^\mu=x^\mu(a^1,a^2,a^3)$. The surface area differential dS_μ is a vector orthogonal to the surface at a, since $T^\mu dS_\mu=0$ for any tangent vector $T^\mu=\sum_{j=1}^{3}\lambda_j\partial x^\mu/\partial a^j$. The reader can verify that the value of a surface integral like (3.2.8) is independent of what coordinates $a^1a^2a^3$ are used to parameterize the surface.

Since dS_μ is manifestly a vector, the "charge" Q is independent of the reference frame used to calculate the integral. However, the ghost of the original reference frame still lives on in the choice of the surface over which we integrated, $x^0=0$. In order to show that Q is really a scalar, we have to show that we get the same result no matter what surface we use.

The original surface was $x^\mu c_\mu=0$, where $c^\mu=(1,0,0,0)$ is a vector along the time axis of the original coordinate system. Another observer will use the surface $x^\mu b_\mu = 0$, where b^μ is a vector along his time axis. We want to show that the corresponding charges,

$$Q_c = \int_{x\cdot c=0} dS_\mu J^\mu, \tag{3.2.10}$$

$$Q_b = \int_{x\cdot b=0} dS_\mu J^\mu \tag{3.2.11}$$

are equal. We will show that $Q_c = Q_b$ by using $\partial_\mu J^\mu = 0$ and Gauss' theorem. To this end we consider the four dimensional volume bounded by the surfaces $x\cdot c=0$, $x\cdot b=0$, and $\mathbf{x}^2 = R^2$ (see Figure 3.1). This volume consists of two disjoint regions, labeled I and II. We will apply Gauss' theorem to the integral

$$0 = \int_{II} d^4x\, \partial_\mu J^\mu - \int_I d^4x\, \partial_\mu J^\mu. \tag{3.2.12}$$

In general, Gauss' theorem tells us that

$$\int_V d^4x\, \partial_\mu F^\mu = \int_{\partial V} dS_\mu F^\mu, \tag{3.2.13}$$

where dS_μ is the outward pointing normal vector to the surface ∂V which bounds V. (Note that this four-dimensional Gauss' theorem is proved in exactly the same way as the corresponding theorem in three dimensional Euclidean space; the metric tensor does not enter the proof at all.) In this particular case we have

$$0 = \int_{\substack{x\cdot b=0\\ \mathbf{x}^2<R^2}} dS_\mu J^\mu - \int_{\substack{x\cdot c=0\\ \mathbf{x}^2<R^2}} dS_\mu J^\mu + \int_{\text{Sides}} dS_\mu J^\mu.$$

Now let the radius R become infinite. As long as J^μ falls off sufficiently rapidly at large distances, we obtain

$$\int_{x\cdot b=0} dS_\mu J^\mu = \int_{x\cdot c=0} dS_\mu J^\mu \tag{3.2.14}$$

as desired.

We have seen that the conserved charge Q associated with a conserved vector current J^μ is a scalar. By making very slight modifications in the proof, the reader can verify the generalization of this theorem: if $J^{\alpha\cdots\beta\mu}(x)$ is an Nth rank tensor field which satisfies the conservation equation

$$\partial_\mu J^{\alpha\cdots\beta\mu} = 0, \tag{3.2.15}$$

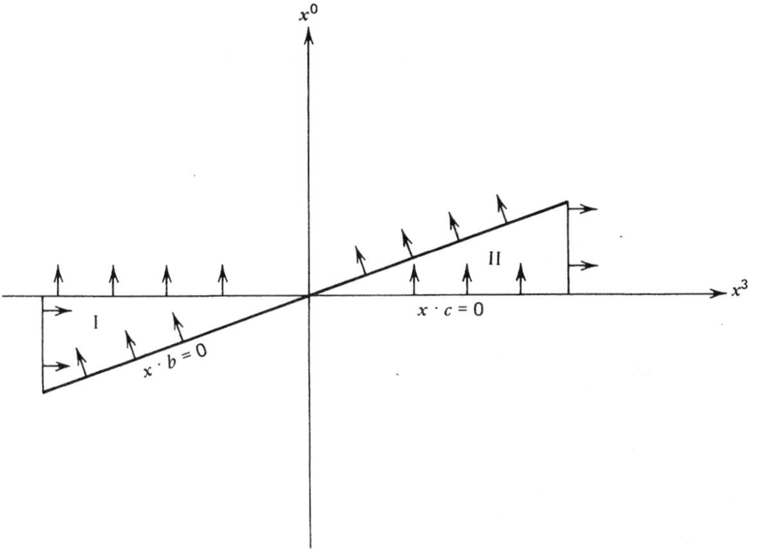

Figure 3.1 Surfaces used for proving that $Q_b = Q_c$. The small arrows represent the direction of the normal vectors dS_μ chosen for the surface integrals. To avoid confusion, we have plotted the covariant components dS_μ of the normal vectors; a figure showing the contravariant components dS^μ would look quite different.

then the corresponding conserved charge

$$Q^{\alpha \cdots \beta} = \int d\mathbf{x} J^{\alpha \cdots \beta 0}(0, \mathbf{x}) \tag{3.2.16}$$

is a tensor of rank $N-1$.

In the next section we discuss an important example of a conserved tensor of rank 1, the total four-momentum of the system P^μ. According to our present results, we may expect that the conservation of four-momentum results from the existence of a tensor field, $T^{\mu\nu}(x)$ which obeys $\partial_\nu T^{\mu\nu} = 0$. We will find that this is indeed the case.

3.3 CONSERVATION OF MOMENTUM AND ENERGY

In the example of a scalar field with an external source we found that when the source $S(x^\mu)$ is independent of the time x^0, then a certain quantity E, constructed from the fields, is conserved. Here we derive a simple and important generalization of this result.

Consider a theory of field components $\phi_1(x)$, $\phi_2(x), \ldots, \phi_n(x)$. The

Lagrangian density $\mathcal{L}(\phi_K, \partial_\mu \phi_K, x)$ can depend on the fields and their derivatives, and also can depend explicitly on x if we include an external source. Suppose, however, that $\mathcal{L}$ does not depend explicitly on one of x^0, x^1, x^2, or x^3—say x^σ:

$$\frac{\partial \mathcal{L}(\phi_K, \partial_\mu \phi_K, x)}{\partial x^\sigma} = 0. \tag{3.3.1}$$

Of course $\mathcal{L}$ still depends implicitly on this coordinate x^σ via its dependence on $\phi(x)$ and $\partial_\mu \phi(x)$:

$$\partial_\sigma \mathcal{L}(\phi_K(x), \partial_\mu \phi_K(x), x) = \sum_{K=1}^{n} \frac{\partial \mathcal{L}}{\partial \phi_K} \partial_\sigma \phi_K + \frac{\partial \mathcal{L}}{\partial(\partial_\mu \phi_K)} \partial_\sigma \partial_\mu \phi_K. \tag{3.3.2}$$

Now construct the following candidate for a conserved current

$$T_\sigma{}^\mu(x) = \delta_\sigma^\mu \mathcal{L} - \sum_{K=1}^{n} (\partial_\sigma \phi_K) \frac{\partial \mathcal{L}}{\partial(\partial_\mu \phi_K)}. \tag{3.3.3}$$

A simple calculation using (3.3.2) gives $\partial_\mu T_\sigma{}^\mu$:

$$\begin{aligned} \partial_\mu T_\sigma{}^\mu = \sum_{K=1}^{n} \Bigg\{ & \frac{\partial \mathcal{L}}{\partial \phi_K}(\partial_\sigma \phi_K) + \frac{\partial \mathcal{L}}{\partial(\partial_\mu \phi_K)}(\partial_\sigma \partial_\mu \phi_K) \\ & - (\partial_\mu \partial_\sigma \phi_K) \frac{\partial \mathcal{L}}{\partial(\partial_\mu \phi_K)} - (\partial_\sigma \phi_K) \partial_\mu \frac{\partial \mathcal{L}}{\partial(\partial_\mu \phi_K)} \Bigg\} \\ = \sum_{K=1}^{n} & (\partial_\sigma \phi_K) \left[\frac{\partial \mathcal{L}}{\partial \phi_K} - \partial_\mu \frac{\partial \mathcal{L}}{\partial(\partial_\mu \phi_K)} \right]. \end{aligned} \tag{3.3.4}$$

As long as the fields obey the equations of motion, the quantities in square brackets are identically zero, so that

$$\partial_\mu T_\sigma{}^\mu = 0. \tag{3.3.5}$$

Thus if the Lagrangian is independent of the coordinate x^σ, the quantity

$$P_\sigma = \int d\mathbf{x} T_\sigma{}^0(x) \tag{3.3.6}$$

is conserved.

The quantity $P^0 = -P_0$ which is conserved when $\mathcal{L}$ is independent of time is called the energy of the system. The quantity $P^1 = P_1$ which is conserved when $\mathcal{L}$ is independent of x^1 is called the one-component of the momentum of the system, and similarly for P^2 and P^3. There is apparently a close connection between these ideas in classical field theory, the corresponding theorem in the classical mechanics of point particles, and the theorem in quantum mechanics that energy is conserved when the Hamiltonian is independent of time and momentum is conserved when the Hamiltonian does not depend on the position of the center of mass of the system. On a less abstract level, we base the identification of P_μ as energy and momentum on the observation to be made in the next chapter that P_μ as calculated from (3.3.3) and (3.3.6) is just what one would expect on the basis of the identification $\mathbf{P}=M\mathbf{v}$ and $E=\frac{1}{2}M\mathbf{v}^2+$(internal energy) for slowly moving systems.

We did not make use of Lorentz invariance in this section. Indeed, in Chapter 7 we use the conserved energy and momentum we have found here in discussing nonrelativistic systems. However, if the Lagrangian is independent of x and is a Lorentz scalar then $T_\sigma{}^\mu(x)$ defined by (3.3.3) will be a tensor. One can give a simple formal proof of this; but it will be more illuminating to see how it works in examples. According to the results of the last section, when $T_\sigma{}^\mu$ is a tensor the quantities P^μ form a four-vector —the "four-momentum" of the system—just as they do for a point particle.

PROBLEMS

1. Calculate $T_\sigma{}^\mu$ for the free scalar field, (3.1.2), and verify directly from the equation of motion (3.1.5) that $\partial_\mu T_\sigma{}^\mu = 0$.
2. Consider the Lagrangian for a scalar field ϕ:

$$\mathcal{L} = \frac{1}{2}\left[(\partial_0\phi)^2 - v^2 \sum_{j=1}^{3} (\partial_j\phi)^2 \right],$$

   where v is a parameter. Show that the equation of motion for ϕ is the wave equation for waves with velocity v. Calculate $T_\sigma{}^\mu$ for this Lagrangian.
3. Consider the Lagrangian (3.1.6). If this external source $S(x)$ is not independent of time, the energy defined in (3.2.1) and (3.2.4) will not be conserved. Argue that the power per unit volume produced by the source is $\partial_\mu \mathcal{E}^\mu(x)$. Find the outgoing wave solution $\phi(x)$ produced by a source function $S(x) = S_0 \cos\omega t\,\delta(\mathbf{x})$ and calculate the time averaged power produced by the source in two ways:

$$\langle P \rangle = \left\langle \int_{|\mathbf{x}|=R} d\mathbf{A}\cdot \boldsymbol{\mathcal{E}} \right\rangle = \left\langle \int d\mathbf{x}\, \partial_\mu \mathcal{E}^\mu \right\rangle$$

   (Note that the instantaneous power produced by this singular source is infinite.)

CHAPTER 4

The Mechanics of Fluids

The simplest class of continuous media is the perfect fluids, that is, fluids for which one can ignore heat conductivity and viscosity. In this chapter we develop and discuss the relativistically covariant formulation of (perfect) fluid mechanics.

For most terrestrial purposes (for example, discussing the flow of air over an airplane wing), the nonrelativistic approximation to the covariant equations of motion is quite adequate. We discuss this approximation in Section 4.3 and, more thoroughly, in Chapter 7. However, we choose to devote most of the chapter to the covariant formulation with an eye toward discussion of the coupling of material media with electromagnetism (Chapters 8 and 10) and with gravity (Chapters 11 and 12). It is a welcome side benefit that the relativistic theory is in many ways simpler than the nonrelativistic theory.

4.1 DESCRIPTION OF FLUID MOTION

The Fields R_a

The object of fluid mechanics is to predict the future motion of a fluid from given initial conditions. But before we can predict future motion we must specify how we will describe any possible motion. One complete description of the motion of a fluid could be obtained by attaching a set of three coordinate labels (R_1, R_2, R_3) to each "droplet" of the fluid.* (See Figure 4.1.) The motion of the fluid could then be specified by giving a function $x^j(\mathbf{R}, t)$, the position at time t of the droplet labeled by $\mathbf{R}$. However we will find it more convenient to specify the motion of the fluid

*In this chapter we consider a fluid that is infinite in extent, so that the range of the material coordinates is $-\infty < R_a < \infty$. This procedure gives us the correct physics inside the fluid, but does not tell us what happens at the surface of a finite body of fluid. We consider finite material bodies and discuss boundary conditions in the next chapter.

by giving the inverse function $R_a(x^j,t)=R_a(x^\mu)$, $a=1,2,3$. The function $R_a(x^j,t)$ tells the labels R_a of the droplet which is at position x^j at time t.

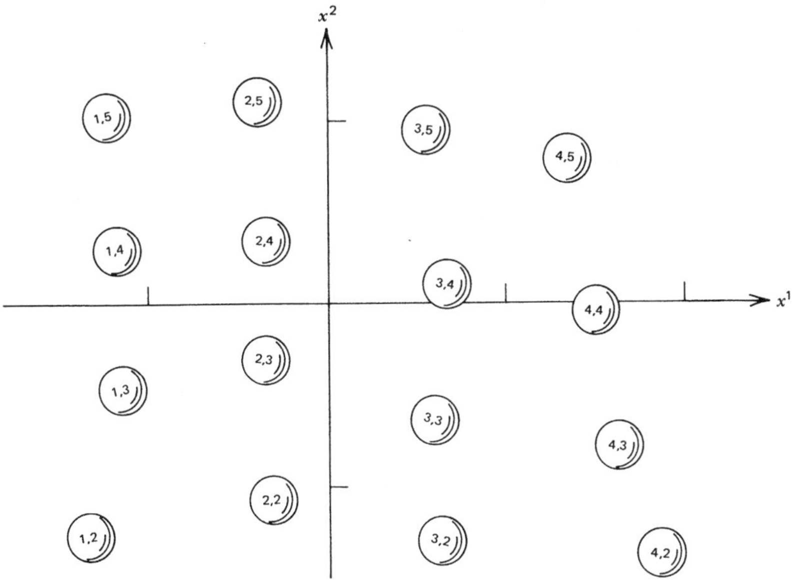

Figure 4.1 Fluid particles in two dimensional space with labels (R_1, R_2). The particles flow past the fixed coordinate axes x^1, x^2.

The labels **R** constitute the "material coordinate system," which flows with the fluid. We can choose this system in any way that suits our fancy. For instance, we might imagine placing the fluid in a rectangular fish tank and marking off a Cartesian coordinate system in the fluid, then pouring it into an experimental apparatus. We might even imagine that the material coordinates **R** are marked on small plastic balls that are carried along by the fluid. In any case, it is evident that during the course of the fluid motion the material coordinate system will not normally be a Cartesian coordinate system. Fortunately the laboratory coordinates x^j which are the independent variables do form a Cartesian system.

The functions $R_a(x)$ will serve as the basic fields describing the fluid. Since the labels R_a on a fluid droplet do not change when observed from different Lorentz frames, the fields $R_1(x)$, $R_2(x)$, $R_3(x)$ are three scalar fields. Our object now is to find equations of motion for these fields.

Let us define $n(\mathbf{R})$ to be the density of the fluid as measured in the material coordinate system: $n(\mathbf{R})dR_1dR_2dR_3$ is the number of atoms (or molecules, or electrons, or whatever) in a parallelepiped of fluid at $\mathbf{R}$ with sides dR_1, dR_2, dR_3 along the R-coordinate axes. Since the coordinate labels R_a move with the fluid, $n(\mathbf{R})$ is fixed for all time once we choose the R-coordinates—assuming that no atoms are created or destroyed.

The Matter Current

Since the number of atoms in our fluid is conserved (by assumption), we may expect that there is a conserved "matter" current $J^\mu(x)$ with

$$\int d^3\mathbf{x}J^0(t,\mathbf{x}) = \text{number of atoms.} \tag{4.1.1}$$

The expression for $J^\mu(x)$ is simple and elegant:

$$\begin{aligned} J^\mu(x) &= n(R_a(x))\epsilon^{\mu\alpha\beta\gamma}(\partial_\alpha R_1(x))(\partial_\beta R_2(x))(\partial_\gamma R_3(x)) \\ &= n(R_a)\epsilon^{\mu\alpha\beta\gamma}\frac{1}{3!}\epsilon_{abc}(\partial_\alpha R_a)(\partial_\beta R_b)(\partial_\gamma R_c). \end{aligned} \tag{4.1.2}$$

To test this ansatz, we first show that $J^0(x)$ is the number of atoms per unit volume as measured in a laboratory coordinate system:

$$\begin{aligned} J^0 &= n(R_a)\epsilon^{0\alpha\beta\gamma}(\partial_\alpha R_1)(\partial_\beta R_2)(\partial_\gamma R_3) \\ &= n(R_a)\epsilon_{ijk}(\partial_i R_1)(\partial_j R_2)(\partial_k R_3) \\ &= n(R_a)\det\left[\frac{\partial R_a}{\partial x^j}\right]. \end{aligned}$$

However, the Jacobean $\det(\partial R_a/\partial x^j)$ is just the factor needed to transform the density $n(R) = dN/d^3R$ measured in the material coordinates into a density dN/d^3x measured in the laboratory coordinates. Secondly, we verify the claim that J^μ is conserved:

$$\begin{aligned} \partial_\mu J^\mu = \epsilon^{\mu\alpha\beta\gamma}\bigg[&\frac{\partial n(R_a)}{\partial R_a}(\partial_\mu R_a)(\partial_\alpha R_1)(\partial_\beta R_2)(\partial_\gamma R_3) + n(\partial_\mu\partial_\alpha R_1)(\partial_\beta R_2)(\partial_\gamma R_3) \\ &+ n(\partial_\alpha R_1)(\partial_\mu\partial_\beta R_2)(\partial_\gamma R_3) + n(\partial_\alpha R_1)(\partial_\beta R_2)(\partial_\mu\partial_\gamma R_3)\bigg]. \end{aligned}$$

Each term in the right hand side of this equation is zero because $\epsilon^{\mu\nu\rho\sigma}$ is antisymmetric. Thus $\partial_\mu J^\mu = 0$. Also, we note that $J^\mu(x)$ is a vector field, so that the total number of atoms $\int d^3\mathbf{x}J^0(x)$ is a scalar, as it must be.

Density and Four-Velocity

Consider a droplet of fluid which at time x^0 is located at position x^j. It will often be important to know what this piece of fluid looks like as viewed in an inertial reference frame which is, at that time, moving along with the droplet. (See Fig. 4.2.) Such a reference frame is called a local rest frame of the fluid at x. (There are many local rest frames at x, related to one another by rotations.) According to observers in a local rest frame the piece of fluid is at rest at the moment of interest. Clearly, questions about the internal constitution of the fluid at x should be referred to such observers.

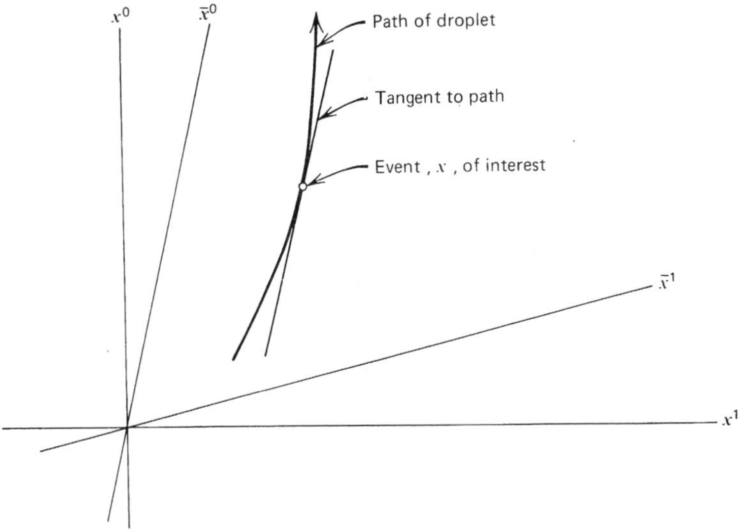

Figure 4.2 Coordinate axes $\bar{x}^1$, $\bar{x}^2$ of a local rest frame of the fluid at the point x.

One such question is, "What is the density of the fluid at x?" Since the density $J^0(x)$ is one component of a four-vector, the answer depends on whom you ask. But the value of the density that is relevant to a description of the internal state of a piece of fluid (as opposed to its state of motion) is the density measured in a local rest frame at x. We call this density $\rho(x)$.

How is $\rho(x)$ related to $J^\mu(x)$? In a local rest frame at x we have $J^k(x)=0$ for $k=1,2,3$. This is clear enough physically: the number of atoms crossing a unit surface area in unit time is zero if the fluid is at rest. It can also be verified from the definition (4.1.2) of J^μ and the observation that $\partial_0 R_a(x)=0$ if the fluid at x is at rest. In a local rest frame we have by definition $J^0(x)=\rho(x)$. Thus

$$-J_\mu(x)J^\mu(x)=\rho(x)^2 \tag{4.1.3}$$

in a local rest frame; but $J_\mu J^\mu$ is a Lorentz scalar, so this equation holds independently of the frame chosen to measure J^μ.

Another important quantity that can be written in terms of J^μ is the four-velocity $u^\mu(x)$ of the fluid at x. The reader will recall from Chapter 1 that the four-velocity of a particle whose path in space-time is $x^\mu(\tau)$ is the tangent vector to the path, normalized to have length 1:

$$u^\mu(\tau)=\frac{dx^\mu}{d\tau}\bigg/\sqrt{-\frac{dx^\nu}{d\tau}\frac{dx_\nu}{d\tau}}\;. \tag{4.1.4}$$

The relation between the four-velocity and the ordinary velocity $v_j = dx^j/dx^0$ is

$$v^j=\frac{u^j}{u^0}\,. \tag{4.1.5}$$

In this chapter we are dealing with a "particle" of fluid at x. Its four-velocity can be found using the observation that in a local rest frame at x the four-velocity is $u^\mu(x)=(1,0,0,0)$. Thus

$$u^\mu(x)=\frac{1}{\rho(x)}J^\mu(x) \tag{4.1.6}$$

in the local rest frame. But since this is a covariant equation, it holds in any reference frame.

By combining (4.1.6) and (4.1.5) we can verify the physically obvious result that the matter current $J^k(x)$ is simply equal to the matter density times the velocity of the fluid:

$$J^k=\rho u^k=\rho u^0\frac{u^k}{u^0}=J^0 v^k. \tag{4.1.7}$$

Entropy

We propose to describe the fluid using the three fields $R_a(x)$. We have seen that the rest frame density ρ can be calculated from $R_a(x)$. But we know from experiment that another variable besides ρ—like the temperature or pressure—must be given in order to specify the internal state of a droplet of fluid. Let us choose to use the entropy per atom, s, as the other thermodynamic variable. (We measure s in a local rest frame of the droplet.)

In this chapter we avoid letting $s(x)$ be a fourth field needed to describe the fluid by considering an "ideal fluid" in which the rest frame entropy of each droplet remains unchanged as the fluid moves along, so that s is a fixed function of R. (Then we can derive a "conservation of entropy" theorem using the conserved current $S^{\mu}=s(R)J^{\mu}$.)

By considering ideal fluids, we neglect entropy increasing effects like heat flow and viscosity. These dissipative effects are considered in Chapter 13.

4.2 FLUID DYNAMICS

Having three fields $R_a(x)$ to describe the fluid, we now need a Lagrangian. In this chapter we are not concerned with how to choose the Lagrangian. We just write down a likely candidate and see how it works. The candidate is recommended by its simplicity*:

$$\mathcal{L}(R_a, \partial_{\mu} R_a) = \mathcal{L}(R_a, \rho). \tag{4.2.1}$$

Here $\mathcal{L}(R_a, \rho)$ can be any function of R_a and the rest density ρ.

We can say the same thing in a more convenient notation by writing

$$\mathcal{L} = -\rho U(\mathcal{V}, R_a), \tag{4.2.2}$$

where

$$\mathcal{V} = 1/\rho \tag{4.2.3}$$

is the volume per atom as measured in a local rest frame. The function U will turn out to be the internal energy per atom of the fluid at R_a when it is compressed to a volume per atom $\mathcal{V}$. We use $\mathcal{V}$ instead of ρ in U because $\mathcal{V}$ is the customary choice of variable in thermodynamics. The possible dependence on R_a is included in order to allow for the possibility that U may depend on the specific entropy $s(R_a)$ and the chemical composition of the fluid at R_a.

*Several action principles to describe relativistic fluid mechanics have appeared in the literature. The original proposal of what was essentially the action given by (4.2.1) was by G. Herglotz, *Ann. Phys.* **36**, 493 (1911). See also A. H. Taub, *Phys. Rev.* **94**, 1468 (1954); K. Tam, *Can. J. Phys.* **44**, 2403 (1966); A. H. Taub, *Commun. Math. Phys.* **15**, 235 (1969); B. Schutz, *Phys. Rev.* **D2**, 2762 (1970); G. A. Maugin, *Ann. Inst. Henri Poincaré* **16**, 133 (1972).

At this point we are in a position to write down the Euler-Lagrange equations of motion:

$$\frac{\partial \mathcal{L}}{\partial R_a} - \partial_\nu \frac{\partial \mathcal{L}}{\partial (\partial_\nu R_a)} = 0. \tag{4.2.4}$$

These equations are apparently three rather complicated second order partial differential equations which determine the motion of the three fields $R_a(x)$. We put aside further discussion of the equations of motion until we have analyzed the energy momentum tensor $T_\mu{}^\nu$ in the next section.

4.3 THE ENERGY-MOMENTUM TENSOR

As long as the fields obey the equations of motion (4.2.4), the energy-momentum tensor

$$T_\mu{}^\nu = g_\mu^\nu \mathcal{L} - (\partial_\mu R_a) \frac{\partial \mathcal{L}}{\partial (\partial_\nu R_a)} \tag{4.3.1}$$

must obey the conservation equation (3.3.5), $\partial_\nu T_\mu{}^\nu = 0$.

In order to understand the physics of $T_\mu{}^\nu$, we will write it out in an explicit and simple form. Straightforward differentiation gives for the second term

$$-(\partial_\mu R_a) \frac{\partial \mathcal{L}}{\partial (\partial_\nu R_a)} = \left[U - \mathcal{V} \frac{\partial U}{\partial \mathcal{V}} \right] (\partial_\mu R_a) \frac{\partial \rho}{\partial (\partial_\nu R_a)}.$$

Recall that $\rho = [-J_\lambda J^\lambda]^{1/2}$, where

$$J^\lambda = n(\mathbf{R}) \frac{1}{3!} \epsilon_{abc} \epsilon^{\lambda\alpha\beta\gamma} (\partial_\alpha R_a)(\partial_\beta R_b)(\partial_\gamma R_c).$$

Thus

$$-(\partial_\mu R_a) \frac{\partial \mathcal{L}}{\partial (\partial_\nu R_a)} = -\left[U - \mathcal{V} \frac{\partial U}{\partial \mathcal{V}} \right] \frac{1}{\rho} J_\lambda$$
$$\times \left\{ \tfrac{1}{2} n \epsilon_{abc} \epsilon^{\lambda\nu\beta\gamma} (\partial_\mu R_a)(\partial_\beta R_b)(\partial_\gamma R_c) \right\}.$$

Consider the factor in braces $\{\ \}$, which we can call $\{F_\mu^{\lambda\nu}\}$. First, notice that this factor vanishes unless $\lambda \neq \nu$, so assume $\lambda \neq \nu$. If $\mu = \nu$, then F equals J^λ. If $\mu = \lambda$, then F equals $-J^\nu$. If $\mu \neq \nu$ and $\mu \neq \lambda$, then in each nonzero term in the sum over β and γ, either β or γ must equal μ; but then the sum over a, b, c vanishes because of the antisymmetry of ϵ_{abc}. Therefore

$$\{F_\mu^{\lambda\nu}\} = g_\mu^\nu J^\lambda - g_\mu^\lambda J^\nu.$$

Thus we have, using $u^\mu = J^\mu/\rho$,

$$-(\partial_\mu R_a)\frac{\partial \mathcal{L}}{\partial(\partial_\nu R_a)} = \left[\rho U - \frac{\partial U}{\partial \mathcal{V}}\right](u_\mu u^\nu + g_\mu^\nu) \tag{4.3.2}$$

and

$$T_\mu{}^\nu = \left[\rho U - \frac{\partial U}{\partial \mathcal{V}}\right]u_\mu u^\nu - g_\mu^\nu \frac{\partial U}{\partial \mathcal{V}}. \tag{4.3.3}$$

Energy and Pressure for Fluid at Rest

What physics does this expression for $T_\mu{}^\nu$ contain? First, let us look at $T^{\mu\nu}$ in a local rest frame of a droplet of fluid. The energy density is

$$T^{00} = \rho U(\mathcal{V}, R_a) \qquad \text{(rest frame)}. \tag{4.3.4}$$

Since T^{00} is the energy per unit volume, $U = T^{00}/\rho$ is the energy per atom of the droplet with label R_a. It includes the rest energy of the atoms.

The momentum density T^{k0} in the rest frame is zero, but the momentum flow T^{kl} is not. We find

$$T^{kl} = -\frac{\partial U}{\partial \mathcal{V}}\delta_{kl} \qquad \text{(rest frame)}. \tag{4.3.5}$$

That is, if we imagine a small surface element with area $d\mathbf{A}$ in the fluid, the fluid on one side of the surface exerts a force $F^k = T^{kl}\,dA_l = -\partial U/\partial \mathcal{V}\,dA_k$ on the fluid that lies on the other side of the surface. This is exactly what we mean when we say that there is a pressure

$$P = -\frac{\partial U}{\partial \mathcal{V}} \tag{4.3.6}$$

in the fluid at point R. We call the substance under discussion a fluid because the only internal forces in it are the pressure forces $T^{kl} = \delta_{kl}P$.

We have been supposing that the specific entropy s of each drop of fluid remains constant as the drop moves along and its volume changes. Thus the partial derivative $\partial U/\partial \mathcal{V}$ is the partial derivative of the internal energy with respect to $\mathcal{V}$ at constant entropy. Therefore, the equation (4.3.6) relating the pressure to the internal energy of the fluid is precisely the first law of thermodynamics as it applies to adiabatic changes of state of a fluid droplet.

Energy and Momentum for a Slowly Moving Fluid

We have found that the momentum tensor in a local rest frame of the fluid is

$$T^{\mu\nu}=\begin{bmatrix} \rho U & 0 & 0 & 0 \\ 0 & P & 0 & 0 \\ 0 & 0 & P & 0 \\ 0 & 0 & 0 & P \end{bmatrix}. \tag{4.3.7}$$

The tensor that reduces to (4.3.7) in the local rest frame is given uniquely by

$$T^{\mu\nu}=(\rho U+P)u^{\mu}u^{\nu}+g^{\mu\nu}P, \tag{4.3.8}$$

which is the momentum tensor (4.3.3) that we obtained directly from the Lagrangian.

We can get a better feeling for this momentum tensor if we write out the nonrelativistic approximation to $T^{\mu\nu}$ and try to understand the physical basis for its various terms. In Chapter 7, we undertake a more thorough investigation of nonrelativistic theories, starting from nonrelativistic Lagrangians. Here we are just interested in the nonrelativistic form of $T^{\mu\nu}$ for the purpose of interpretation, since Newtonian mechanics is more familiar than covariant mechanics.

To derive the nonrelativistic approximation to $T^{\mu\nu}$, we assume that all velocities of interest are small compared to 1 (that is, the speed of light). Thus we can approximate $u^{\mu}=(1,\mathbf{v})[1-\mathbf{v}^2]^{-1/2}$ by $u^{\mu}\cong(1,\mathbf{v})$ and $\rho=J^0[1-\mathbf{v}^2]^{1/2}$ by $\rho\cong J^0$. We must also make this approximation in the energy U. We write

$$U(\mathcal{V},R_a)=m(R_a)+\tilde{U}(\mathcal{V},R_a), \tag{4.3.9}$$

where $m(R_a)$ is the mass per atom: the energy per atom measured with the fluid at rest at some standard density. Then $\tilde{U}(\mathcal{V},R_a)$ is the energy added by compressing the fluid to the density $\mathcal{V}$. Now, $\tilde{U}$ is much smaller than m in ordinary situations; $\tilde{U}/m$ is roughly equal to the square of a typical atomic velocity. Thus we assume

$$\frac{\tilde{U}}{m}\sim(\text{velocity})^2\ll 1. \tag{4.3.10}$$

To find the nonrelativistic approximation to $T^{\mu\nu}$ we simply expand $T^{\mu\nu}$ in powers of $\mathbf{v}^2$ and $(\tilde{U}/m)$ and keep only the lowest order terms.

For the momentum density T^{k0} we find

$$T^{k0}\cong(mJ^0)v^k. \tag{4.3.11}$$

This is just the mass density times the velocity, as we might have expected.

The momentum current is

$$T^{kl} \cong mJ^0 v^k v^l + \delta_{kl} P. \tag{4.3.12}$$

The first term represents a momentum density $mJ^0 v^k$ being carried along with the fluid at velocity v^l. The second term gives the momentum transfer due to the pressure. In conclusion, the nonrelativistic equation representing conservation of momentum is

$$\frac{\partial}{\partial t}(mJ^0 v^k) + \frac{\partial}{\partial x^l}(mJ^0 v^k v^l + \delta_{kl} P) \cong 0. \tag{4.3.13}$$

Now let us look at energy conservation. To lowest order, we find

$$\begin{aligned} T^{00} &\cong mJ^0, \\ T^{0l} &\cong mJ^0 v^l. \end{aligned} \tag{4.3.14}$$

However, the conservation equation

$$\frac{\partial}{\partial t} mJ^0 + \frac{\partial}{\partial x^l}(mJ^0 v^l) = 0 \tag{4.3.15}$$

is really a statement of conservation of mass. We know independently that (rest) mass is exactly conserved in our model because the current $J_m^\mu(x) = m(R_a(x))J^\mu(x)$ obeys $\partial_\mu J_m^\mu = 0$ for any motion of the fluid, whether or not $R_a(x)$ obeys the equations of motion. Thus to see the real content of the energy conservation equation $\partial_\mu T^{0\mu} = 0$ we need to expand $T^{0\mu}$ to one more order:

$$T^{00} \cong mJ^0 + \tfrac{1}{2} m\mathrm{v}^2 J^0 + \tilde{U} J^0, \tag{4.3.16}$$

$$T^{0l} \cong mJ^0 v^l + \left(\tfrac{1}{2} m\mathrm{v}^2 + \tilde{U}\right) J^0 v^l + P v^l. \tag{4.3.17}$$

Now we can recognize in T^{00} the density of kinetic energy $\frac{1}{2} m\mathrm{v}^2 J^0$ and the density of internal energy $\tilde{U} J^0$. In the energy current T^{0l}, the second term represents the energy density $(\frac{1}{2} m\mathrm{v}^2 + \tilde{U}) J^0$ being carried along with the fluid with the velocity v^l. The last term Pv^l represents the rate at which the fluid on one side of an imaginary surface is doing work on the fluid on the other side of the surface because of the pressure.

When we subtract the mass conservation equation $\partial_\mu J_m^\mu = 0$ from the covariant energy conservation equation we obtain

$$0 \cong \frac{\partial}{\partial t}\left[\left(\tfrac{1}{2} m\mathrm{v}^2 + \tilde{U}\right) J^0\right] + \frac{\partial}{\partial x^l}\left[\left(\tfrac{1}{2} m\mathrm{v}^2 + \tilde{U}\right) J^0 v^l + P v^l\right], \tag{4.3.18}$$

which represents nonrelativistic energy conservation.

Energy and Momentum for a Fast Moving Fluid

There is a sense in which the relativistic expression (4.3.8) is so elegant that any further discussion of it can only muddy the waters, but there are two points worthy of special mention. The first is that $T^{\mu\nu} = T^{\nu\mu}$. Thus the energy current T^{0k} equals the momentum density T^{k0}. In later chapters we will have more to say about the symmetry of the momentum tensor.

Second, the energy density T^{00} contains a term $(u^0u^0 - 1)P$ which is proportional to the pressure. Since $u^0u^0 - 1 \sim v^2$ for $v \ll 1$, this term does not appear in the nonrelativistic expression for the energy density. Its presence is entirely a relativistic effect.

4.4 THE INTERNAL ENERGY FUNCTION

We have seen how the dynamics of a perfect fluid is determined once the Lagrangian $\mathcal{L} = -\rho U(\mathcal{V}, R_a)$ is known. But how are we to know $U(\mathcal{V}, R_a)$? This question falls in the domain of statistical mechanics and thermodynamics rather than classical field theory; nevertheless, a few examples may be helpful here.

For a homogenous fluid, the internal energy per atom is a function of the volume per atom, $\mathcal{V}$, and of the entropy per atom s. The entropy may be a function of the material coordinates R_a, but since we are not allowing irreversible processes, this function $s(R_a)$ does not change in time. Thus $U(\mathcal{V}, R_a)$ has the form $U(\mathcal{V}, R_a) = U(\mathcal{V}, s(R_a))$.

In a particular problem, the function $s(R_a)$ would be determined from the initial conditions. If the initial temperature rather than the initial entropy were given, one would compute $s(R_a)$ using $T = \partial U/\partial s$.

For some simple fluids the internal energy function $U(\mathcal{V}, s)$ can be computed from statistical mechanics. Consider, for example, an ideal Fermi gas composed of noninteracting particles of mass m, spin j. We suppose that the particle velocities (measured in the rest frame of the gas) are small, so that the particle energies can be approximated by $E = m + \frac{1}{2}m\mathrm{v}^2$. One then finds from statistical mechanics that $U(\mathcal{V}, s)$ has the form*

$$U(\mathcal{V}, s) = m + \mathcal{V}^{-2/3} F(s). \tag{4.4.1}$$

The function $F(s)$ cannot be written in closed form in terms of well known functions. In the high temperature limit ($s \to \infty$) it is

$$F(s) \cong \left[\frac{3\pi}{m} (2j+1)^{-2/3} e^{-5/3} \right] e^{(2/3)(s/k)}, \tag{4.4.2}$$

*See K. Huang, *Statistical Mechanics* (Wiley, New York, 1963), Chapter 11.

where k is Boltzmann's constant. In the low temperature limit ($s \to 0$),

$$F(s) \cong \left[\frac{3\pi}{10m} \left(\frac{6\pi^2}{2j+1} \right)^{2/3} \right] \left(1 + \frac{5}{3\pi^2} \left(\frac{s}{k} \right)^2 \right). \qquad (4.4.3)$$

If the particles in the last example are very energetic, one can calculate $U(\mathcal{V},s)$ by approximating the particle energies by $E = |\mathbf{p}|$, where $\mathbf{p}$ is the particle momentum. This ultrarelativistic Fermi gas model finds important astrophysical applications in describing the electrons in a white dwarf star and the neutrons in a neutron star. The internal energy function is*

$$U = \mathcal{V}^{-1/3} G(s), \qquad (4.4.4)$$

$$G(s) \cong \left[3 \left(\frac{\pi^2}{2j+1} \right)^{1/3} e^{-4/3} \right] e^{(1/3)(s/k)} \quad \text{for} \quad \frac{s}{k} \gg 1, \qquad (4.4.5)$$

$$G(s) \cong \left[\frac{27\pi^2}{32(2j+1)} \right]^{1/3} \left[1 + \frac{2}{3\pi^2} \left(\frac{s}{k} \right)^2 \right] \quad \text{for} \quad \frac{s}{k} \ll 1. \qquad (4.4.6)$$

For more complicated fluids, a calculation of $U(\mathcal{V},s)$ from first principles is impractical. Instead, one must rely on experimental results to reconstruct $U(\mathcal{V},s)$. For instance if we are interested in the behavior of water near a particular temperature T_0 and pressure P_0, we can use a Taylor expansion for $U(\mathcal{V},s)$ about the corresponding point $\mathcal{V}_0, S_0$:

$$U(\mathcal{V},s) = M_0 + T_0(s - s_0) - P_0(\mathcal{V} - \mathcal{V}_0)$$
$$+ \tfrac{1}{2}A_0(s - s_0)^2 + B_0(s - s_0)(\mathcal{V} - \mathcal{V}_0) + \tfrac{1}{2}C_0(\mathcal{V} - \mathcal{V}_0)^2. \qquad (4.4.7)$$

For $T_0 = 293$ K (room temperature) and $P_0 = 1.01 \times 10^6$ erg/cm^3 (atmospheric pressure), the constants in (4.4.7) are measured to be†

$$\begin{aligned} M_0 &= 18.0 \text{ g/mole}, \\ \mathcal{V}_0 &= 18.0 \text{ cm}^3\text{/mole}, \\ s_0 &= \text{unknown}, \\ A_0 &= 3.93 \times 10^{-7} \text{ mole K}^2\text{/erg}, \\ B_0 &= -1.78 \text{ mole K/cm}^3, \\ C_0 &= 1.22 \times 10^9 \text{ mole erg/cm}^6. \end{aligned} \qquad (4.4.8)$$

*Huang, *loc. cit.*

†The constants in (4.4.8) are derived from data in M. W. Zemanski, *Heat and Thermodynamics*, 4th ed. (McGraw-Hill, New York, 1957), Table 13.2.

Of course, the determination of A_0, B_0, C_0 from experimental data is rather indirect (see Problem 1).

Notice that $C_0 = -\partial P/\partial \mathcal{V}$ is very large compared with $P/\mathcal{V}$: $-\mathcal{V}P^{-1}\partial P/\partial \mathcal{V} \sim 2\times 10^4$. Thus a 1% increase in the pressure on a drop of water decreases its volume by only 0.00005 %. This is the reason that water is often treated in practical problems as if it were incompressible. We discuss the principle of stationary action for "incompressible" materials toward the end of the next chapter.

4.5 USE OF MOMENTUM CONSERVATION EQUATIONS AS EQUATIONS OF MOTION

We have seen that the equation $\partial_\nu T^{\mu\nu} = 0$ representing conservation of momentum and energy are a consequence of the Euler-Lagrange equations of motion for the fluid. However, the physics of the momentum equations is more transparent than that of the Euler-Lagrange equations. Thus it is fortunate that the three momentum conservation equations are equivalent to the three Euler-Lagrange equations, and hence may be used as the equations of motion.

To see why the two sets of equations are equivalent, let us write out $\partial_\nu T_k{}^\nu$, repeating the steps in the proof of momentum conservation in Chapter 3:

$$\begin{aligned}
\partial_\nu T_k{}^\nu &= \partial_\nu \left[g_k^\nu \mathfrak{L} - (\partial_k R_a) \frac{\partial \mathfrak{L}}{\partial(\partial_\nu R_a)} \right] \\
&= (\partial_k R_a) \frac{\partial \mathfrak{L}}{\partial R_a} + (\partial_k \partial_\nu R_a) \frac{\partial \mathfrak{L}}{\partial(\partial_\nu R_a)} \\
&\quad - (\partial_\nu \partial_k R_a) \frac{\partial \mathfrak{L}}{\partial(\partial_\nu R_a)} - (\partial_k R_a) \partial_\nu \frac{\partial \mathfrak{L}}{\partial(\partial_\nu R_a)} \\
&= (\partial_k R_a) \left\{ \frac{\partial \mathfrak{L}}{\partial R_a} - \partial_\nu \frac{\partial \mathfrak{L}}{\partial(\partial_\nu R_a)} \right\}.
\end{aligned} \tag{4.5.1}$$

This equation implies that $\partial_\nu T_k{}^\nu = 0$ whenever the Euler-Lagrange equations

$$\frac{\partial \mathfrak{L}}{\partial R_a} - \partial_\nu \frac{\partial \mathfrak{L}}{\partial(\partial_\nu R_a)} = 0$$

are satisfied. But as long as the transformation $R_a(t, \mathbf{x})$ from the laboratory coordinates x^k to the fluid coordinates R_a is nonsingular, the matrix

$M_{ka} = \partial_k R_a$ will have an inverse, so that we can rewrite (4.5.1) as

$$\frac{\partial \mathcal{L}}{\partial R_a} - \partial_\nu \frac{\partial \mathcal{L}}{\partial(\partial_\nu R_a)} = (M^{-1})_{ak} \partial_\nu T_k{}^\nu. \tag{4.5.2}$$

Thus the Euler-Lagrange equations are satisfied whenever $\partial_\nu T_k{}^\nu = 0$.

This derivation is apparently quite general. It applies to the mechanics of all of the materials we will discuss. But we will need the Euler-Lagrange equations as equations of motion for fields other than R_a when we discuss electromagnetism and gravity.

4.6 THE CIRCULATION THEOREM

A homogeneous fluid has a remarkable invariance property. Imagine two identical beakers of fluid. Stir the fluid in one beaker and then bring it to rest again, without changing the fluid's density or adding any entropy. Then you will be unable to experimentally distinguish the stirred fluid from the unstirred fluid.

We use this invariance property of fluids to prove an important conservation theorem about fluid motion. This theorem is a special application of a general theorem, called Noether's theorem, which connects invariance properties and conservation laws. (Noether's theorem is stated and proved in its full generality in Chapter 9.)

We begin with a mathematical statement of the invariance property just described. Consider a fluid of uniform composition and uniform specific entropy s. In such a homogeneous fluid the internal energy per atom depends only on the volume per atom $\mathcal{V}$ and not on the material coordinates R_a:

$$\mathcal{L}(R_a, \partial_\mu R_a) = -\rho U(\mathcal{V}). \tag{4.6.1}$$

Without loss of generality, we may suppose that the material coordinates R_a have been chosen in such a way that the density $n(\mathbf{R}) \equiv dN/d^3\mathbf{R}$ is a constant, n.

Now suppose that another material coordinate system $\overline{R}_a$ is laid out in the fluid. The new coordinates $\overline{R}_a$ are then functions $\overline{R}_a = F_a(\mathbf{R})$ of the old. (We may imagine that the new coordinates are chosen after stirring the fluid.) Let the coordinates $\overline{R}_a$ be such that the density $dN/d^3\overline{\mathbf{R}}$ measured in the $\overline{\mathbf{R}}$ system is still n; that is,

$$\det\left(\frac{\partial \overline{R}_a}{\partial R_b}\right) = 1. \tag{4.6.2}$$

The Lagrangian is invariant under this change of field variables:

$$\mathcal{L}(\partial_\mu \bar{R}_a(x)) = \mathcal{L}(\partial_\mu R_a(x)) \tag{4.6.3}$$

where $\bar{R}_a(x) = F_a(\mathbf{R}(x))$. To see this, recall that $\mathcal{L}$ depends only on $\mathcal{V} = [-J_\mu J^\mu]^{-1/2}$, so we have only to show that $\bar{\mathcal{V}} = \mathcal{V}$. Consider J^μ in a local rest frame of the material, in which $\bar{J}^k = J^k = 0$; then we need only show that $\bar{J}^0 = J^0$. Using (4.1.2) and the condition (4.6.2) we find

$$\bar{J}^0 = n \det\left(\frac{\partial \bar{R}_a}{\partial x^j}\right) = n \det\left(\frac{\partial \bar{R}_a}{\partial R_b}\right) \det\left(\frac{\partial R_b}{\partial x^j}\right) = J^0.$$

This proves (4.6.3).

We can now use the invariance property (4.6.3) to prove the circulation theorem. Suppose that the two material coordinate systems are nearly identical,

$$\bar{R}_a = R_a + \epsilon \xi_a(\mathbf{R}), \tag{4.6.4}$$

where ϵ is considered infinitesimal. To first order in ϵ, the requirement (4.6.2) is

$$\frac{\partial \xi_a}{\partial R_a} = 0. \tag{4.6.5}$$

The invariance property (4.6.3) says, again to first order in ϵ,

$$\begin{aligned} 0 &= \mathcal{L}(\partial_\mu R_a + \epsilon \partial_\mu \xi_a) - \mathcal{L}(\partial_\mu R_a) \\ &= \epsilon \frac{\partial \mathcal{L}}{\partial(\partial_\mu R_a)}(\partial_\mu \xi_a). \end{aligned}$$

This can be rewritten as

$$0 = \epsilon \partial_\mu J_\xi^\mu - \epsilon \xi_a \left(\partial_\mu \frac{\partial \mathcal{L}}{\partial(\partial_\mu R_a)}\right). \tag{4.6.6}$$

where

$$J_\xi^\mu = \frac{\partial \mathcal{L}}{\partial(\partial_\mu R_a)} \xi_a. \tag{4.6.7}$$

Since $\mathcal{L}$ does not depend on R_a the Euler-Lagrange equations of motion are

$$0 = \partial_\mu \frac{\partial \mathcal{L}}{\partial(\partial_\mu R_a)}. \tag{4.6.8}$$

Suppose now that the fields $R_a(x)$ obey these equations of motion. Then $\partial_\mu J_\xi^\mu = 0$, so the quantity

$$Q_\xi(t) = \int d\mathbf{x} J_\xi^0(\mathbf{x}, t) \tag{4.6.9}$$

is conserved. This is a remarkably powerful conservation law because there is one conserved quantity for each displacement function $\xi_a(\mathbf{R})$ which obeys the divergence condition $\partial \xi_a / \partial R_a = 0$.

The content of the conservation law can be displayed in an intuitively appealing form by choosing for the displacement function $\xi_a(\mathbf{R})$ a function that displaces the material coordinates around a loop drawn in the fluid. Let this loop be described by

$$R_a = \Omega_a(\tau) \qquad 0 \leqslant \tau \leqslant 1,$$

$$\Omega_a(0) = \Omega_a(1).$$

Let $\xi_a(\mathbf{R})$ be the function

$$\xi_a(\mathbf{R}) = -(mn)^{-1} \int_0^1 d\tau \frac{d\Omega_a}{d\tau} \delta^3(\mathbf{R} - \Omega(\tau)),$$

where $\delta^3(\mathbf{R} - \Omega)$ is a Dirac delta function*. This function $\xi_a(R)$ obeys the divergence condition (4.6.5):

$$\begin{aligned}
\frac{\partial \xi_a}{\partial R_a} &= -(mn)^{-1} \int_0^1 d\tau \frac{d\Omega_a}{d\tau} \frac{\partial}{\partial R_a} \delta^3(\mathbf{R} - \Omega(\tau)) \\
&= (mn)^{-1} \int_0^1 d\tau \frac{d}{d\tau} \delta^3(\mathbf{R} - \Omega(\tau)) \\
&= (mn)^{-1} \left[\delta^3(\mathbf{R} - \Omega(1)) - \delta^3(\mathbf{R} - \Omega(0)) \right] \\
&= 0.
\end{aligned}$$

The conserved quantity corresponding to this displacement function is

$$Q(t) = -(mn)^{-1} \int_0^1 d\tau \int d\mathbf{x} \frac{\partial \mathcal{L}}{\partial(\partial_0 R_a)} \frac{\partial \Omega_a}{d\tau} \delta^3(\mathbf{R} - \Omega(\tau)).$$

It is a straightforward exercise to bring this expression into tractable form. The $\mathbf{x}$ integral can be done by changing integration variables from $\mathbf{x}$ to $\mathbf{R}$

*For our purpose, $\delta^3(\mathbf{R} - \Omega)$ could be any function that is sharply peaked at $\mathbf{R} - \Omega = 0$ and obeys $\int d\mathbf{R} \delta(\mathbf{R} - \Omega) = 1$.

and using the delta function:

$$Q(t) = -(mn)^{-1}\int_0^1 d\tau \frac{d\Omega_a}{d\tau}\left[\frac{\partial(\mathbf{x})}{\partial(\mathbf{R})}\frac{\partial \mathcal{L}}{\partial(\partial_0 R_a)}\right]_{\mathbf{x}=\mathbf{x}(\Omega(\tau))}.$$

Now we can insert a factor $\delta_{ab} = (\partial R_a/\partial x^j)(\partial x^j/\partial R_b)$:

$$Q(t) = -m^{-1}\int_0^1 \frac{\partial x^j}{\partial R^b}\frac{\partial \Omega_b}{d\tau}d\tau\left[n\frac{\partial(\mathbf{R})}{\partial(\mathbf{x})}\right]^{-1}\frac{\partial \mathcal{L}}{\partial(\partial_0 R_a)}(\partial_j R_a).$$

The first three factors in the intrgrand can be written as $(d\mathbf{x}/d\tau)d\tau = d\mathbf{x}$. The Jacobean factor is $n\,\partial(\mathbf{R})/\partial(\mathbf{x}) = J^0 = \rho u^0$. The rest of the integrand was evaluated in (4.3.2):

$$-\frac{\partial \mathcal{L}}{\partial(\partial_0 R_a)}(\partial_j R_a) = \rho[U + P\mathcal{V}]u^0 u^j.$$

Thus we have

$$Q(t) = \oint dx^j m^{-1}[U + P\mathcal{V}]u^j \tag{4.6.10}$$

Here $\oint dx^j \cdots$ signifies the line integral around the loop. The loop moves with the fluid. The conserved quantity Q is called the (relativistic) circulation around the loop.

In the case of ordinary terrestrial fluids, Q is well approximated by its nonrelativistic limit. We set $u^j = v^j[1 - v^2]^{-1/2} \cong v^j$ and $U + P\mathcal{V} \cong m$, as we did in Section 4.3 in discussing $T^{\mu\nu}$. This gives

$$Q(t)_{\text{nonrel}} = \oint d\mathbf{x}\cdot\mathbf{v}. \tag{4.6.11}$$

One can also derive the nonrelativistic version of the circulation theorem by first making the nonrelativistic approximation in the Lagrangian, as we will do in Chapter 7, then deriving the conservation of Q_{nonrel} as an exact consequence of the nonrelativistic equations of motion.

The present proof of the circulation theorem also establishes the theorem under the more general hypothesis that the specific entropy $s(\mathbf{R})$ is *not* constant throughout the fluid, but that the loop under consideration lies in a surface of constant s.

4.7 SOUND WAVES

Of all the dynamically possible motions of the fluid, the simplest is that in which a uniform fluid just sits there and does nothing. In that situation, we can choose the fluid coordinates R_a to be identical with the laboratory coordinates: $R_a(t,\mathbf{x})=x^a$. Also, because the fluid is uniform, we have $n(\mathbf{R})=n$, $U(\mathcal{V},R_a)=U(\mathcal{V})$.

What happens if this uniform fluid is now disturbed slightly? To find out, we suppose that

$$R_a(t,\mathbf{x})=x^a+\phi_a(t,\mathbf{x}), \tag{4.7.1}$$

where $\phi_a(x)$ is a small deviation from the simple motion $\mathbf{R}=\mathbf{x}$. Then we write the equations of motion as equations for ϕ, using the assumed smallness of ϕ to justify neglecting terms of order ϕ^2, ϕ^3, etc.

We will use the momentum conservation equations $\partial_\mu T^{k\mu}=0$ as the equations of motion, so we will need expressions for T^{k0} and T^{kl} to lowest order in ϕ. The calculation is quite straightforward. Using $J^\mu = n\epsilon^{\mu\alpha\beta\gamma}(\partial_\alpha R_1)(\partial_\beta R_2)(\partial_\gamma R_3)$ we find

$$\begin{aligned} J^0 &\cong n(1+\partial_a\phi_a), \\ J^k &\cong -n(\partial_0\phi_k), \\ \rho &\cong \sqrt{-J_\mu J^\mu} = n(1+\partial_a\phi_a), \end{aligned} \tag{4.7.2}$$

Thus, using ρ_0 for n and $\mathcal{V}_0$ for n^{-1},

$$\begin{aligned} T^{k0} &= (\rho U(\mathcal{V})+P(\mathcal{V}))\frac{J^k J^0}{\rho^2} \\ &\cong -(\rho_0 U(\mathcal{V}_0)+P(\mathcal{V}_0))(\partial_0\phi_k), \end{aligned} \tag{4.7.3}$$

$$\begin{aligned} T^{kl} &= (\rho U(\mathcal{V})+P(\mathcal{V}))\frac{J^k J^l}{\rho^2}+\delta_{kl}P(\mathcal{V}) \\ &\cong \delta_{kl}\left(P(\mathcal{V}_0)-\left(\frac{dP}{d\mathcal{V}}\right)_{\mathcal{V}=\mathcal{V}_0}\mathcal{V}_0\,\partial_a\phi_a\right) \end{aligned} \tag{4.7.4}$$

Thus the approximate equation of motion is

$$\begin{aligned} 0 = &-\left[\rho_0 U(\mathcal{V}_0)+P(\mathcal{V}_0)\right](\partial_0\partial_0\phi_k) \\ &-\mathcal{V}_0 P'(\mathcal{V}_0)(\partial_k\partial_a\phi_a). \end{aligned} \tag{4.7.5}$$

This is the differential equation for sound waves in the fluid. Its simplest solutions are traveling plane waves

$$\phi_a(t,\mathbf{x}) = \phi_a e^{i(\mathbf{k}\cdot\mathbf{x}-\omega t)} \tag{4.7.6}$$

with

$$[\rho_0 U_0 + P_0]\omega^2\phi_j = -\mathcal{V}_0 P'(\mathcal{V}_0)k_j k_a \phi_a. \tag{4.7.7}$$

Apparently the only solutions with $\omega \neq 0$ are "longitudinal" waves in which the vibration direction is the same as the direction of propagarion: $\phi_j = (\text{const})k_j$. In this case ω is related to k by $(\rho_0 U_0 + P_0)\omega^2 = -\mathcal{V}_0 P_0' k^2$. Thus the speed of propagation of these waves is

$$v^2 = \frac{\omega^2}{k^2} = \frac{-\mathcal{V}(dP/d\mathcal{V})}{\rho U + P}. \tag{4.7.8}$$

In normal circumstances the energy per atom U is well approximated by the average atomic mass m, and $\rho m \gg P$. In this approximation we have

$$v^2 \cong \frac{+\rho(dP/d\rho)}{\rho m} = \frac{dP}{d(\rho m)}. \tag{4.7.9}$$

That is, v^2 is the derivative of the pressure with respect to the mass density (evaluated at constant entropy per unit mass.) The relativistic generalization of this familiar law can be obtained by noting that

$$U + \mathcal{V}P = U - \mathcal{V}\frac{dU}{d\mathcal{V}} = U + \rho\frac{dU}{d\rho} = \frac{d}{d\rho}(\rho U)$$

so

$$v^2 = \frac{dP/d\rho}{d(\rho U)/d\rho} = \frac{dP}{d(\rho U)}. \tag{4.7.10}$$

That is, v^2 is the derivative of the pressure with respect to the energy density.

If, instead of longitudinal waves, we choose transverse waves with $\mathbf{k}\cdot\boldsymbol{\phi} = 0$ we find that $\omega = 0$. This means that the stationary fluid solution $R_a = x^a$ is not really stable against perturbations of the form $\phi_a(t,\mathbf{x}) = f(t)\phi_a e^{i\mathbf{k}\cdot\mathbf{x}}$ with $\mathbf{k}\cdot\boldsymbol{\phi} = 0$. The most general time dependence $f(t)$ of such a disturbance is $f(t) = \alpha + vt$ instead of $f(t) = \alpha e^{-i\omega t}$. In real fluids, of course, such free flowing motions are eventually damped out by viscous forces which we are ignoring in this chapter.

PROBLEMS

1. Use the data (4.4.8) for water at room temperature and atmospheric pressure to compute:
   (a) The velocity of sound.
   (b) The volume expansivity $\beta = \frac{1}{\mathcal{V}}\left(\frac{\partial \mathcal{V}}{\partial T}\right)_{P=\text{const.}}$
   (c) The isothermal compressibility $k=(1/\rho)(\partial\rho/\partial P)_{T=\text{const}}$.
   (d) The heat capacity at constant pressure

$$C_P = T\left(\frac{\partial s}{\partial T}\right)_{P=\text{const}}.$$

CHAPTER 5

The Mechanics of Elastic Solids

We have seen how the equations of motion of a fluid can be derived and interpreted (and applied to the simple problem of sound waves). In this chapter we use similar methods to discuss a wider class of materials that includes fluids as a special case. These materials are usually called elastic materials. The word elastic implies the absence of such irreversible processes as heat flow and the permanent deformation of a copper wire when it is bent. Thus we seek to describe the mechanical behavior of simple solids like steel, salt crystals, rubber and Jello, but do not try to include putty or a loaf of bread.

5.1 THE LAGRANGIAN AND THE DEFORMATION MATRIX

We can describe the kinematically possible motions of a solid by using the three scalar fields $R_a(x)$ that we used to describe fluids. What we need is a different Lagrangian. Let us look at a general class of Lagrangians and see what sorts of dynamics it describes.

Suppose that the Lagrangian depends on the fields R_a and their first derivatives $\partial_\mu R_a$, but not on higher derivatives. Since we want a covariant theory, suppose also that $\mathcal{L}$ is a Lorentz scalar. Then the form of the dependence of $\mathcal{L}$ on the three vectors $\partial_\mu R_1, \partial_\mu R_2, \partial_\mu R_3$ is limited: $\mathcal{L}$ can depend only on the six scalar products

$$G_{ab} \equiv (\partial_\mu R_a)(\partial^\mu R_b). \qquad (5.1.1)$$

Thus we consider Lagrangians of the general form*

$$\mathcal{L}(G_{ab}, R_a).$$

*G. Herglotz first proposed the variational principle $0 = \delta \int dx\, \mathcal{L}(G_{ab}, R_a)$, using slightly different variables, in 1911; see G. Herglotz, *Ann. Phys.* **36**, 493 (1911). For a more recent variational principle for elastic solids (or elastic dielectric solids) see H. G. Schöpf, *Ann. Phys.* **9**, 301 (1962); Y. Yu, *J. Accoust. Soc. Am.* **36**, 111 (1964); R. A. Grot, *J. Math Phys.* **11**, 109 (1970); G. A. Maugin and A. C. Eringen, *J. Math. Phys.* **13**, 1777 (1972).

In Chapter 6 we look at some important special cases from this class of Lagrangians.

The Deformation Matrix

Before we begin the discussion of the equations of motion and the momentum tensor, it is important to understand what information about the state of the material is contained in the matrix G_{ab}. Consider G_{ab} in a local rest frame at the point x. In such a frame, $\partial_0 R_a = 0$ at x, so that

$$G_{ab} = \frac{\partial R_a}{\partial x^k}\frac{\partial R_b}{\partial x^k}. \tag{5.1.2}$$

The material coordinates of the "particle" of material located at the position $\mathbf{x}$ at the time t in question are $R_a(t,\mathbf{x})$. The material coordinates of particles located at neighboring positions are given approximately by the derivatives $\partial R_a/\partial x^k$ at x:

$$R_a(t,\mathbf{x}+\delta\mathbf{x}) \sim R_a(t,x) + \frac{\partial R_a}{\partial x^k}\delta x^k.$$

In this sense the derivatives $\partial R_a/\partial x^k$ may be said to describe the displacement of the material near the point x.

Apparently G_{ab} is determined from $\partial R_a/\partial x^k$; but not all of the information in $\partial R_a/\partial x^k$ can be recovered from G_{ab}. Suppose that we deform the material from $R_a(\mathbf{y}) = y^a$ to $R_a(\mathbf{y}) = F_a(\mathbf{y})$, then rotate it with a rotation $y^l = R_{lk}x^k$. The result is $R_a(x^k) = F_a(R_{lk}x^k)$, so that

$$\frac{\partial R_a}{\partial x^k} = \frac{\partial F_a}{\partial y^l}R_{lk}. \tag{5.1.3}$$

But the final rotation has no effect on G_{ab}:

$$G_{ab} = \frac{\partial F_a}{\partial y^l}R_{lk}R_{jk}\frac{\partial F_b}{\partial y^j} = \frac{\partial F_a}{\partial y^l}\frac{\partial F_b}{\partial y^l}. \tag{5.1.4}$$

Conversely, if we know G_{ab} at a point we can recover $\partial R_a/\partial x_k$ up to an unknown final rotation. Since $\mathbf{G}$ has the form $\mathbf{G} = \mathbf{M}\mathbf{M}^T$, it is a symmetric matrix and it is positive (that is, all of its eigenvalues are positive). Thus it has a positive symmetric square root $\sqrt{\mathbf{G}}$ such that $\sqrt{\mathbf{G}}\,\sqrt{\mathbf{G}} = \mathbf{G}$. Let us show that $M_{ak} = \partial R_a/\partial x^k$ has the form

$$\mathbf{M} = \sqrt{\mathbf{G}}\,\mathbf{R},$$

where $\mathbf{R}$ is a rotation matrix. To prove this assertion, define $\mathbf{R} = \sqrt{\mathbf{G}}^{\,-1}\mathbf{M}$

and calculate $\mathbf{RR}^T$:

$$\mathbf{RR}^T = \sqrt{\mathbf{G}}^{-1}\mathbf{MM}^T\sqrt{\mathbf{G}}^{-1} = \sqrt{\mathbf{G}}^{-1}\mathbf{G}\sqrt{\mathbf{G}}^{-1} = \mathbf{1}.$$

Furthermore, $\det\mathbf{R} = [\det\mathbf{G}]^{-1/2}\det\mathbf{M} > 0$ as long as $\det\mathbf{M} > 0$. Thus $\mathbf{R}$ is indeed a rotation matrix.

In summary, G_{ab} contains precisely that information about the local displacement of the material which is not affected by a final rotation.

If we imagine that the material is made of "atoms" connected by "springs," the matrix $\mathbf{G}$ tells us how much the springs have been stretched. The distance between an atom with coordinates R_a and a neighboring atom with coordinates $R_a + \delta R_a$ is determined directly in terms of the inverse $\mathbf{G}^{-1}$ of $\mathbf{G}$:

$$(\delta\mathbf{x})^2 = \frac{\partial x^k}{\partial R_a}\delta R_a \frac{\partial x^k}{\partial R_b}\delta R_b$$

$$= (\mathbf{G}^{-1})_{ab}\delta R_a \delta R_b. \tag{5.1.5}$$

For this reason, $\mathbf{G}$ is often called the deformation matrix. (The reader should be warned that several closely related matrices go under the name of "deformation matrix" or "deformation tensor" in the literature. The present matrix $\mathbf{G}$ was introduced by G. Piola* at the early date of 1833, but its inverse, called Green's deformation tensor, is more popular now. Readers familiar with Riemanian geometry will recognize that the matrix elements G_{ab} are the contravariant components of the metric tensor in the material coordinate system.)

The Deformation Matrix and the Density

We can connect this discussion of the deformation matrix with our previous discussion of fluids by noting that the density of the material in a local rest frame, $\rho = \sqrt{-J_\mu J^\mu}$ is simply related to the determinant of $\mathbf{G}$. In a local rest frame we find

$$\rho = n\det\left(\frac{\partial R_a}{\partial x^j}\right) = n\sqrt{\det\frac{\partial R_a}{\partial x^j}\frac{\partial R_b}{\partial x^j}},$$

or

$$\rho = n\sqrt{\det\mathbf{G}}. \tag{5.1.6}$$

Since both sides of this equation are scalars, the equality holds in any

*G. Piola, 1833, *La meccanica de'corpi naturalamente estesi trattata col calcolo delle variazioni; Opusc. mat. fis. di deversi autori,* Vol. 1 (Giusti, Milano, 1833), pp. 201–236.

frame. Thus the fluid discussed in the preceding chapter is a special case of the class of materials discussed here: if $\mathcal{L}(G_{ab}, R_a)$ depends only on $\det\mathbf{G}$ and R_a, we recover the case of the fluid.

Some Examples

We can get a better feeling for the deformation matrix by working through the calculation of $\mathbf{G}$ for some simple deformations of a two dimensional solid.

1. Uniform compression:

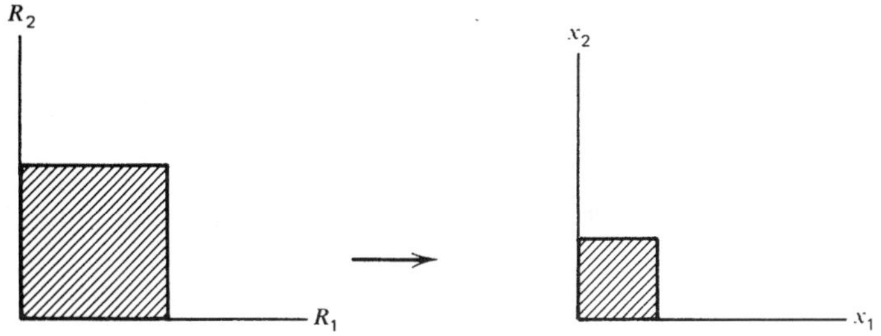

$$\begin{pmatrix} R_1 \\ R_2 \end{pmatrix} = \begin{pmatrix} e^{+\omega} & 0 \\ 0 & e^{+\omega} \end{pmatrix} \begin{pmatrix} x^1 \\ x^2 \end{pmatrix},$$

$$\mathbf{G} = \begin{pmatrix} e^{+2\omega} & 0 \\ 0 & e^{+2\omega} \end{pmatrix}.$$

2. Squeeze along R_1 axis with no compression:

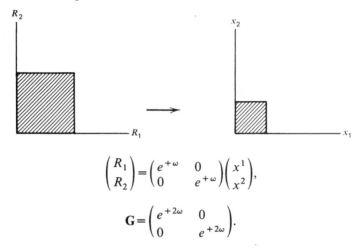

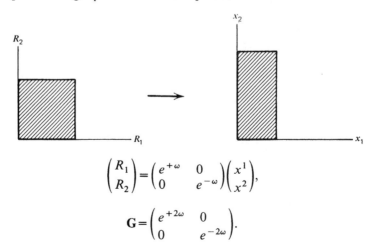

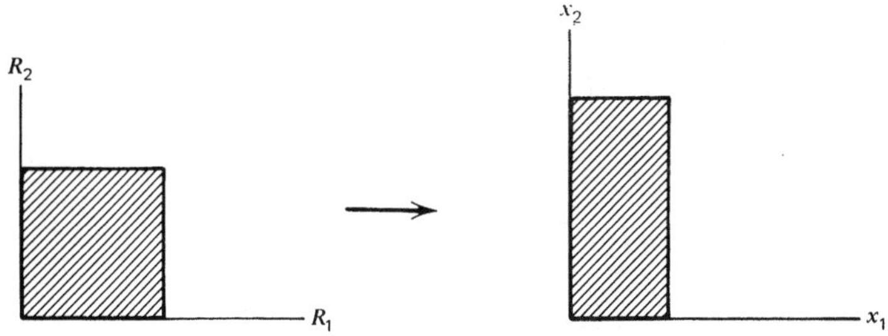

$$\begin{pmatrix} R_1 \\ R_2 \end{pmatrix} = \begin{pmatrix} e^{+\omega} & 0 \\ 0 & e^{-\omega} \end{pmatrix} \begin{pmatrix} x^1 \\ x^2 \end{pmatrix},$$

$$\mathbf{G} = \begin{pmatrix} e^{+2\omega} & 0 \\ 0 & e^{-2\omega} \end{pmatrix}.$$

3. Squeeze along axis inclined at an angle Θ, with no compression:

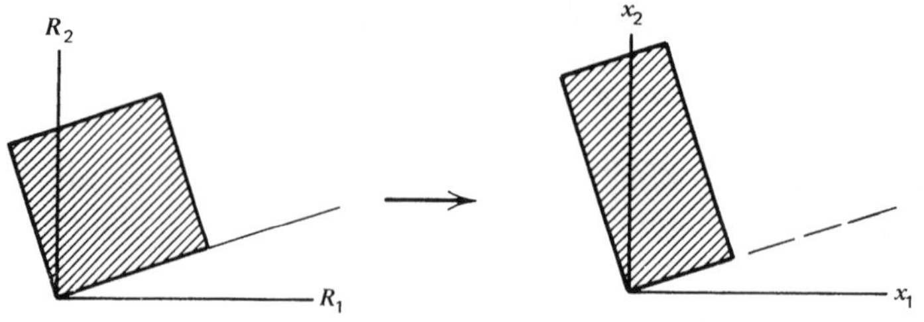

$$\begin{pmatrix} R_1 \\ R_2 \end{pmatrix} = \begin{pmatrix} \cos\theta & -\sin\theta \\ \sin\theta & \cos\theta \end{pmatrix} \begin{pmatrix} e^{+\omega} & 0 \\ 0 & e^{-\omega} \end{pmatrix} \begin{pmatrix} \cos\theta & \sin\theta \\ -\sin\theta & \cos\theta \end{pmatrix} \begin{pmatrix} x^1 \\ x^2 \end{pmatrix},$$

$$\mathbf{G} = \begin{pmatrix} \cosh 2\omega + \sinh 2\omega \cos 2\theta & \sinh 2\omega \sin 2\theta \\ \sinh 2\omega \sin 2\theta & \cosh 2\omega - \sinh 2\omega \cos 2\theta \end{pmatrix}.$$

4. Shear along the R_1 axis:

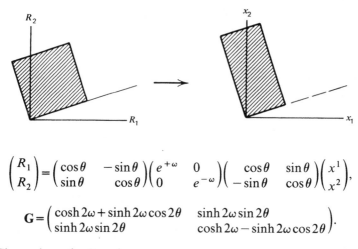

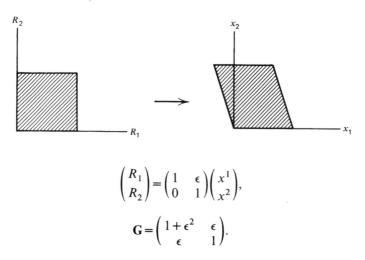

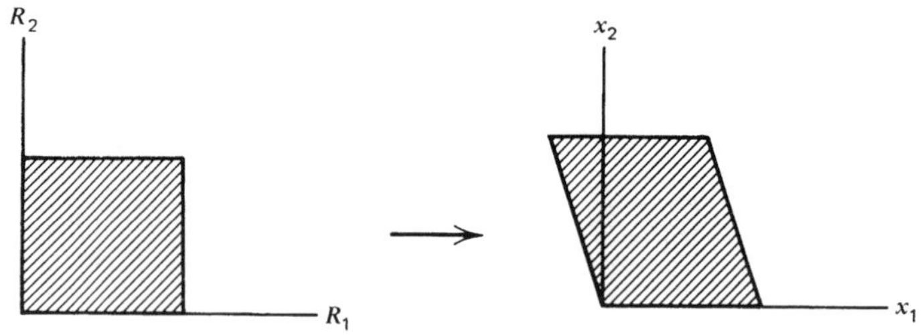

$$\begin{pmatrix} R_1 \\ R_2 \end{pmatrix} = \begin{pmatrix} 1 & \epsilon \\ 0 & 1 \end{pmatrix} \begin{pmatrix} x^1 \\ x^2 \end{pmatrix},$$

$$\mathbf{G} = \begin{pmatrix} 1+\epsilon^2 & \epsilon \\ \epsilon & 1 \end{pmatrix}.$$

Note that det $\mathbf{G}=1$ in Example 4, indicating that the density of the material is not changed when it is sheared. The strain matrix produced by this shear can also be produced by a properly chosen compressionless squeeze as in Example 3. (To be precise, the proper choice is $\cot 2\theta = \sinh\omega = \epsilon/2$). The functions $R_a(\mathbf{x})$ are different in these two cases, even though the deformation matrices are the same. In Example 3, the matrix relating $\mathbf{R}$ and $\mathbf{x}$ is symmetric; it is the square root of $\mathbf{G}$. The matrix relating $\mathbf{R}$ and $\mathbf{x}$ in Example 4 is related to $\sqrt{\mathbf{G}}$ by a final rotation.

5.2 THE MOMENTUM TENSOR

We come now to the dynamics of the elastic material. The Euler-Lagrange equations of motion corresponding to a Lagrangian of the form $\mathcal{L}(G_{ab}, R_a)$ are

$$\frac{\partial \mathcal{L}}{\partial R_a} - 2\partial_\mu\left[\frac{\partial \mathcal{L}}{\partial G_{ab}}\partial^\mu R_b\right] = 0. \tag{5.2.1}$$

(Here and in the following, we assume that the dependence of $\mathcal{L}$ on the symmetric matrix $\mathbf{G}$ is written in a symmetric fashion, so that $\partial \mathcal{L}/\partial G_{ab} = \partial \mathcal{L}/\partial G_{ba}$.)

As we found when we discussed the special case of fluid dynamics, the physics of energy-momentum conservation is more transparent than the physics of the Euler-Lagrange equations. Furthermore, the three momentum conservation equations $\partial_\nu T^{k\nu} = 0$ in any fixed reference frame are equivalent to the three Euler-Lagrange equations (5.2.1). (The proof is the same as that given in Section 4.5.) Therefore we concentrate our attention on the energy-momentum tensor $T^{\mu\nu}$.

The expression we obtain for $T^{\mu\nu}$ is easiest to interpret if we write $\mathcal{L}$ in the form

$$\mathcal{L}(G_{ab}, R_a) = -\rho U(G_{ab}, R_a), \tag{5.2.2}$$

where $\rho = n\sqrt{\det \mathbf{G}} = \sqrt{-J_\mu J^\mu}$ is the rest frame density. It is a straightforward matter to calculate $T^{\mu\nu}$ from the general formula $T_\mu{}^\nu = g_\mu^\nu \mathcal{L} - (\partial_\mu R_a)\partial \mathcal{L}/\partial(\partial_\nu R_a)$ (see Section 4.3). One finds

$$T^{\mu\nu} = \rho U u^\mu u^\nu + 2\rho(\partial^\mu R_a)(\partial^\nu R_b)\frac{\partial U}{\partial G_{ab}}, \tag{5.2.3}$$

where $u^\mu = J^\mu/\rho$ is the four-velocity of the material.

In a local rest frame of the material the energy density is ρU. Thus U can be identified as the internal energy per atom. There is no energy flow or momentum density in the local rest frame. The momentum flow is*

$$T^{kl} = 2\rho\frac{\partial U}{\partial G_{ab}}(\partial_k R_a)(\partial_l R_b) \qquad \text{(rest frame).} \tag{5.2.4}$$

One can think of this as a momentum flow tensor $\bar{T}_{ab} = 2\rho\partial U/\partial G_{ab}$ in the material coordinate system, transformed to the laboratory coordinates by

*In books on elasticity theory or continuum mechanics, one encounters the "stress tensor" S^{kl}, which is identical to the momentum flow tensor T^{lk} in a material at rest, except for sign: $S^{kl} = -T^{lk}$.

means of the matrix $\partial R_a/\partial x^k$, but this interpretation need be only suggestive. What is important is that the stress in the material is proportional to the rate of variation of the internal energy as the deformation G_{ab} is changed. As an example, we recall that the pressure in a fluid is $-\partial U/\partial \mathcal{V}$.

Energy Conservation

The fact that the rest frame momentum flow tensor is proportional to $\partial U/\partial G_{ab}$ is directly related to conservation of energy. Suppose that $T^{\mu\nu}$ had the form

$$T^{\mu\nu} = \rho U(G_{ab}, R_a)u^\mu u^\nu + (\partial^\mu R_a)(\partial^\nu R_b)\overline{T}_{ab}, \tag{5.2.5}$$

but that $\overline{T}_{ab}$ were unknown. We might then consider the equation $u_\mu \partial_\nu T^{\mu\nu} = 0$, which expresses energy conservation in a local rest frame. Using the form (5.2.5) and the kinematic results $\partial_\nu(\rho u^\nu)=0$, $u_\mu \partial_\nu u^\mu = \frac{1}{2}\partial_\nu(u_\mu u^\mu)=0$ and $u_\mu \partial^\mu R_a = 0$, this equation reads

$$\begin{aligned} 0 &= -\rho u^\nu \partial_\nu U + u^\mu(\partial_\nu \partial_\mu R_a)(\partial^\nu R_a)\overline{T}_{ab} \\ &= -\rho u^\nu \partial_\nu U + \tfrac{1}{2}(u^\mu \partial_\mu G_{ab})\overline{T}_{ab}. \end{aligned} \tag{5.2.6}$$

Using the chain rule to differentiate U, and remembering that $u^\nu \partial_\nu R_a = 0$, we would have

$$0 = \left[\tfrac{1}{2}\overline{T}_{ab} - \rho \frac{\partial U}{\partial G_{ab}}\right] u^\nu \partial_\nu G_{ab}. \tag{5.2.7}$$

Thus we would be led to identify $\overline{T}_{ab}$ with $2\rho\,\partial U/\partial G_{ab}$ on the basis of energy conservation.

5.3 BOUNDARY CONDITIONS

The action principle presented above and in the preceding chapter is applicable to materials that fill the whole of space. It is a simple matter to extend the formalism to cover the mechanics of finite material bodies. In this section we consider bodies that are free to float through space and bodies whose boundary surfaces are rigidly fixed or are constrained to slide over another surface. The relevant boundary conditions really need no derivation, since they can be written down immediately on the basis of "common sense" or "physical intuition." Nevertheless it is interesting to see how the correct boundary conditions arise from the principle of stationary action.

Free Boundary

Consider first a solid body that is free to float in space (as in the one dimensional example in Section 2.2). We may suppose that the surface of the body is described by an equation of the form $F(R_a)=0$, with $F(R_a)>0$ defining the inside of the body. The action that describes the dynamics of such a body consists of the usual Lagrangian density integrated over the interior of the body only:

$$A=-\int d^4x\Theta(F(R_a(x)))\rho U(G_{ab},R_a). \tag{5.3.1}$$

Here $\Theta(\tau)$ is the unit step function: $\Theta(\tau)=0$ for $\tau<0$, $\Theta(\tau)=1$ for $\tau>0$.

When we calculate the variation of this action corresponding to a variation $\delta R_a(x)$ of the fields, surface terms will arise both from the variation of the region of integration as $R_a(x)$ is varied and also from integration by parts. These surface terms can be treated expeditiously by using the relation

$$\frac{d}{d\tau}\Theta(\tau)=\delta(\tau), \tag{5.3.2}$$

where $\delta(\tau)$ is the Dirac delta function. We find, after an integration by parts,

$$\begin{aligned}\delta A=&-\int d^4x\Theta(F)\left\{\frac{\partial(\rho U)}{\partial R_a}-\partial_\nu\frac{\partial(\rho U)}{\partial(\partial_\nu R_a)}\right\}\delta R_a\\ &-\int d^4x\left[\delta(F)\frac{\partial F}{\partial R_a}\delta R_a\rho U\right.\\ &\left.-\delta(F)\frac{\partial F}{\partial R_b}(\partial_\nu R_b)\frac{\partial(\rho U)}{\partial(\partial_\nu R_a)}\delta R_a\right].\end{aligned} \tag{5.3.3}$$

The first term leads to the usual equations of motion (5.2.1) in the interior of the body and need not concern us here. Because of the factor $\delta(F)$, the second term contains contributions only from the surface $F(R_a(x))=0$ in space-time swept out by the surface of the body. After a little algebra, it can be written

$$\delta A_{\text{surface}}=\int d^4x\left[\delta(F)\frac{\partial F}{\partial R_c}\partial_\nu R_c\right]2\rho(\partial^\nu R_b)\frac{\partial U}{\partial G_{ab}}\delta R_a. \tag{5.3.4}$$

This is an integral over the surface $F=0$. It can be rewritten in the more familiar notation for a surface integral that employs the differential surface area vector dS_ν, defined in (3.2.9). Using the relation

$$\int d^4x\,\delta(F)(\partial_\nu F)\cdots=\int_{F(R(x))=0} dS_\nu\cdots \tag{5.3.5}$$

(see Problem 2), we have

$$\delta A_{\text{surface}}=\int dS_\nu 2\rho(\partial^\nu R_b)\frac{\partial U}{\partial G_{ab}}\delta R_a. \tag{5.3.6}$$

From the principle of stationary action, we know that $\delta A=0$ for all δR. By choosing $\delta R=0$ except near a point on the surface, we conclude that

$$2\rho(\partial^\nu R_b)\frac{\partial U}{\partial G_{ab}}dS_\nu=0$$

on the surface. We multiply this equation by $(\partial^\mu R_a)$ and add to it the kinematic identity

$$\rho U u^\mu u^\nu dS_\nu=0,$$

which is true because u^ν lies in the surface, while dS_ν is the normal vector to that surface. We obtain

$$0=\left[\rho U u^\mu u^\nu+2\rho(\partial^\mu R_a)(\partial^\nu R_b)\frac{\partial U}{\partial G_{ab}}\right]dS_\nu$$

or

$$0=T^{\mu\nu}dS_\nu \tag{5.3.7}$$

on the surface.

This is a sensible boundary condition. In a local rest frame of a particle of material at the surface of the body, $dS_0=0$ and dS_k is a vector normal to the surface, so (5.3.7) says that no momentum flows into the body.

The alert reader may have noted that the boundary condition (5.3.7) could have been derived with less fuss by incorporating the factor $\Theta(F(R_a))$ into $n(R_a)$, the density of the material in the R-coordinate system. Then the action

$$A=-\int d^4x\rho U(G_{ab},R_a),$$

with $\rho=n\sqrt{\det \mathbf{G}}$, is the same as (5.3.1). But now no special derivation is needed to take the boundary into account; the results of our previous derivations are correct as they stand. We have only to remember the implicit delta functions when we interpret the results. For instance, the momentum tensor $T^{\mu\nu}$ contains an implicit factor $\Theta(F)$; thus the momentum conservation equation contains a delta function:

$$0=\partial_\nu(\Theta(F)T^{\mu\nu})=\Theta(F)\partial_\nu T^{\mu\nu}+\delta(F)(\partial_\nu F)T^{\mu\nu}.$$

Inside the material, this is just the differential equation $\partial_\nu T^{\mu\nu}=0$. But on the surface it implies the boundary condition

$$T^{\mu\nu}(\partial_\nu F)=0,$$

which is the same as (5.3.7).

In those cases, to be discussed presently, where the boundary is not completely free to move, momentum is not conserved and this simplified derivation of the boundary conditions does not work.

Clamped Boundary

Suppose that we are interested in the motion of a material body whose boundary is attached to an unyielding support. The supporting structure may either be stationary or may execute some prescribed motion. For instance, one might glue a rectangular piece of rubber into a closed steel box and ask for the motion of the rubber when the box is shaken.

We again let the surface of the material be defined by an equation $F(R_a)=0$, with $F(R_a)>0$ being the interior of the material. Similarly, the surface of the supporting material is described by an equation $g(t,\mathbf{x})=0$. The boundary condition is that $R_a(x)$ is a prescribed function on the boundary:

$$R_a(x)=G_a(x) \qquad \text{when} \quad g(x)=0, \tag{5.3.8}$$

where the function $G_a(x)$ maps the surface $g(x)=0$ onto the surface $F(R_a)=0$.

The variational principle that determines the motion of the material is still represented by the equation

$$0=-\delta\int d^4x\Theta(F)\rho U(G_{ab},R_a).$$

But now we must modify the meaning of this equation in order to allow for the constraint (5.3.8). The modification required may be familiar to the reader from a study of classical mechanics. We require that the actual motion of the material, $R_a(x)$, satisfy the constraint (5.3.8). We also require that the variation δA of the action be zero when the fields are varied from the actual motion to some slightly different motion $R_a(x)+\delta R_a(x)$ which also satisfies the constraint. Thus the variation must satisfy the constraint

$$\delta R_a(x)=0 \qquad \text{when} \quad g(x)=0. \tag{5.3.9}$$

As in the "free boundary" case, the variational equation is

$$\begin{aligned}0&=\delta A\\&=-\int d^4x\Theta(F)\left\{\frac{\partial(\rho U)}{\partial R_a}-\partial_\nu\frac{\partial(\rho U)}{\partial(\partial_\nu R_a)}\right\}\delta R_a\\&\quad+\int_{\text{boundary}} dS_\nu 2\rho(\partial^\nu R_b)\frac{\partial U}{\partial G_{ab}}\delta R_a.\end{aligned} \tag{5.3.10}$$

The first term leads to the usual equations of motion inside the material. The surface term, which led to momentum conservation at a free boundary, is now zero for all allowed variations δR_a, since $\delta R_a=0$ at the boundary. Thus there are no new boundary conditions other than the condition (5.3.8) we imposed to begin with.

In this problem, momentum is conserved inside the body, but not on the boundary. The "physical" reason for this failure of momentum conservation is that the unyielding support attached to the material can absorb momentum. The "mathematical" reason is that the support spoils the invariance of the problem under translations in space-time.

Sliding Boundary

The final boundary condition we investigate is intermediate between the first two. We may imagine again the block of rubber in the steel box, but this time the surface of the rubber is not glued to the surface of the box. Instead, the two surfaces can slide over one another without friction. Another example of such a "sliding" boundary condition is the flow of a nonviscous fluid in a pipe. The surface of the fluid is free to move along the inside surface of the pipe, but the fluid cannot penetrate the pipe. We also assume that no bubble can form between the fluid and the pipe.

To state the boundary condition precisely, we again let $F(R_a)=0$ be the surface of the material and $g(t,\mathbf{x})=0$ be the surface of the supporting vessel. The boundary condition is that these surfaces coincide:

$$F(R_a(x))=0 \qquad \text{when} \quad g(x)=0. \tag{5.3.11}$$

Thus the condition on the allowed variations δR_a is

$$\frac{\partial F}{\partial R_a}\delta R_a=0 \tag{5.3.12}$$

at the boundary.

What does the principle of stationary action tell us in this case? From (5.3.10) we find

$$\delta R_a\left\{2\rho(\partial^\nu R_b)\frac{\partial U}{\partial G_{ab}}dS_\nu\right\}=0 \tag{5.3.13}$$

at the boundary, for all δR_a satisfying $\delta R_a(\partial F/\partial R_a)=0$. The only way the three component vector $V_a=\{2\rho(\partial^\nu R_b)(\partial U/\partial G_{ab})dS_\nu\}$ can be orthogonal to all vectors δR_a which are orthogonal to $\partial F/\partial R_a$ is for V_a to point in the same direction as $\partial F/\partial R_a$:

$$2\rho(\partial^\nu R_b)\frac{\partial U}{\partial G_{ab}}dS_\nu=\alpha(x)\frac{\partial F}{\partial R_a}, \tag{5.3.14}$$

where α is an undetermined scalar. If we multiply this equation by $(\partial^\mu R_a)$ and add to it the identity $\rho U u^\mu u^\nu dS_\nu=0$, we obtain an equation involving $T^{\mu\nu}$ at the boundary:

$$T^{\mu\nu}dS_\nu=\bar{\alpha}(x)dS^\mu, \tag{5.3.15}$$

where $\bar{\alpha}(x)$ is another undetermined scalar.

The physical content of this boundary condition is clear. Since there is no friction between the material and the supporting vessel, the force exerted by the material on the vessel must be directed in the direction orthogonal to the surface of the vessel.

We have discussed examples in which one type of boundary condition applies to the whole surface of the material, but it should be apparent that different parts of the surface can obey different boundary conditions. The cases of a block sliding down an inclined plane or a steel beam with its ends clamped in a fixed position come to mind.

In addition, we have not exhausted the possible types of boundary conditions. For instance, the body may be subject to a pressure P_0 from its environment but be otherwise free to move. Such an external pressure can be incorporated as an extra term in the action,

$$\bar{A}=\int d^4x\,\theta(-F(R_a(x)))P_0.$$

The action $A+\bar{A}$ leads to a conserved momentum tensor that includes the desired extra term, $\bar{T}^{\mu\nu}=g^{\mu\nu}P_0$, outside the body.

More generally, one may ask that the body be subject to prescribed external forces on its boundary. In most situations of physical interest, such forces are derived from a (possibly time dependent) potential energy function. If this is the case, the potential can be incorporated into the action by subtracting it from the Lagrangian. If this is not the case, then there is no action, there are only equations of motion.

5.4 INCOMPRESSIBLE MATERIALS

Some materials, such as solid rubber, are easy to bend, but hard to compress. In this section we develop an approximation that treats these materials as being impossible to compress.

To say that the material is hard to compress means that the internal energy U increases sharply as the density ρ is changed from some equilibrium density $\rho_0(R_a)$. We may imagine, for instance, that U contains a term proportional to $k(\rho-\rho_0)^2$, where the "spring constant" k is very large. Let us, then, consider the Lagrangian

$$\mathfrak{L} = -\rho\left[U(G_{ab}, R_a) + \tfrac{1}{2}k\left(\frac{\rho}{\rho_0}-1\right)^2\right], \tag{5.4.1}$$

and examine the behavior of the material in the limit $k\to\infty$.

The momentum tensor derived from the Lagrangian (5.4.1) is

$$T^{\mu\nu} = \rho\left[U + \tfrac{1}{2}k\left(\frac{\rho}{\rho_0}-1\right)^2\right]u^\mu u^\nu$$
$$+2\rho(\partial^\mu R_a)(\partial^\nu R_b)\frac{\partial U}{\partial G_{ab}} + (g^{\mu\nu}+u^\mu u^\nu)\lambda(x),$$

where we have defined

$$\lambda(x) = k\frac{\rho^2}{\rho_0}\left(\frac{\rho}{\rho_0}-1\right). \tag{5.4.3}$$

The factor λ appears in $T^{\mu\nu}$ in the role of a pressure [see (4.3.8)]. Where the material is slightly compressed ($\rho>\rho_0$), $\lambda(x)$ is positive and tends to dilate it; conversely, where the material is slightly decompressed, $\lambda(x)$ is negative and pulls it back in. During any physical motion of the material, the magnitude of λ will be comparable to that of other pressures or momentum currents in the body. Thus (ρ/ρ_0-1) will be small:

$$\left(\frac{\rho}{\rho_0}-1\right) = 0\left(\frac{1}{k}\right). \tag{5.4.4}$$

In the limit $k\to\infty$, we can therefore replace the equation of motion $\partial_\nu T^{\mu\nu}=0$ by the equations

$$\rho=\rho_0 \tag{5.4.5}$$

$$\partial_\nu T_I^{\mu\nu}=0, \tag{5.4.6}$$

where

$$T_I^{\mu\nu} = \rho_0 U u^\mu u^\nu + 2\rho_0(\partial^\mu R_a)(\partial^\nu R_b)\frac{\partial U}{\partial G_{ab}}$$

$$+(g^{\mu\nu} + u^\mu u^\nu)\lambda(x). \tag{5.4.7}$$

We have increased the number of equations by one, but we have also added one more field—the "pressure" $\lambda(x)$.

A Variational Principle

Can the equations of motion (5.4.5) and (5.4.6) for an incompressible material be derived from a variational principle? Indeed they can. We have only to consider $\lambda(x)$ to be one of the fields along with $R_a(x)$, and use the Lagrangian

$$\mathcal{L}_I = -\rho U(G_{ab}, R_a) - \lambda\left(\frac{\rho}{\rho_0} - 1\right). \tag{5.4.8}$$

Since the Lagrangian does not contain $\partial_\mu \lambda$, the equation of motion that arises from varying $\lambda(x)$ is simply

$$0 = \frac{\partial \mathcal{L}}{\partial \lambda} = -\left(\frac{\rho}{\rho_0} - 1\right). \tag{5.4.9}$$

The momentum tensor derived from $\mathcal{L}_I$ is

$$T^{\mu\nu} = \rho U u^\mu u^\nu + 2\rho(\partial^\mu R_a)(\partial^\nu R_b)\frac{\partial U}{\partial G_{ab}}$$

$$+(g^{\mu\nu} + u^\mu u^\nu)\frac{\rho}{\rho_0}\lambda - \lambda\left(\frac{\rho}{\rho_0} - 1\right) g^{\mu\nu}$$

When we set $\rho = \rho_0$ on account of (5.4.9), we obtain precisely the expression (5.4.7) for the momentum tensor.

In treating our material as "incompressible" we are artificially imposing the condition $\rho = \rho_0$. Thus we may reasonably suppose that we need only know how $U(G_{ab}, R_a)$ behaves on the surface $n\sqrt{\det \mathbf{G}} = \rho_0$ in order to solve the equations of motion. But this supposition is not obviously true, since it seems that we need to know the form of U slightly off of this surface in order to calculate $\partial U / \partial G_{ab}$. To show that the values of U off of the surface $\rho = \rho_0$ are irrelevant, consider using another function $U(G_{ab}, R_a)$

which differs from U for $\rho \neq \rho_0$:

$$\bar{U}(G_{ab}, R_a) = U(G_{ab}, R_a) + \frac{1}{\rho} F(G_{ab}, R_a)\left(\frac{\rho}{\rho_0} - 1\right). \tag{5.4.10}$$

The revised Lagrangian is

$$\bar{\mathcal{L}}_{\mathrm{I}} = -\rho U - [\lambda + F]\left(\frac{\rho}{\rho_0} - 1\right). \tag{5.4.11}$$

This differs from $\mathcal{L}_{\mathrm{I}}$ only by the replacement of $\lambda(x)$ by $\lambda(x) + F(G_{ab}, R_a)$. Thus to every solution $R_a(x)$, $\lambda(x)$ of the original problem $\delta A = 0$, there corresponds a solution R_a, λ of the new problem $\delta A = 0$ given by

$$\bar{R}_a(x) = R_a(x) \tag{5.4.12}$$

$$\bar{\lambda}(x) = \lambda(x) - F(G_{ab}(x), R_a(x)).$$

The effect of the change in U for $\rho \neq \rho_0$ is entirely absorbed into the "pressure" $\lambda(x)$, and does not affect the motion of the material, $R_a(x)$.

Incompressibility as a Constraint

There is another way of formulating Hamilton's principle for incompressible materials which is of some interest. We demand that the action

$$A = \int d^4x\, \mathcal{L}(R_a, \partial_\mu R_a), \tag{5.4.13}$$

with $\mathcal{L} = -\rho U$, be made stationary subject to the constraint

$$\frac{\rho}{\rho_0} - 1 = 0 \tag{5.4.14}$$

on the admissible fields $R_a(x)$. Let us call this variational problem the constrained problem. We would like to show that a solution of the "unconstrained problem" formulated in the preceding subsection solves the constrained problem.

In seeking solutions to the constrained problem, we would write

$$0 = \delta A_C = \int d^4x \frac{\delta A_C}{\delta R_a(x)} \delta R_a(x), \tag{5.4.15}$$

where $\delta A_C / \delta R_a(x)$ is a shorthand notation for

$$\frac{\delta A_C}{\delta R_a(x)} = \frac{\partial \mathcal{L}}{\partial R_a} - \partial_\mu \frac{\partial \mathcal{L}}{\partial(\partial_\mu R_a)}. \tag{5.4.16}$$

Ordinarily, we would conclude that $\delta A_C/\delta R_a(x)=0$, but we cannot do so here because the constraint (5.4.14) does not allow us to choose $\delta R_a(x)$ freely. Since both R_a and $R_a+\delta R_a$ must satisfy the constraint, the variation $\delta R_a(x)$ must satisfy

$$\frac{\partial}{\partial R_a}\left(\frac{\rho}{\rho_0}-1\right)\delta R_a+\frac{\partial}{\partial(\partial_\mu R_a)}\left(\frac{\rho}{\rho_0}-1\right)\partial_\mu \delta R_a=0. \tag{5.4.17}$$

There is a standard technique for dealing with such problems.* One multiplies the constraint function (ρ/ρ_0-1) at x^μ by a "Lagrange undetermined multiplier" $\lambda(x^\mu)$, integrates this product over all x^μ, and subtracts the result from the action. This gives a revised action

$$A_U=\int dx\left[\mathcal{L}_C-\lambda(x)\left(\frac{\rho}{\rho_0}-1\right)\right]. \tag{5.4.18}$$

Then we require that $\delta A_U=0$ for all (unconstrained) variations of the fields $R_a(x)$ and of the multipliers $\lambda(x)$. This is precisely the unconstrained problem formulated earlier.

Let us suppose that $(R_a(x), \lambda(x))$ is a solution to the unconstrained problem. We will show that $R_a(x)$ solves the constrained problem. First, the constraint equation (5.4.14) for $R_a(x)$ is satisfied, since $\delta A_U=0$ when $\lambda(x)$ is varied. To show that $\delta A_C=0$ when δR_a satisfies (5.4.17), we take any such δR_a and calculate δA_U:

$$0=\delta A_U=\int dx\Bigg\{\frac{\delta A_C}{\delta R_a(x)}\delta R_a(x)$$

$$-\lambda(x)\left[\frac{\partial}{\partial R_a}\left(\frac{\rho}{\rho_0}-1\right)\delta R_a+\frac{\partial}{\partial(\partial_\mu R_a)}\left(\frac{\rho}{\rho_0}-1\right)\partial_\mu \delta R_a\right]\Bigg\}.$$

The quantity in square brackets is zero because of (5.4.17), so that we are left with

$$0=\int dx\frac{\delta A_C}{\delta R_a(x)}\delta R_a(x)=\delta A_C.$$

Thus the constrained problem is solved.†

*See I. M. Gelfand and S. V. Fomin, *Calculus of Variations* (Prentice-Hall, New York, 1963).

†A proof that a solution of the constrained problem gives a solution of the unconstrained problem would be much harder. See Problem 3 and Naum I. Akhiezer, *The Calculus of Variations* (Blaisdell Publishing Co., Waltham, Mass. 1962), p. 117.

Of course, the constrained problem is of only academic interest, since the only way we can solve it is by converting it to the unconstrained problem. However, the analogue of the constrained problem is of great importance in the nonrelativistic theory of a rigid body. There the constraints so restrict the motion of the body that there are only six degrees of freedom left—only the position of the center of mass of the body and the three Euler angles are free to vary. Therefore one can write the action as a function of these six variables and obtain six ordinary differential equations of motion for these variables as functions of time.*

PROBLEMS

1. Verify equations (5.2.3) and (5.3.4).
2. Let the surface $F(x)=0$ be described parametically by $x^\mu = X^\mu(a^1,a^2,a^3)$ and let $H^\nu(x)$ be any function. Show that the integral $\int d^4x\delta(F(x))(\partial_\nu F)H^\nu(x)$ equals the surface integral

$$\int_{F(x)=0} dS_\nu H^\nu(x) \equiv -\int d^3a\epsilon_{\nu\alpha\beta\gamma}\frac{\partial x^\alpha}{\partial a^1}\frac{\partial x^\beta}{\partial a^2}\frac{\partial x^\gamma}{\partial a^3}H^\nu(X(\mathbf{a})).$$

(Notice that the first form of the integral is manifestly invariant under changes of parameters a, while the second form is manifestly invariant under changes of the function F describing the surface.)

3. Consider the following constrained variational problem: let $A_C = \int d^4x\,\mathcal{L}(R_a,\partial_\mu R_a)$ and require $\delta A_C=0$ when $R_a(x)$ and $R_a(x)+\delta R_a(x)$ satisfy the constraint $F(R_a(x),x)=0$. Compare this to the unconstrained problem $0=\delta A_U=\delta\int d^4x[\mathcal{L}(R_a,\partial_\mu R_a)-\lambda(x)F(R_a(x),x)]$ where $R_a(x)$ and $\lambda(x)$ are varied freely. Let $R_a(x)$ be a solution of the constrained problem. Prove that there exists a $\lambda(x)$ such that $(R_a(x),\lambda(x))$ solves the unconstrained problem. (*Hint*: since F does not depend on $\partial_\mu R_a$, you can find $\lambda(x)$ by solving an *algebraic* equation that arises from $\delta A_U=0$ for $\delta R_a(x)=\alpha(x)(\partial F/\partial R_a)$, $\alpha(x)$ arbitrary.)

*See Herbert Goldstein, *Classical Mechanics* (Addison-Wesley, Reading, Mass., 1950), p. 93, and Problem 2 in Chapter 7.

CHAPTER 6

Special Types of Solids

In the preceding chapter we discussed a general class of models for materials, based on the Langrangian $\mathfrak{L} = -\rho U(G_{ab}, R_a)$. Now we are ready to consider some important special cases: "Hooke's law" materials, isotropic materials, and isotropic Hooke's law materials.

6.1 HOOKE'S LAW

The Strain Matrix

Imagine a solid, say, a crystal, in its natural unbent state. That is, each little piece of crystal is in its lowest energy state, with all its atoms lined up in a regular pattern. While the crystal is in this natural state, we lay down the material coordinate labels R_a so that the material coordinates form a Cartesian coordinate system with the same scale of length as the laboratory coordinate system. The material coordinate axes can be aligned in some convenient way with respect to the crystal lattice. (See Figure 6.1.)

With such a choice of material coordinates R_a, we have $R_a(t,\mathbf{x}) = x^a$ for the unbent crystal when it is at rest in the laboratory and properly oriented. Thus

$$G_{ab} = (\partial_\mu R_a)(\partial^\mu R_a) = \delta_{ab} \tag{6.1.1}$$

whenever the crystal is in its natural unbent state.

In view of (6.1.1), it will be convenient to define a matrix

$$s_{ab} = \tfrac{1}{2}(\delta_{ab} - G_{ab}). \tag{6.1.2}$$

Then $s_{ab} = 0$ represents the natural state of the crystal, and s_{ab} tells how much the crystal has been squeezed or twisted from its natural state. We will call s_{ab} the "strain matrix."

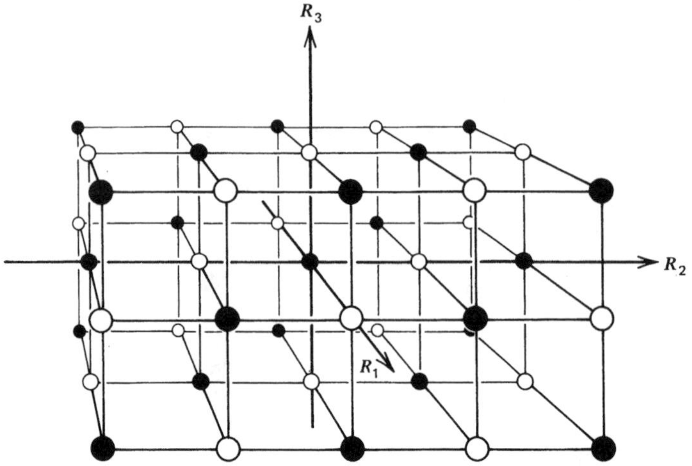

Figure 6.1 A convenient choice of internal coordinates R_a for a NaCl crystal.

The Hooke's Law Lagrangian

The internal energy per atom, U, can be regarded as a function of s_{ab} instead of G_{ab} if we please. Our plan in this section is to approximate U by the lowest order terms in the Taylor expansion of U in powers of s_{ab}. Since $s_{ab}=0$ is the lowest energy state, the terms in the expansion linear in s_{ab} are zero. Thus we must keep the quadratic terms. For the rest of this section we neglect the cubic and higher order terms. This gives

$$U=m+\frac{1}{2n}C_{abcd}s_{ab}s_{cd} \tag{6.1.3}$$

(The factor $(2n)^{-1}$, with n the density in the material coordinate system, is conventional.)

We thus obtain a definite model for the behavior of elastic materials which can be regarded as an approximation to the general model, valid when the strain s is small. The condition $|s_{ab}|\ll 1$ does not limit us to studying small displacements of the material from its natural state (such as sound waves). For instance, a long thin crystal can be bent into the shape of a pretzel, so that $R_a(t,\mathbf{x})-x^a$ and $\partial R_a/x^j-\delta_{aj}$ are large, but s_{ab} is everywhere small. In principle, we can calculate the forces required to so bend the crystal.

The equation (6.1.3) defining the model is apparently a generalization of

Hooke's law, which says that the internal energy of a spring is proportional to the square of its displacement from a relaxed position.

Many authors define the strain matrix s_{ab} which occurs in Hooke's law as $\bar{s}_{ab}=\frac{1}{2}([G^{-1}]_{ab}-\delta_{ab})$. This definition is natural if one is using $x^j(R_a,t)$ as the fields, but is not very natural from the viewpoint of this book. Since $\bar{s}_{ab}\cong s_{ab}$ for small strain, the difference between the two definitions is not very important.

The Elastic Constants

The mechanical properties of a Hooke's law material are determined by the constants m and C_{abcd}. The constant m is the internal energy per atom of the solid in its natural state, including the binding energy. Thus it is the mass per atom as measured by weighing a chunk of the material.

The constants C_{abcd} are called the elastic constants of the solid. At first sight there appear to be $(3)^4=81$ of them. However, we can assume that C_{abcd} obeys the symmetry requirements

$$C_{abcd}=C_{bacd}=C_{abdc}=C_{cdab}. \tag{6.1.4}$$

If it did not, we could replace C_{abcd} with

$$\begin{aligned}\overline{C}_{abcd}=\tfrac{1}{8}(&C_{abcd}+C_{bacd}+C_{abdc}+C_{badc}\\&+C_{cdab}+C_{dcab}+C_{cdba}+C_{dcba})\end{aligned}$$

without changing U. Since a 3×3 symmetric matrix has 6 independent components and a 6×6 symmetric matrix has 21 independent components, only 21 of elastic constants C_{abcd} are independent.

The elastic constants are further restricted by the requirement that $s_{ab}=0$ be a state of stable equilibrium—that is, a minimum of U. Thus we demand that

$$C_{abcd}s_{ab}s_{cd}>0 \tag{6.1.5}$$

for all symmetric matrices s_{ab}.

The elastic constants of some typical crystals are displayed in Table 6.1. Note that the symmetries of the crystal can limit the number of independent elastic constants. For simple cubic crystals like sodium chloride there are only 3 independent constants; for monoclinic crystals like sodium thiosulphate there are 13. The values of $(2n)^{-1}C_{abcd}$ are interesting: they are the "spring constants" in the expression (6.1.3) for the potential energy per atom. For instance, the value corresponding to $C_{1111}=5\times 10^{11}$ erg/cm^3 for sodium chloride is $(2n)^{-1}C_{1111}=3$ eV/atom.

Table 6.1 Elastic Constants of Some Crystals (10^{11} erg/cm^3)[a]

	Diamond	NaCl	$Na_2S_2O_3$
C_{1111}	95	5.0	3.3
C_{2222}	(95)	(5.0)	3.0
C_{3333}	(95)	(5.0)	4.6
C_{2323}	43	1.3	0.6
C_{3131}	(43)	(1.3)	1.1
C_{1212}	(43)	(1.3)	0.6
C_{1122}	39	1.3	1.8
C_{1133}	(39)	(1.3)	1.8
C_{1123}	(0)	(0)	(0)
C_{1131}	(0)	(0)	0.3
C_{1112}	(0)	(0)	(0)
C_{2233}	(39)	(1.3)	1.7
C_{2223}	(0)	(0)	(0)
C_{2231}	(0)	(0)	1.0
C_{2212}	(0)	(0)	(0)
C_{3323}	(0)	(0)	(0)
C_{3331}	(0)	(0)	−0.7
C_{3312}	(0)	(0)	(0)
C_{2331}	(0)	(0)	(0)
C_{2312}	(0)	(0)	−0.3
C_{3112}	(0)	(0)	(0)

[a]Values in parentheses are not independent, but are related to the preceding data by crystal symmetries. The data are from S. Bhagavantam, *Proc. Ind. Acad. Sci.*, **16A**, 79 (1955). See also R. Hearmon, *Rev. Mod. Phys.*, **18**, 409 (1946).

We have been speaking as if m and C_{abcd} were constant throughout the material. But there is nothing inconsistent in allowing m and C_{abcd} to be functions of R_a in order to describe a piece of material of nonuniform composition.

The Momentum Tensor

How is the momentum tensor of an elastic solid related to the strain matrix s_{ab} and the elastic constants? From the general form (5.2.3) for $T^{\mu\nu}$ and our expression (6.1.3) for U we learn that

$$T^{\mu\nu} = \rho U u^\mu u^\nu - \frac{\rho}{n}(\partial^\mu R_a)(\partial^\nu R_b)C_{abcd}s_{cd}. \tag{6.1.6}$$

In a local rest frame of a piece of material, the momentum flow is

$$T^{kl} = -\frac{\rho}{n}(\partial_k R_a)(\partial_l R_b)C_{abcd}s_{cd}. \tag{6.1.7}$$

Thus it is said "the stress is proportional to the strain." It is also proportional to ρ and some factors $\partial R_a/\partial x^k$ which can be thought of as transforming the components of the momentum tensor from the material coordinate system to the laboratory coordinate system.

We will return to the momentum tensor (6.1.6) in order to discuss sound waves in a Hooke's law solid later in this chapter.

6.2 ISOTROPIC MATERIALS

In general, the internal energy U is a more or less arbitrary function of the six independent components of the deformation matrix G_{ab}. If the material is isotropic the situation improves somewhat: U is a function of only three deformation variables. Let us see how this comes about, and what these variables are.

First, we should make clear the meaning of "isotropic." We begin by considering the deformation matrix $G_{ab}=(\partial_\mu R_a)(\partial^\mu R_b)$. Suppose that we used different material coordinates $\overline{R}_a$, related to R_a by a rotation:

$$\overline{R}_a = \mathcal{R}_{ab} R_b. \tag{6.2.1}$$

Then the deformation matrix defined with the new coordinates is

$$\overline{G}_{ab} = \mathcal{R}_{ac} \mathcal{R}_{bd} G_{cd}. \tag{6.2.2}$$

That is, G_{ab} transforms like a tensor under rotations of the material coordinate system.

Consider now a homogeneous material, so that U is independent of R_a. Suppose, in addition, that there are no preferred directions in the material. Then it cannot make any difference how we choose the directions of the material coordinate axes:

$$U(\mathcal{R}_{ac} \mathcal{R}_{bd} G_{cd}) = U(G_{ab}). \tag{6.2.3}$$

We call such a material "isotropic." Rubber, steel, Jello, and air are examples of isotropic materials.

Without making things any more difficult for ourselves, we can drop the requirement that the material be homogeneous by considering $U(G_{ab}, R_a)$ such that

$$U(\mathcal{R}_{ac} \mathcal{R}_{bd} G_{cd}, R_a) = U(G_{ab}, R_a). \tag{6.2.4}$$

This equation defines what we call a "locally isotropic" material.

It is easy to make up some functions $g(\mathbf{G})$ of G_{ab} that are invariant under rotations in the sense of (6.2.3). Three conventional choices are

$$g_{\mathrm{I}}(\mathbf{G}) = G_{aa},$$

$$g_{\mathrm{II}}(\mathbf{G}) = \tfrac{1}{2}\left[(G_{aa})^2 - (G_{ab}G_{ba})\right], \tag{6.2.5}$$

$$g_{\mathrm{III}}(\mathbf{G}) = \det\mathbf{G}.$$

The reader should verify that these functions are indeed invariant (see Problem 4). Any function $U(g_{\mathrm{I}}, g_{\mathrm{II}}, g_{\mathrm{III}}; R_a)$ of these invariants will satisfy (6.2.4) and will thus describe a locally isotropic material.

Are there any other invariants that are not simply functions of g_{I}, g_{II}, and g_{III}? To answer this question, look at the equation that determines the eigenvalues λ of the symmetric matrix $\mathbf{G}$: $0 = \det(\mathbf{G} - \lambda\mathbf{1})$. When written out in detail this equation reads

$$0 = -\lambda^3 + g_{\mathrm{I}}(\mathbf{G})\lambda^2 - g_{\mathrm{II}}(\mathbf{G})\lambda + g_{\mathrm{III}}(\mathbf{G}). \tag{6.2.6}$$

Thus when we know g_{I}, g_{II}, and g_{III} we can solve (6.2.6) for the three eigenvalues of $\mathbf{G}$. But the "diagonalization" theorem of linear algebra tells us that if we know the eigenvalues $\lambda_1 \lambda_2, \lambda_3$ of $\mathbf{G}$, then we know $\mathbf{G}$ up to an unknown rotation $\mathfrak{R}$:

$$\mathbf{G} = \mathfrak{R}\begin{pmatrix} \lambda_1 & 0 & 0 \\ 0 & \lambda_2 & 0 \\ 0 & 0 & \lambda_3 \end{pmatrix}\mathfrak{R}^T.$$

Finally, if we know $\mathbf{G}$ up to a rotation, then we can calculate the value $g(\mathbf{G})$ for any invariant function we may have had in mind. Thus g_{I}, g_{II}, g_{III} provide a complete set of independent invariant functions of $\mathbf{G}$.

What is the form of $T^{\mu\nu}$ for a locally isotropic material? To find out, we assume that the internal energy has the form $U(g_{\mathrm{I}}, g_{\mathrm{II}}, g_{\mathrm{III}}; R_a)$ and calculate $T^{\mu\nu}$, beginning with the general form (5.2.3) for elastic materials. In the expression that results, all of the indices $a, b, c \cdots$ on the fields R_a are contracted against one another. Thus some simplification can be achieved by using an "inside out" deformation tensor

$$C_{\mu\nu} = (\partial_\mu R_a)(\partial_\nu R_a). \tag{6.2.7}$$

This tensor is called Cauchy's deformation tensor; it is a "purely spacelike" tensor in that $C_{\mu 0} = C_{0\mu} = 0$ in a local rest frame of a piece of material. The calculation just described is straightforward. The result is reasonably simple:

$$
\begin{aligned}
T^{\mu\nu} &= \rho U u^{\mu} u^{\nu} \\
&\quad + 2\rho\left[\frac{\partial U}{\partial g_{\rm I}} + C^{\lambda}_{\lambda}\frac{\partial U}{\partial g_{\rm II}}\right] C^{\mu\nu} \\
&\quad - 2\rho\frac{\partial U}{\partial g_{\rm II}} C^{\mu\lambda} C^{\nu}_{\lambda} \\
&\quad + 2\rho g_{\rm III}\frac{\partial U}{\partial g_{\rm III}}\left[g^{\mu\nu} + u^{\mu}u^{\nu}\right].
\end{aligned} \tag{6.2.8}
$$

Note that we recover the momentum tensor (4.3.3) for a fluid if U is a function of R_a and $g_{\rm III} = \rho^2/n^2$ only.

6.3 ISOTROPIC HOOKE'S LAW MATERIALS

The simplest kind of solid is an isotropic solid which obeys Hooke's law. To study such a solid, we can begin with the Hooke's law internal energy

$$U = m + \frac{1}{2n} C_{abcd} s_{ab} s_{cd}. \tag{6.3.1}$$

Under a rotation of the material coordinates, the strain matrix s_{ab} transforms like a tensor. If the expression for U is to remain invariant, the array of elastic constants must obey

$$\mathcal{R}_{a'a}\,\mathcal{R}_{b'b}\,\mathcal{R}_{c'c}\,\mathcal{R}_{d'd}\,C_{a'b'c'd'} = C_{abcd} \tag{6.3.2}$$

for all rotation matrices $\mathcal{R}$. We can make up such an invariant tensor C_{abcd} using Kroneker deltas (see Section 1.1). There are two obvious choices that satisfy the symmetry requirements (6.1.4): $C_{abcd} = \delta_{ab}\delta_{cd}$ and $C_{abcd} = \delta_{ac}\delta_{bd} + \delta_{ad}\delta_{bc}$. The most general invariant C_{abcd} is a linear combination of these (see Problem 8):

$$C_{abcd} = \lambda\delta_{ab}\delta_{cd} + \mu\left[\delta_{ac}\delta_{bd} + \delta_{ad}\delta_{bc}\right]. \tag{6.3.3}$$

With this choice for C_{abcd}, the internal energy is

$$U = m + \frac{1}{2n}\left[\lambda(s_{aa})^2 + 2\mu s_{ab}s_{ab}\right]. \tag{6.3.4}$$

In order for $U - m$ to be positive for all strains s_{ab}, the constant μ and the combination $(\lambda + \frac{2}{3}\mu)$ must both be positive. This can best be seen by writing s_{ab} in the form $s_{ab}=\frac{1}{3}s\delta_{ab}+r_{ab}$, where $s=s_{aa}$ and $r_{aa}=0$; then $U-m$ takes the form

$$U-m=\frac{1}{2n}\left[(\lambda+\tfrac{2}{3}\mu)s^2+2\mu r_{ab}r_{ab}\right].$$

This expression is positive if $\mu>0$ and $(\lambda+\frac{2}{3}\mu)>0$. Conversely, if $(\lambda+\frac{2}{3}\mu)$ were negative, we could make $U-m$ negative by choosing $s_{ab}=\delta_{ab}$; and if μ were negative, we could make $U-m$ negative by choosing for s_{ab} any matrix with $s_{aa}=0$.

The constants λ and μ that characterize the solid are called Lame constants. The Lame constants of a few materials are given in Table 6.2.

Table 6.2 Values of the Lame constants for some materials.[a]

Substance	λ ($\times 10^{11}$ erg/cm^3)	μ ($\times 10^{11}$ erg/cm^3)
Copper	10	4.8
Brass	8.7	3.7
Tungsten carbide steel	17	22
Flint glass	3.7	3.2
Fused quartz	1.6	3.1

[a]Data from G. W. C. Kaye and T. H. Laby, *Tables of Physical and Chemical Constants* (Longman, London, 1963).

The momentum tensor for the isotropic Hooke's law solid can be found by substituting (6.3.3) into (6.1.6):

$$T^{\mu\nu}=\rho Uu^{\mu}u^{\nu}-\frac{\rho}{n}(\partial^{\mu}R_a)(\partial^{\nu}R_b)$$
$$\times\left[\lambda\delta_{ab}s_{cc}+2\mu s_{ab}\right]. \qquad (6.3.5)$$

One can also write this momentum tensor by using the Cauchy deformation tensor, as we did for the general isotropic elastic solid in the preceding section, but no great advantage results.

6.4 SOUND PROPAGATION IN A HOOKE'S LAW SOLID

What happens if a Hooke's law solid is perturbed slightly from its equilibrium state $R_a(t,\mathbf{x})=x^a$? To find out, we define the displacement from equilibrium ϕ:

$$R_a(t,\mathbf{x})=x^a+\phi_a(t,\mathbf{x}). \qquad (6.4.1)$$

We stipulate that ϕ is small and write out the equations of motion for ϕ, neglecting terms of order ϕ^2 and higher.

It is convenient to use the momentum conservation equations $\partial_\mu T^{i\mu}=0$ as equations of motion, so we need to write out $T^{i\mu}$:

$$T^{i\mu}=\rho\left(m+\frac{1}{2n}C_{abcd}s_{ab}s_{cd}\right)u^i u^\mu$$
$$-\frac{\rho}{n}(\partial^i R_a)(\partial^\mu R_b)C_{abcd}s_{cd}. \qquad (6.4.2)$$

The quantities u^i and s_{cd} are zero if $\phi=0$, so we evaluate them to first order in ϕ; everything else on the right hand side of (6.4.2) can then be evaluated at $\phi=0$. Straightforward calculation shows that

$$u^i \cong -(\partial_0\phi_i),$$
$$s_{cd}\cong -\tfrac{1}{2}(\partial_c\phi_d+\partial_d\phi_c). \qquad (6.4.3)$$

Thus

$$T^{i0}\cong -nm(\partial_0\phi_i),$$
$$T^{ij}\cong C_{ijlm}(\partial_l\phi_m). \qquad (6.4.4)$$

This gives the equation of motion for the perturbation field $\phi(t,\mathbf{x})$:

$$\partial_0\partial_0\phi_i=\frac{1}{mn}C_{ijlm}\partial_j\partial_l\phi_m. \qquad (6.4.5)$$

A plane wave

$$\phi_i(t,\mathbf{x})=\phi_i e^{i(\mathbf{k}\cdot\mathbf{x}-\omega t)} \qquad (6.4.6)$$

is a solution of the equation of motion, provided that ω, $\mathbf{k}$, and ϕ are related by

$$\omega^2\phi_i=\frac{1}{mn}C_{ijlm}k_j k_l\phi_m. \qquad (6.4.7)$$

Evidently, sound propagation in an anisotropic crystal is more complicated than in a fluid. For a given wave vector $\mathbf{k}$, (6.4.7) is an eigenvalue equation of the form

$$\omega^2\phi_i=M(\mathbf{k})_{im}\phi_m.$$

In general, we find three eigenvalues ω^2 and three corresponding eigenvectors $\boldsymbol{\phi}$, none of which need be either in the direction of $\mathbf{k}$ or perpendicular to it. However, since $M_{im}=M_{mi}$, the eigenvectors will be perpendicular to one another.

Are the eigenvalues ω^2 always positive, so that $\phi_j(t,\mathbf{x})$ represents a sound wave and not an exponentially growing disturbance? The eigenvalues of the matrix $M(\mathbf{k})_{im}$ will be positive if $\phi_i M(\mathbf{k})_{im}\phi_m > 0$ for all vectors ϕ_i; that is, if

$$(k_i\phi_j + k_j\phi_i)C_{ijlm}(k_l\phi_m + k_m\phi_l) > 0. \tag{6.4.8}$$

But (6.4.8) is guaranteed to hold because of the positivity property (6.1.5) of C_{ijlm}: $s_{ij}C_{ijlm}s_{lm} > 0$ for all symmetric s_{lm}. This positivity requirement was necessary to ensure that the state of zero strain was the minimum energy state of a piece of crystal.

Sound propagation in an isotropic Hooke's law solid is much simpler. The relation (6.4.7) relating ω, $\mathbf{k}$, and $\boldsymbol{\phi}$ becomes

$$\omega^2\boldsymbol{\phi} = \frac{\lambda+\mu}{mn}\mathbf{k}(\mathbf{k}\cdot\boldsymbol{\phi}) + \frac{\mu}{mn}(\mathbf{k}\cdot\mathbf{k})\boldsymbol{\phi}. \tag{6.4.9}$$

For a given wave vector $\mathbf{k}$, one eigenvector $\boldsymbol{\phi}$ points in the direction of $\mathbf{k}$. The corresponding eigenvalue ω^2 is $\omega^2 = [(\lambda+2\mu)/m]\mathbf{k}^2$. Thus the speed of longitudinal sound waves is

$$v_{\text{long}} = \sqrt{\frac{\lambda+2\mu}{mn}}\ . \tag{6.4.10}$$

The other two eigenvectors $\boldsymbol{\phi}$ are orthogonal to $\mathbf{k}$; the corresponding eigenvalue ω^2 is $\omega^2=[\mu/mn]\mathbf{k}^2$. Thus the speed of transverse sound waves is

$$v_{\text{trans}} = \sqrt{\frac{\mu}{mn}}\ . \tag{6.4.11}$$

In seismology the longitudinal sound waves in the earth's crust are called P waves and transverse waves are called S waves. The measured velocities of P and S waves in wet granite are about $v_P = 5.5$ km/sec and $v_S = 3.0$ km/sec.*

*C. H. Scholz, L. R. Sylees, and Y. P. Aggarwal, *Science*, **181**, 803 (1973). This article reports on the usefulness of measured changes in the ratio v_P/v_S as a predictor of earthquakes.

PROBLEMS

1. The structure of sodium chloride is shown in Figure 6.1. Show that the symmetries of this crystal imply that only three of the elastic constants are independent, as indicated in Table 6.1.
2. Sodium thiosulphate is a monoclinic crystal. This means that the crystal lattice has lattice points at $\mathbf{R}=n_1\mathbf{a}+n_2\mathbf{b}+n_3\mathbf{c}$ where n_1, n_2, n_3 are integers and $\mathbf{a}=(a,0,0)$, $\mathbf{b}=(0,b,0)$, $\mathbf{c}=(c\cos\beta,0,c\sin\beta)$ with $a\neq b\neq c$ and $\beta\neq\pi/2$. Supposing that the crystal has the full symmetry of its crystal lattice, show that there are only 13 independent elastic constants, as indicated in Table 6.1.
3. Diamond has the following structure. Take $a=3.56\times10^{-8}$ cm as the unit of length. Then there are carbon atoms at the corners of a cube of length 1, $\mathbf{R}=(0,0,0)$, $(1,0,0)$, $(1,1,0)$, $(0,1,0)$, $(0,0,1)$, $(1,0,1)$, $(1,1,1)$, $(0,1,1)$; centered on the faces of the cube at $\mathbf{R}=(\frac{1}{2},\frac{1}{2},0)$, $(\frac{1}{2},\frac{1}{2},1)$, $(\frac{1}{2},0,\frac{1}{2})$, $(\frac{1}{2},1,\frac{1}{2})$, $(0,\frac{1}{2},\frac{1}{2})$, $(1,\frac{1}{2},\frac{1}{2})$; and inside the cube at $\mathbf{R}=(\frac{1}{4},\frac{1}{4},\frac{1}{4})$, $(\frac{3}{4},\frac{3}{4},\frac{1}{4})$, $(\frac{1}{4},\frac{3}{4},\frac{3}{4})$, $(\frac{3}{4},\frac{1}{4},\frac{3}{4})$. This pattern repeats itself: if an atom is located at $\mathbf{R}$, then atoms will be located at $\mathbf{R}+(n_1,n_2,n_3)$ for all integers n_1, n_2, n_3. Show that there are only three independent elastic constants for diamond, as indicated in Table 6.1.
4. Prove that the functions $g_{\mathrm{I}}, g_{\mathrm{II}}, g_{\mathrm{III}}$ defined in (6.2.5) are invariant under rotations.
5. Find expressions giving $g_{\mathrm{I}}, g_{\mathrm{II}}, g_{\mathrm{III}}$ as functions of the eigenvalues $\lambda_1, \lambda_2, \lambda_3$ of $\mathbf{G}$.
6. How are the invariants $g_{\mathrm{I}}, g_{\mathrm{II}}, g_{\mathrm{III}}$ of $\mathbf{G}$ related to the corresponding invariants of the inverse matrix $\mathbf{G}^{-1}$?
7. Find expressions giving $g_{\mathrm{I}}(\mathbf{G})$, $g_{\mathrm{II}}(\mathbf{G})$, $g_{\mathrm{III}}(\mathbf{G})$ as functions of the Cauchy deformation tensor $C_{\mu\nu}$.
8. Show that an invariant tensor C_{abcd} with the symmetry properties $C_{abcd}=C_{bacd}=C_{abdc}=C_{cdab}$ must have the form $C_{abcd}=\lambda\delta_{ab}\delta_{cd}+\mu[\delta_{ac}\delta_{bd}+\delta_{ad}\delta_{bc}]$. (*Hint*: see Problem 9 in Chapter 1.)
9. Calculate the deformation energy per unit of undeformed volume, $W\equiv n(U-m)$, in an isotropic Hooke's law solid that is uniformly compressed: $R_a=e^{\omega}x^a$. Justify the name "modulus of compression" for the quantity $R=\lambda+\frac{2}{3}\mu$.
10. A cylindrical rod made of an isotropic Hooke's law material is given a small twist: $R_1=\cos\theta x^1+\sin\theta x^2$, $R_2=-\sin\theta x^1+\cos\theta x^2$, $R_3=x^3$ with $\theta=\omega x^3$. Calculating to first order in ω, find the deformation energy density $W=n(U-m)$. Is the interior of the rod in mechanical

equilibrium? What forces must be applied to the surface of the rod to maintain the rod in the twisted state.

11. A cylindrical rod made of an isotropic Hooke's law material is stretched by means of a force F applied to the ends of the rod. The force is applied uniformly over the end surfaces in the direction normal to these surfaces. No forces are applied to the side surface. What is the resulting shape of the rod? (In principle this is a problem in partial differential equations, but the problem can be more easily solved by using your physical intuition to guess the form of the answer.) Find the limiting form of your answer for small F and relate it to "Young's modulus," $E=\mu(3\lambda+2\mu)/(\lambda+\mu)$, and "Poisson's ratio," $\nu=\frac{1}{2}\lambda/(\lambda+\mu)$.

12. Calculate the deformation energy density $W=n(U-m)$ for an isotropic Hooke's law solid that is subjected to a shear along the R_1-axis: $R_1=x^1+\omega x^2$, $R_2=x^2$, $R_3=x^3$. Why is μ often called the "shear modlus"?

13. A regular tetrahedron made of an isotropic Hooke's law material is thrown into the ocean. When it comes to rest on the bottom, it is subjected to the pressure P of the water (we neglect the previous pressure of the air). What is the final shape and size of the tetrahedron?

14. Find the natural vibration frequencies of a solid sphere made out of an isotropic Hooke's law material.

15. A spherical shell with inner radius r_1 and outer radius r_2 is made out of an isotropic Hooke's law material. The shell is turned inside out by cutting a small hole in it and pushing the material through the hole. Write down an ordinary differential equation with boundary conditions which determines the new inner and outer radii of the shell.

CHAPTER 7

The Nonrelativistic Approximation

In Chapter 4 we calculated the nonrelativistic approximation to the momentum tensor of a fluid, with the limited goal of getting a better feeling for the physics of $T^{\mu\nu}$. Now we look more deeply into the nonrelativistic theory of elastic materials. We begin by finding the nonrelativistic Lagrangian as an approximation to the Lorentz invariant Lagrangian. From then on, we will treat the nonrelativistic theory as a separate, self-consistent theory.

7.1 THE LAGRANGIAN

We exploit two features of ordinary terrestrial experiments in order to derive an approximate Lagrangian for elastic materials. The first feature is that nothing but light moves with anything near the speed of light. Thus

$$\frac{\partial R_a}{\partial x^0} \ll 1. \tag{7.1.1}$$

To discuss the second feature we write

$$U(G_{ab}, R_a) = m(R_a) + \tilde{U}(G_{ab}, R_a). \tag{7.1.2}$$

Here $m(R_a)$ is the energy per atom of material in some standard "undeformed" state. That is, the mass per atom obtained by weighing a chunk of material in the standard state. The remaining function $\tilde{U}(G_{ab}, R_a)$ is then the energy required to deform the material. Under ordinary conditions one can change the energy of a piece of material only by a very tiny fraction by deforming it. Thus the second feature is

$$\frac{\tilde{U}}{m} \ll 1. \tag{7.1.3}$$

Our procedure is to expand the Lagrangian $\mathfrak{L} = -\rho U$ in powers of

$(\partial_0 R_a)^2$ and $\tilde{U}/m$ and keep only the leading terms.* For the invariant density ρ we can write $\rho=[-J_\mu J^\mu]^{1/2}$ and recall that the density $J^0 = n \det[\partial R_a/\partial x^j]$ contains no factor of $\partial_0 R_a$, while $\mathbf{J}$ is proportional to $\partial_0 R_a$. Thus

$$\rho=\sqrt{(J^0)^2-\mathbf{J}^2}=J^0\left(1-\tfrac{1}{2}\left(\frac{\mathbf{J}}{J^0}\right)^2+\cdots\right)$$

$$=J^0(1-\tfrac{1}{2}\mathbf{v}^2+\cdots), \tag{7.1.4}$$

where $\mathbf{v}=\mathbf{J}/J^0$ is the velocity of the material. For U we can use $U = m(R_a)+\tilde{U}(G_{ab},R_a)$. The leading contribution to $\tilde{U}(G_{ab},R_a)$ is obtained by neglecting the term $-(\partial_0 R_a)(\partial_0 R_b)$ in the argument G_{ab}. Thus

$$U(G_{ab},R_a)=m(R_a)+\tilde{U}(\tilde{G}_{ab},R_a)+\cdots \tag{7.1.5}$$

where

$$\tilde{G}_{ab}=(\partial_k R_a)(\partial_k R_b). \tag{7.1.6}$$

When we insert (7.1.4) and (7.1.5) into $\mathcal{L}=-\rho U$ we obtain

$$\mathcal{L}=-J^0\left[1-\tfrac{1}{2}\mathbf{v}^2+\cdots\right]\left[m(R_a)+\tilde{U}(\tilde{G}_{ab},R_a)+\cdots\right]. \tag{7.1.7}$$

We drop the higher order terms indicated by dots in (7.1.7).

We can also drop the term $-mJ^0$. Although it is much larger than the other two terms, it does not contribute anything to the equations of motion. To see this, note that the conserved mass current, mJ^μ, can be written as the divergence of a tensor quantity: $mJ^\mu=\partial_\nu(mJ^\nu x^\mu)$. Thus when the fields $R_a(x)$ are varied, the variation of $\int d^4x\, mJ^0$ is always zero:

$$\delta\int d^4x\, mJ^0=\int d^4x\,\partial_\nu\left[\delta(mJ^\nu x^0)\right]=0.$$

Therefore this term does not contribute to the variation δA of the action.

We now have the nonrelativistic Lagrangian we sought:

$$\mathcal{L}=\tfrac{1}{2}mJ^0\mathbf{v}^2-J^0\tilde{U}(\tilde{G}_{ab},R_a). \tag{7.1.8}$$

Note that $\mathcal{L}$ has the form

$$\mathcal{L}=(\text{kinetic energy})-(\text{potential energy})$$

familiar from the nonrelativistic mechanics of point particles. The kinetic energy is "$\tfrac{1}{2}MV^2$," and the potential energy is a function of the

*If we had not chosen units in which $c=1$, this procedure would amount to formally expanding $\mathcal{L}$ in powers of c^{-1}.

coordinates $\mathbf{x}(\mathbf{R},t)$ of the particles of material, but does not depend on their velocities.

7.2 ENERGY AND MOMENTUM

In Chapter 3 we defined energy and momentum currents

$$T_\mu{}^\nu = \delta_\mu{}^\nu \mathcal{L} - (\partial_\mu R_a)\frac{\partial \mathcal{L}}{\partial(\partial_\nu R_a)} \tag{7.2.1}$$

with

$$\text{energy} = -\int d\mathbf{x}\, T_0{}^0,$$

$$\text{momentum} = P^j = \int d\mathbf{x}\, T_j{}^0. \tag{7.2.2}$$

When the fields R_a obey the Euler-Lagrange equations of motion, this energy and momentum are conserved: $\partial_\nu T_\mu{}^\nu = 0$. This follows from the fact that $\mathcal{L}$ does not depend explicitly on t or $\mathbf{x}$; the proof does not involve the transformation properties of $\mathcal{L}$ under Lorentz transformations. Of course, since $\mathcal{L}$ is not a Lorentz scalar, $T_\mu{}^\nu$ will not be a tensor under Lorentz transformations. We return to the transformation properties of $T_\mu{}^\nu$ in the next section.

The computation of $T_\mu{}^\nu$ using (7.2.1) and the Lagrangian (7.1.8) is the sort of calculation that is worth doing as an exercise (Problem 4), but not worth reproducing in a book. It is somewhat more involved than the calculation of $T_\mu{}^\nu$ for a Lorentz invariant Lagrangian because the cases $T_0{}^0$, $T_0{}^j$, $T_j{}^0$, $T_j{}^k$ must be considered separately. We simply state the results.

The momentum density is simply the mass density times the velocity:

$$T^{j0} \equiv T_j{}^0 = mJ^0 v^j. \tag{7.2.3}$$

The momentum current contains a kinetic term equal to the momentum density times the velocity of the material, plus a "stress" term that tells how internal forces arise from the deformation of the material:

$$T^{jk} \equiv T_j{}^k = mJ^0 v^j v^k + 2J^0(\partial_j R_a)(\partial_k R_b)\frac{\partial \tilde{U}}{\partial \tilde{G}_{ab}}. \tag{7.2.4}$$

Note that the nonrelativistic momentum current for an elastic material is the same as the relativistic momentum current as long as the material is at rest.

We can, if we like, use the three momentum conservation equations $\partial_k T^{jk} = 0$ as equations of motion in place of the three Euler-Lagrange equations. The proof that these two sets of equations are equivalent is the same as that given in Section 4.5.

The energy density computed using (7.2.1) is just "kinetic energy" plus "potential energy":

$$T^{00} = -T_0{}^0 = \tfrac{1}{2} m J^0 \mathrm{v}^2 + J^0 \tilde{U}. \tag{7.2.5}$$

The energy current is

$$\begin{aligned} T^{0k} = -T_0{}^k = &\left(\tfrac{1}{2} m J^0 \mathrm{v}^2 + J^0 \tilde{U}\right) v^k \\ &+ 2J^0 (\partial_j R_a)(\partial_k R_b) \frac{\partial \tilde{U}}{\partial \tilde{G}_{ab}} v^j. \end{aligned} \tag{7.2.6}$$

The first term arises from the fact that each piece of material carries an energy per atom $\frac{1}{2}\mathrm{mv}^2 + \tilde{\mathrm{U}}$ along with it at velocity v^k. The second term represents the rate at which one piece of material is doing work on the piece of material ahead of it (see Section 4.3).

Nonrelativistic versus Relativistic Dynamics

The use of the nonrelativistic equations of motion produces the correct covariant results in problems of the types considered in the preceding chapter. For static problems, the nonrelativistic approximation is no approximation at all. When we considered sound waves, the assumption that the perturbation of the material from its uniform state was small implied that $(\partial_0 R_a)^2 \ll 1$ and $(\tilde{U}/m) \ll 1$. Thus we got the nonrelativistic result even though we started from the Lorentz invariant Lagrangian.

However, the nonrelativistic Lagrangian gives only approximately covariant results for the propagation of sound in a deformed material, since the momentum density in a moving stressed material contains terms proportional to energy of deformation times velocity and stress times velocity (see (4.7.3)); these terms are not included in the nonrelativistic approximation. Finally, of course, nonrelativistic dynamics differs radically from covariant dynamics when the material is moving at large velocities.

7.3 GALILEAN INVARIANCE

By making the approximation that all velocities are small we have lost the invariance under Lorentz transformations of the theory of elastic materials. We have gained instead invariance under "Galilean" transformations. The

Galilean transformations include rotations of the coordinates and also "Galilean boosts" $(t,x^j)\to(\bar{t},\bar{x}^j)$ with $\bar{t}=t$, $\bar{x}^j=x^j+V^jt$. This boost is a transformation to a coordinate system whose origin is moving with a constant velocity $-V^j$. The most general Galilean transformation of the coordinate system is a combination of a rotation and a boost:

$$\bar{t}=t,$$

$$\bar{x}^j=\mathcal{R}_{jk}x^k+V^jt. \tag{7.3.1}$$

The easiest way to verify that the nonrelativistic equations of motion are covariant under Galilean transformations is to examine the nonrelativistic action,

$$A[R]=\int d^4x\left(\tfrac{1}{2}mJ^0\mathbf{v}^2-J^0\tilde{U}(\tilde{G}_{ab})\right). \tag{7.3.2}$$

The notation $A[R]$ here indicates the dependence of A on the three functions $R_a(x^\mu)$. If we apply the formula (7.3.2) to calculate an action using the transformed coordinates $\bar{x}^\mu$ given in (7.3.1), we obtain the transformed action $\bar{A}[R]=A[\bar{R}]$, where $\bar{R}_a(\bar{x})=R_a(x)$. The quantities mJ^0, $\tilde{G}_{ab}$, transform like scalars (for example, $\bar{J}^0(\bar{x})=J^0(x)$); but v^j transforms into $\bar{v}^j(\bar{x})=\mathcal{R}_{jk}v^k(x)+V^j$. Thus the difference between the original and transformed action is

$$\bar{A}[R]-A[R]=\int d^4x\tfrac{1}{2}mJ^0\left(2V^j\mathcal{R}_{jk}v^k+V^kV^k\right).$$

Since $J^0v^j=J^j$, this is

$$\bar{A}[R]-A[R]=V^j\mathcal{R}_{jk}\int d^4xmJ^k$$

$$+\tfrac{1}{2}\mathbf{V}^2\int d^4xmJ^0. \tag{7.3.3}$$

But the conserved mass current mJ^μ is equal to the divergence $\partial_\nu(mJ^\nu x^\mu)$, as we observed in Section 7.1. Thus $\delta(\bar{A}[R]-A[R])$ is zero for any variation $\delta R_a(x)$ of the fields which vanishes outside of some bounded region of space-time. Therefore the requirement that $\delta\bar{A}$ be zero is exactly the same as the requirement that δA be zero: the transformed equations of motion are equivalent to the original equations of motion.

To see precisely how Galilean covariance of the equations of motion works, and to find out how objects like $T^{\mu\nu}$ transform under Galilean transformations, we construct a bit of formal apparatus similar to that used to discuss Lorentz invariance.

Let us call a set of physical quantities $a^\mu=(a^0,a^1,a^2,a^3,a^4)$ a "five-vector" if it transforms under a Galilean transformation (7.3.1) according to the linear transformation

$$\begin{aligned}\bar{a}^0&=a^0\\ \bar{a}^j&=\mathcal{R}_{jk}a^k+V^ja^0\\ \bar{a}^4&=a^4+V^ja^j+\tfrac{1}{2}\mathbf{V}^2a^0.\end{aligned}\tag{7.3.4}$$

That is, $\bar{a}^\mu=M^\mu{}_\nu a^\nu$ where

$$M^\mu{}_\nu=\begin{bmatrix}1&0&0&0&0\\ V^1&\mathcal{R}_{11}&\mathcal{R}_{12}&\mathcal{R}_{13}&0\\ V^2&\mathcal{R}_{21}&\mathcal{R}_{22}&\mathcal{R}_{23}&0\\ V^3&\mathcal{R}_{31}&\mathcal{R}_{32}&\mathcal{R}_{33}&0\\ \tfrac{1}{2}\mathbf{V}^2&V^1&V^2&V^3&1\end{bmatrix}.\tag{7.3.5}$$

Five-vectors are familiar objects. An important example is formed from the mass, momentum, and kinetic energy $E=\frac{1}{2}m\mathbf{v}^2$ of a free particle:

$$\mathcal{P}^\mu=(m,P^1,P^2,P^3,E).$$

Another example is the differential operator $\mathcal{D}^\mu$ defined by

$$\mathcal{D}^\mu=\left(0,\frac{\partial}{\partial x^1},\frac{\partial}{\partial x^2},\frac{\partial}{\partial x^3},-\frac{\partial}{\partial t}\right).$$

There is an invariant scalar product that can be formed between five-vectors. The reader can easily verify that the product

$$A^\mu B^\nu \mathcal{G}_{\mu\nu}=A^jB^j-A^0B^4-A^4B^0\tag{7.3.6}$$

is invariant under Galilean transformations.* Note that we adopt the convention in this section of summing Greek indices from 0 to 4. A useful example of this invariant inner product is the differential operator

$$\frac{1}{m}\mathcal{G}_{\mu\nu}\mathcal{P}^\mu\mathcal{D}^\nu=\frac{\partial}{\partial t}+v^j\frac{\partial}{\partial x^j}\qquad v^j=\frac{P^j}{m}.\tag{7.3.7}$$

*We might have defined the Galilean transformations abstractly as consisting of those linear transformations of five-vectors that leave the inner product (7.3.6) invariant and also leave the zero component of five-vectors unchanged. The interested reader can verify by transforming to components (a^+,a^1,a^2,a^3,a^-) with $a^\pm=(a^0\pm a^4)/\sqrt{2}$ that the Galilean transformations form a subgroup of the Lorentz group acting on a space with four spacelike and one timelike dimensions. See J. B. Kogut and D. E. Soper, *Phys. Rev.* **D1**, 2901 (1970).

When applied to a scalar field $\Phi(t,\mathbf{x})$, this differential operator gives the time derivative of the value of Φ as observed by a particle with momentum $\mathcal{P}^\mu$:

$$\frac{d}{dt}\phi(t,C^j+v^jt)=\left[\left(\frac{\partial}{\partial t}+v^j\frac{\partial}{\partial x^j}\right)\phi(t,x^j)\right]_{x^j=C^j+v^jt}.$$

The differential operator (7.3.7) is often called the substantial time derivative.

There are some five-vectors whose components a^0, a^1, a^2, a^3 are of physical interest but whose component a^4 plays no role in physics. For example, the coordinates $x^0=t$, x^1, x^2, x^3 form the first four components of a five-vector, but the fifth component $x^4=\mathbf{x}^2/2t$ has no use that I know of. Another important five-vector with an irrelevant fifth component is the matter density J^0 and current J^1, J^2, J^3, which we have used extensively in the previous chapters. (The fifth component $J^5=\mathbf{J}^2/2J^0$ is, if you will, the density of kinetic energy of the material divided by the mass per atom.) Note that the equation representing conservation of the number of atoms,

$$0=\frac{\partial}{\partial t}J^0+\frac{\partial}{\partial x^k}J^k=\mathcal{G}_{\mu\nu}\,\mathcal{D}^\mu\,\mathcal{J}^\nu,$$

is Galilean invariant, in addition to being Lorentz invariant.

We have seen that (mass, momentum, energy) forms a five-vector and that (density, current) forms the first four components of a five-vector. Thus it is natural to think that the mass, momentum, and energy densities and their corresponding currents form a five-tensor:

$$\mathcal{T}^{\mu\nu}=\begin{bmatrix} mJ^0 & mJ^1 & mJ^2 & mJ^3 & * \\ T^{10} & T^{11} & T^{12} & T^{13} & * \\ T^{20} & T^{21} & T^{22} & T^{23} & * \\ T^{30} & T^{31} & T^{32} & T^{33} & * \\ T^{00} & T^{01} & T^{02} & T^{03} & * \end{bmatrix}. \tag{7.3.8}$$

(The components $\mathcal{T}^{\mu 4}$ are "irrelevant" and so have been indicated by asterisks.) According to this ansatz, $\mathcal{T}^{\mu\nu}$ should transform under Galilean transformations (7.3.1) according to

$$\bar{\mathcal{T}}^{\mu\nu}=M^\mu{}_\alpha M^\nu{}_\beta\,\mathcal{T}^{\alpha\beta}, \tag{7.3.9}$$

where $M^\mu{}_\alpha$ is the transformation matrix defined in (7.3.5). This can be directly checked using our expressions (7.2.3) to (7.2.6) for $\mathcal{T}^{\mu\nu}$.

The transformation law can also be verified in greater generality by considering the components $\mathcal{T}^{\mu\nu}$ to be averages of microscopic quantities and using the known transformation properties of the momenta of atoms, the force between two atoms, etc. (See Problem 1.)

The transformation law (7.3.9) is of practical importance for finding the components of $\mathcal{T}^{\mu\nu}$ for a moving material when the components are known in a reference frame in which the material is instantaneously at rest. We can also use it to verify trivially that the nonrelativistic equations of motion for an elastic material are Galilean invariant. We need only note that the equations

$$\sum_{\mu=0}^{3} \frac{\partial}{\partial x^{\mu}}(mJ^{\mu})=0,$$

$$\sum_{\mu=0}^{3} \frac{\partial}{\partial x^{\mu}} T^{j\mu}=0,$$

$$\sum_{\mu=0}^{3} \frac{\partial}{\partial x^{\mu}} T^{0\mu}=0,$$

representing conservation of mass, momentum, and energy, can be compactly summarized in the manifestly covariant form

$$\mathcal{G}_{\mu\nu}\, \mathcal{D}^{\mu}\, \mathcal{T}^{\alpha\nu}=0. \tag{7.3.10}$$

PROBLEMS

1. Consider the following atomic model for a nonrelativistic solid. Atoms numbered $N=1,2,3,\ldots,10^{23}$ have masses m_N and positions $\mathbf{x}_N(t)$; they interact with one another via conservative forces

$$F^{j}_{M\to N}=-\frac{\partial}{\partial x^{j}_{N}} V_{NM}\left(|\mathbf{x}_N-\mathbf{x}_M|\right).$$

We can define the mass density in this model by

$$mJ^{0}(x^{\mu})=\sum_{N} m_N \delta\left(\mathbf{x}-\mathbf{x}_N(t)\right),$$

where $\delta(\mathbf{x})$ is a Dirac delta function. (If we prefer a nice smooth mass density, we can use a "smeared out" delta function $\delta_\lambda(\mathbf{x})=(\sqrt{\pi}\,\lambda)^{-3}\exp(-\mathbf{x}^2\lambda^{-2})$, where the length λ is small, but much larger than atomic dimensions.) We define in a similar fashion

$$mJ^k(x^\mu)=\sum_N m_N \dot{x}_N^k \delta(\mathbf{x}-\mathbf{x}_N),$$

$$T^{k0}(x^\mu)=\sum_N m_N \dot{x}_N^k \delta(\mathbf{x}-\mathbf{x}_N),$$

$$T^{kl}(x^\mu)=\sum_N m_N \dot{x}_N^k \dot{x}_N^l \delta(\mathbf{x}-\mathbf{x}_N)$$

$$+\sum_{N<M} F^k_{M\to N}\left[x_N^l - x_M^l\right]$$

$$\times\int_0^1 d\sigma\, \delta\left(\mathbf{x}-\left[\sigma \mathbf{x}_N+(1-\sigma)\mathbf{x}_M\right]\right),$$

$$T^{00}(x^\mu)=\sum_N \tfrac{1}{2} m_N \dot{\mathbf{x}}_N^2 \delta(\mathbf{x}-\mathbf{x}_N)$$

$$+\tfrac{1}{2}\sum_{N\neq M} V_{NM}(|\mathbf{x}_N-\mathbf{x}_M|)\,\delta(\mathbf{x}-\mathbf{x}_N),$$

$$T^{0l}(x^\mu)=\sum_N \tfrac{1}{2} m_N \dot{\mathbf{x}}_N^2 \dot{x}_N^l \delta(\mathbf{x}-\mathbf{x}_N)$$

$$+\tfrac{1}{2}\sum_{N\neq M} V_{NM}(|\mathbf{x}_N-\mathbf{x}_M|)\,\dot{x}_N^l\,\delta(\mathbf{x}-\mathbf{x}_N)$$

$$+\tfrac{1}{2}\sum_{N\neq M} (\mathbf{F}_{M\to N}\cdot \dot{\mathbf{x}}_N)\left(x_N^l - x_M^l\right),$$

$$\times\int_0^1 d\sigma\, \delta\left(\mathbf{x}-\left[\sigma \mathbf{x}_N+(1-\sigma)\mathbf{x}_M\right]\right).$$

Using this model, $\mathbf{F}=M\mathbf{a}$, and the definitions given, verify the mass, momentum, and energy conservation laws $\partial_\mu(mJ^\mu)=0$, $\partial_\mu T^{\nu\mu}=0$. Also verify the transformation law (7.3.9) for mJ^μ, $T^{\mu\nu}$.

2. Discuss the behavior of $\tilde{U}(\tilde{G}_{ab}, R_a)$ for a body which is nearly "rigid," $\tilde{G}_{ab}\cong\delta_{ab}$. Derive an approximation to the equations of motion appropriate for a "rigid body" and show how these equations of motion can be obtained from a variational problem in which rigidity is imposed as a constraint. (See Section 5.4.)

Show that when the rigidity constraint is imposed, only six variables are left free to vary. Use Hamilton's principle to find the equations of motion for these variables.

3. What is wrong with the idea of a relativistic rigid body, $\tilde{G}_{ab} = \delta_{ab}$? *Hint*: Consider the distance between points on a rotating rigid phonograph record.
4. Verify (7.2.3), (7.2.4), (7.2.5), and (7.2.6).
5. Consider the mechanics of the earth's atmosphere, viewed as a perfect nonrelativistic fluid with a gravitational potential energy density $-mg\frac{r_E}{|\mathbf{x}|}J^0$. Let $\mathbf{x}$ be the coordinate vector in a reference system rotating along with the earth with angular velocity $\boldsymbol{\Omega}$. Show that the Lagrangian describing this system is

$$\mathcal{L} = \tfrac{1}{2}mJ^0(\mathbf{v}+\boldsymbol{\Omega}\times\mathbf{x})^2 - J^0U(J^0) + mg\frac{r_E}{|\mathbf{x}|}J^0.$$

Find the conserved energy density and energy flow that arise because of the invariance of $\mathcal{L}$ under time translations. Argue directly from Newton's laws, including the centrifugal and coriolis forces, that your energy is indeed conserved.

6. Write the Lagrangian for a nonrelativistic Hooke's law solid, then define $\phi_a(x) = R_a(x) - x^a$, and expand the Lagrangian to second order in ϕ_a to obtain a Lagrangian appropriate for the description of small vibrations. Show that the lowest order equation of motion for ϕ_a is the same as that obtained in (6.4.5).

CHAPTER 8

The Electromagnetic Field

If any one theory serves as a model for "classical field theory," it is Maxwell's theory of the electromagnetic field. Electrodynamics is useful for describing important macroscopic phenomena, as are the theories of continuum mechanics that we have discussed up to now. But it is also a fundamental theory. The contiuum description of matter breaks down when pushed to a length scale of about 10^{-8}m, where the atomic structure of matter becomes important. But electrodynamics seems to be valid down to distances of at least 10^{-16}m. The only change needed at small distances is a switch from classical electrodynamics to quantum electrodynamics: the classical observables $\mathbf{E}(x)$, $\mathbf{B}(x)$ are replaced by quantum mechanical operators $\mathbf{E}(x)$, $\mathbf{B}(x)$. It is interesting that this quantum version of electrodynamics also serves as the primary model for quantum field theory.

8.1 FIELDS AND POTENTIALS

It is safe to suppose that the reader already knows a good deal about electrodynamics, although he may not be familiar with the manifestly Lorentz covariant form of the equations.

The electric field $\mathbf{E}(x)$ and the magnetic field $\mathbf{B}(x)$ are generally defined by writing the equation of motion for a point particle with charge q, momentum P^μ, and velocity $\mathbf{v}$, exposed to the fields $\mathbf{E}$ and $\mathbf{B}$. This is usually called the Lorentz force equation:

$$\frac{d\mathbf{P}}{dt} = q\mathbf{E} + q\mathbf{v} \times \mathbf{B}. \tag{8.1.1}$$

Multiplying by dt gives

$$dP^i = qE_i dt + q\epsilon_{ijk} dx^j B_k.$$

In time dt the change in the energy of the particle is $dP^0 = (d\mathbf{P}/dt) \cdot d\mathbf{x}$.

That is,

$$dP^0 = qE_i\,dx^i.$$

These two equations can be combined into a single equation,

$$dP^\mu = qF^{\mu\nu}dx_\nu \tag{8.1.2}$$

if we let

$$F^{\mu\nu} = \begin{bmatrix} 0 & E_1 & E_2 & E_3 \\ -E_1 & 0 & B_3 & -B_2 \\ -E_2 & -B_3 & 0 & B_1 \\ -E_3 & B_2 & -B_1 & 0 \end{bmatrix}. \tag{8.1.3}$$

Since dx_ν is an arbitrary (timelike) four-vector and dP^μ is a four-vector, the quantities $F^{\mu\nu}$ must form a tensor. This antisymmetric tensor field $F^{\mu\nu}(x)$ is called the electromagnetic field.

The equations of motion for $F^{\mu\nu}$ were discovered in a long series of experiments by Coulomb, Ampere, Faraday, and Henry. Maxwell unified their results and added a crucial term in 1865*. His equations, in modern form, are

$$\partial_\nu \epsilon^{\mu\nu\rho\sigma} F_{\rho\sigma} = 0, \tag{8.1.4}$$

$$\partial_\nu F^{\mu\nu} = \mathcal{J}^\mu. \tag{8.1.5}$$

Here $\mathcal{J}^\mu$ is the electric current: $\mathcal{J}^0$ is the charge density and $\mathcal{J}^k$ is the density of electric current.

It is often useful (and, for our purposes, necessary) to take advantage of the homogeneous Maxwell equation (8.1.4) by introducing a four-vector potential $A^\mu(x)$ such that

$$F_{\mu\nu} = \partial_\mu A_\nu - \partial_\nu A_\mu. \tag{8.1.6}$$

Apparently (8.1.4) is automatically satisfied when $F^{\mu\nu}$ has this form. Conversely, the validity of (8.1.4) implies the existence of a potential A^μ.†

*J. C. Maxwell, *Trans. Roy. Soc. Lond.* **155** (1865).

†This is a special case of a general theorem in the theory of differential forms; see, for example, H. Flanders, *Differential Forms* (Academic Press, New York, 1963). The application to the potential A_μ in electrodynamics is discussed in any electrodynamics text, for example, J. D. Jackson, *Classical Electrodynamics* (Wiley, New York, 1998, Third edition).

If $\bar{A}^{\mu}$ is related to A^{μ} by adding the gradient of a scalar field,

$$\bar{A}^{\mu}(x) = A^{\mu}(x) + \partial^{\mu}\Lambda(x), \tag{8.1.7}$$

then $\bar{A}^{\mu}$ and A^{μ} evidently lead to the same field $F^{\mu\nu}$. The converse of this statement is also true.† The transformation (8.1.7) is called a gauge transformation, or a change of gauge. When the basic equations (8.1.2) and (8.1.5) of electrodynamics are written in a form involving A^{μ}, they are unchanged by a gauge transformation, even though A^{μ} is changed.

8.2 THE LAGRANGIAN

It is possible to reformulate the content of Maxwell's equations as a variational equation $\delta\int d^4x\, \mathcal{L} = 0$. The Lagrangian is very simple:

$$\mathcal{L} = -\tfrac{1}{4}F_{\mu\nu}F^{\mu\nu} + \mathcal{J}_{\mu}A^{\mu}. \tag{8.2.1}$$

We regard $A^{\mu}(x)$ as the field to be varied and consider $F_{\mu\nu}$ to be merely a shorthand way of writing $\partial_{\mu}A_{\nu} - \partial_{\nu}A_{\mu}$. For the moment, the electric current $\mathcal{J}_{\mu}(x)$ may be thought of as a prescribed external source for the electric field. (We will shortly substitute an expression involving matter fields $R_a(x)$ for $\mathcal{J}_{\mu}$.)

We calculate from (8.2.1)

$$\frac{\partial\mathcal{L}}{\partial(\partial_{\nu}A_{\mu})} = F^{\mu\nu}, \quad \frac{\partial\mathcal{L}}{\partial A_{\mu}} = \mathcal{J}^{\mu}.$$

Thus the Euler-Lagrange equations of motion for A_{μ} are

$$\partial_{\nu}F^{\mu\nu} = \mathcal{J}^{\mu}. \tag{8.2.2}$$

This is just the part of Maxwell's equations that is not automatically satisfied by writing $F_{\mu\nu} = \partial_{\mu}A_{\nu} - \partial_{\nu}A_{\mu}$.

8.3 THE ELECTRIC FIELD COUPLED TO MATTER

Consider a piece of material that carries with it an electric charge per atom q, for instance a glass rod that has been rubbed with a piece of fur. This material produces an electric field, and will radiate electromagnetic waves if it is accelerated. Futhermore, if the material is placed in an electromagnetic field it will experience forces; indeed, each part of the material exerts a force on every other part because of the charge carried by the material. In this section we try to describe these interactions using Hamilton's principle. We do not treat currents that flow through the material until the last chapter, since such current flows involve irreversible phenomena.

To begin, we need an expression for the electric current $\mathcal{J}^{\mu}(x)$ carried by the material. We let the charge per atom of material be a function $q(R_a)$ that may depend on the material coordinates R_a. In a local rest frame of a piece of material, the charge density is $\mathcal{J}^0 = qJ^0$, since J^0 is the number of atoms per unit volume. The charge is attached to the material, so that there is no flow of charge $\mathcal{J}^i$ in this frame. Thus the four-current is

$$\mathcal{J}^{\mu} = q(R_a)J^{\mu} \tag{8.3.1}$$

in a local rest frame. Since this is a covariant equation, it remains true in any frame.

The electric current $\mathcal{J}^{\mu}$ defined by (8.3.1) is automatically conserved, since $\partial_{\mu}J^{\mu}=0$ and $(\partial_{\mu}R_a)J^{\mu}=0$.

Now we are prepared to write a Lagrangian. If matter and the electromagnetic field did not interact, the Lagrangian would be $\mathcal{L} = \mathcal{L}_M + \mathcal{L}_E$ where

$$\mathcal{L}_M = -\rho U(G_{ab}, R_a),$$

$$\mathcal{L}_E = -\tfrac{1}{4}F_{\mu\nu}F^{\mu\nu}. \tag{8.3.2}$$

We saw in the preceding section that to produce Maxwell's equations as the equations of motion for A_{μ}, we must add an interaction term

$$\mathcal{L}_I = \mathcal{J}_{\mu}A^{\mu}. \tag{8.3.3}$$

Thus we choose

$$\mathcal{L} = \mathcal{L}_M + \mathcal{L}_E + \mathcal{L}_I. \tag{8.3.4}$$

The Euler-Lagrange equations of motion for A_{μ} are $\partial_{\nu}F^{\mu\nu} = \mathcal{J}^{\mu}$, as we wished. As usual it is convenient to use the momentum conservation equations as equations of motion for R_a. The definition (3.3.3) gives

$$T_{\mu}{}^{\nu} = g_{\mu}{}^{\nu}\mathcal{L} - (\partial_{\mu}R_a)\frac{\partial \mathcal{L}}{\partial(\partial_{\nu}R_a)} - (\partial_{\mu}A_{\alpha})\frac{\partial \mathcal{L}}{\partial(\partial_{\nu}A_{\alpha})}.$$

After some routine calculations we find*

$$T^{\mu\nu} = T_M{}^{\mu\nu} + T_E{}^{\mu\nu} + T_I{}^{\mu\nu}, \tag{8.3.5}$$

*See Problem 1. The reader may be familiar with another momentum tensor in electrodynamics; we discuss this other momentum tensor in Chapters 9 and 10.

where

$$T_M{}^{\mu\nu} = \rho U u^\mu u^\nu + 2\rho(\partial^\mu R_a)(\partial^\nu R_b)\frac{\partial U}{\partial G_{ab}}, \tag{8.3.6}$$

$$T_E{}^{\mu\nu} = (\partial^\mu A_\sigma)F^{\nu\sigma} - \tfrac{1}{4}g^{\mu\nu}F_{\alpha\beta}F^{\alpha\beta}, \tag{8.3.7}$$

$$T_I{}^{\mu\nu} = A^\mu \mathcal{J}^\nu. \tag{8.3.8}$$

We can compute $\partial_\nu T_E{}^{\mu\nu} + \partial_\nu T_I{}^{\mu\nu}$ by using Maxwell's equations:

$$\begin{aligned}\partial_\nu (T_E^{\mu\nu} + T_I^{\mu\nu}) &= -(\partial^\mu \partial_\nu A_\sigma)F^{\sigma\nu} - (\partial^\mu A_\sigma)\mathcal{J}^\sigma \\ &\quad + (\partial^\mu \partial^\beta A^\alpha)F_{\alpha\beta} + (\partial_\nu A^\mu)\mathcal{J}^\nu \\ &= -F^{\mu\nu}\mathcal{J}_\nu.\end{aligned}$$

Thus the momentum conservation equation $\partial_\nu T^{\mu\nu} = 0$ amounts to

$$\partial_\nu T_M^{\mu\nu} = F^{\mu\nu}\mathcal{J}_\nu. \tag{8.3.9}$$

What is the physics behind (8.3.9)? We note that $T_M^{\mu\nu}$ is the momentum tensor of the matter alone. In the absence of an electromagnetic field, the equation $\partial_0 T_M^{\mu 0} = -\partial_k T_M^{\mu k}$ says that the rate of increase of the momentum P^μ contained in a small volume of space is equal to the rate at which momentum is flowing into the volume due to the motion of the material and forces within the material. The extra term $F^{\mu\nu}\mathcal{J}_\nu$ that has now been added to the equation apparently represents an extra contribution to the rate of increase of the momentum of the matter. It is the rate (per unit time and per unit volume) at which momentum is being transferred from the electromagnetic field to the matter. Thus (8.3.9) is simply the analogue for continuous materials of the Lorentz force law (8.1.2) for point particles.

We can in fact derive (8.3.9) from the Lorentz force law by noting that, in a local rest frame of the matter, the average force per unit volume is $\mathbf{f} = \mathcal{J}^0\mathbf{E}$. Thus $\partial_\nu T_M^{\mu\nu} = \mathcal{J}^0 F^{0\mu} = \mathcal{J}_0 F^{\mu 0} = \mathcal{J}_\nu F^{\mu\nu}$, which is just (8.3.9).

It is interesting to note that by adding one term, $A_\mu \mathcal{J}^\mu$, to the noninteracting Lagrangian we obtained equations describing *both* the effect of charged matter on the electromagnetic field and the effect of the field on charged matter. We choose the interaction term so as to obtain Maxwell's equations and we got the Lorentz force equation for free.

PROBLEMS

1. Verify that the momentum tensor corresponding to the Lagrangian (8.3.4) is given by (8.3.5).

2. Show that the action $A = \int dx(-\frac{1}{4}F_{\mu\nu}F^{\mu\nu} + \mathcal{J}_\mu A^\mu)$, where $\mathcal{J}_\mu(x)$ is a prescribed external current, is invariant under gauge transformations if and only if the current $\mathcal{J}_\mu$ is conserved.
3. Show that Maxwell's equation (8.2.2) can hold only if the current $\mathcal{J}_\mu$ is conserved.
4. Although the Lorentz force $-\partial_\nu[T_E^{\mu\nu} + T_I^{\mu\nu}] = F^{\mu\nu}\mathcal{J}_\nu$ is gauge invariant, the electromagnetic momentum tensor $T_E^{\mu\nu} + T_I^{\mu\nu}$ is not. Show, however, that the total momentum contained in the electromagnetic field, $P^\mu = \int d\mathbf{x}(T_E^{\mu 0} + T_I^{\mu 0})$, is invariant under changes of gauge as long as the fields and the gauge function $\Gamma(x)$ fall off fast enough as $|\mathbf{x}| \to \infty$.

CHAPTER 9

Further General Properties of Field Theories

We discussed some of the general methods and theorems of classical field theory in Chapter 3. This framework was adequate for our development of mechanics of elastic materials and for an introduction to electrodynamics, but a further discussion of the framework is now necessary.

9.1 NOETHER'S THEOREM

We found in Chapter 3 that if the Lagrangian depends only on the fields and their derivatives, but not on x^μ, then a certain tensor current $T^{\mu\nu}$ is conserved: $\partial_\nu T^{\mu\nu}=0$. This result is a special case of an important general theorem due to E. Noether,* which we explore here.

To state the theorem in a general setting, let us suppose that the fields $\phi_J(x)$, $J=1,2,\dots,N$, depend on coordinates x^μ, $\mu=1,2,\dots,M$. The Lagrangian can depend on x^μ directly, on the fields, and on the derivatives of the fields up to some finite order,

$$\mathfrak{L}=\mathfrak{L}(\phi_J,\partial_\mu\phi_J,\partial_\mu\partial_\nu\phi_J,\dots;x^\mu). \tag{9.1.1}$$

The variational equation is

$$0=\delta\int dx\,\mathfrak{L}=\int dx\frac{\delta A}{\delta\phi_J(x)}\delta\phi_J(x), \tag{9.1.2}$$

where $\delta A/\delta\phi_J(x)$ is the "variational derivative of A with respect to $\phi_J(x)$":

$$\frac{\delta A}{\delta\phi_J(x)}=\frac{\partial\mathfrak{L}}{\partial\phi_J}-\partial_\mu\frac{\partial\mathfrak{L}}{\partial(\partial_\mu\phi_J)}+\partial_\mu\partial_\nu\frac{\partial\mathfrak{L}}{\partial(\partial_\mu\partial_\nu\phi_J)}-\cdots. \tag{9.1.3}$$

*Noether's theorem was proved by E. Noether, *Nachr. Ges. Wiss. Göttingen*, 171 (1918). Earlier versions of the theorem may be found in G. Hamel, *Z. Math. Phys.* **50**, 1 (1904), and G. Hergoltz, *Ann. Phys.* **36**, 493 (1911).

Transformations

Noether's theorem demonstrates the existence of certain conserved currents when the action is left invariant by a set of transformations of the coordinates and the fields. The transformations are labeled by a continuous parameter ε. Let us call the *original* coordinates and fields $\bar{x}^{\mu}$, $\bar{\phi}_J$ and the transformed coordinates and fields x^{μ}, ϕ_J. Then the set of transformations is specified by giving two functions* that relate x, ϕ to $\bar{x}$, $\bar{\phi}$:

$$x^{\mu}=X^{\mu}(\bar{x};\varepsilon),$$

$$\phi_J(x)=\Phi_J\big(\bar{\phi}(\bar{x}),x;\varepsilon\big). \tag{9.1.4}$$

At $\varepsilon=0$ we assume that the transformation is the identity transformation: $X^{\mu}(\bar{x},0)=\bar{x}^{\mu}$, $\Phi_J(\bar{\phi},\bar{x};0)=\bar{\phi}_J$.

As an example, consider a rotation of the coordinate system in a two dimensional space through an angle ε,

$$X^1(\bar{x};\varepsilon)=\cos\varepsilon\ \bar{x}^1+\sin\varepsilon\ \bar{x}^2,$$

$$X^2(\bar{x};\varepsilon)=-\sin\varepsilon\ \bar{x}^1+\cos\varepsilon\ \bar{x}^2.$$

If ϕ_1 and ϕ_2 are components of a vector field, we have also

$$\Phi_1(\phi,X;\varepsilon)=\cos\varepsilon\ \bar{\phi}_1+\sin\varepsilon\ \bar{\phi}_2,$$

$$\Phi_2(\phi,X;\varepsilon)=-\sin\varepsilon\ \bar{\phi}_2+\cos\varepsilon\ \bar{\phi}^2.$$

Invariance

What does it mean for the action to be left invariant by the transformation (9.1.4)? The original action is

$$\bar{A}=\int d\bar{x}\,\mathcal{L}\left(\bar{\phi}_J(\bar{x}),\frac{\partial\bar{\phi}_J}{\partial\bar{x}^{\mu}},\ldots;\bar{x}^{\mu}\right). \tag{9.1.5}$$

If we use the same function $\mathcal{L}$ to calculate an action using the transformed

*The transformations are assumed to be invertible, and all of the functions involved are assumed to have an appropriate number of continuous derivations.

coordinates and fields we get

$$A(\varepsilon)=\int dx\, \mathcal{L}\left(\phi_J(x), \frac{\partial \phi_J}{\partial x^\mu}, \ldots; x^\mu\right). \tag{9.1.6}$$

If $A(\varepsilon)$ is equal to $\bar{A} \equiv A(0)$ for all ε and for all fields $\phi_J(x)$, then we say that the action is invariant under the set of transformations.*

Apparently we need to investigate the conditions under which $dA(\varepsilon)/d\varepsilon = 0$. The variation of $A(\varepsilon)$ corresponding to a small variation $\delta\varepsilon$ of ε is

$$\delta A(\varepsilon) \equiv \frac{dA}{d\varepsilon}\delta\varepsilon = \int dx\, \delta\mathcal{L}, \tag{9.1.7}$$

where

$$\delta\mathcal{L} = \left(\frac{\partial \mathcal{L}}{\partial \varepsilon}\right)_{x=\text{const}} \delta\varepsilon. \tag{9.1.8}$$

A sufficient condition for δA to be zero is that $\delta\mathcal{L}$ be the divergence of some quantity δB^μ,

$$\delta\mathcal{L} = \partial_\mu \delta B^\mu, \tag{9.1.9}$$

for all fields $\phi(x)$. The invariance condition (9.1.9) is taken as the given condition in Noether's theorem.

Variation of $\mathcal{L}$ When $\mathcal{L}$ Is a Scalar Density

In most applications of Noether's theorem the invariance of the action is obvious because the Lagrangian transforms as a scalar field (or, more precisely, as a "scalar density") under the transformations in question.† In these applications, δB^μ has the simple form

$$\delta B^\mu = -\mathcal{L}\frac{\partial X^\mu}{\partial \varepsilon}\delta\varepsilon. \tag{9.1.10}$$

To see how this comes about, let us first state precisely what is meant by

*The reader may object that the integrals (9.1.5) and (9.1.6) over the entire space may not converge. In that case the equation $A(\varepsilon) = A$ does not make much sense. However in Hamilton's principle one is only concerned with the difference between $A(\varepsilon)$ calculated with fields $\phi_J(x)$ and $A'(\varepsilon)$ calculated with fields $\phi_J(x) + \delta\phi_J(x)$, where $\delta\phi_J(x)$ is zero outside of a bounded region. The integral for $A'(\varepsilon) - A(\varepsilon)$ is always convergent.

†The one exception is general relativity. See (12.2.5).

a scalar density. We define

$$L(x,\varepsilon)=\mathfrak{L}\left(\phi_J(x),\frac{\partial\phi_J}{\partial x^\mu},\dots;x^\mu\right),$$

$$\bar{L}(\bar{x})=\mathfrak{L}\left(\bar{\phi}_J(\bar{x}),\frac{\partial\bar{\phi}_J}{\partial\bar{x}^\mu},\dots;\bar{x}^\mu\right). \tag{9.1.11}$$

If

$$L(x,\varepsilon)=\det\left(\frac{\partial\bar{x}^\alpha}{\partial x^\beta}\right)\bar{L}(\bar{x}), \tag{9.1.12}$$

then one says that $\mathfrak{L}$ transforms as a scalar density. (The word "density" refers to the factor $\det(\partial\bar{x}^\alpha/\partial x^\beta)$; in most examples in this book this factor equals 1.)

In the example given earlier of a theory in two dimensions with rotations as the set of transformations, one verifies immediately that the Lagrangian

$$\mathfrak{L}(\phi_J,x^\mu)=(\phi_1)^2+(\phi_2)^2+\phi_1x^1+\phi_2x^2$$

transforms as a scalar density under rotations.

It is easy to see that the action is invariant if $\mathfrak{L}$ transforms as a scalar density. The difference between $A(\varepsilon)$ and $\bar{A}$ is

$$A(\varepsilon)-\bar{A}=\int dx\,L(x,\varepsilon)-\int d\bar{x}\,\bar{L}(\bar{x}).$$

By making a change of variable in the first integral one obtains

$$A(\varepsilon)-\bar{A}=\int d\bar{x}\left[\det\left(\frac{\partial x^\alpha}{\partial\bar{x}^\beta}\right)L(X(\bar{x},\varepsilon),\varepsilon)-\bar{L}(\bar{x})\right].$$

If $\mathfrak{L}$ is a scalar density under the transformation in question, the integrand vanishes identically, so $A(\varepsilon)-\bar{A}=0$.

If the Lagrangian does transform as a scalar density, we can obtain

(9.1.10) by differentiating (9.1.12) with respect to ε, holding $\bar{x}^\mu$ fixed*:

$$\frac{\partial L(x,\varepsilon)}{\partial x^\mu}\frac{\partial x^\mu}{\partial \varepsilon}+\frac{\partial L(x,\varepsilon)}{\partial \varepsilon}=-\det\left(\frac{\partial \bar{x}^\alpha}{\partial x^\beta}\right)\frac{\partial \bar{x}^\nu}{\partial x^\mu}\frac{\partial^2 x^\mu}{\partial \bar{x}^\nu \partial \varepsilon}\bar{L}(\bar{x})$$

$$=-\det\left(\frac{\partial \bar{x}^\alpha}{\partial x^\beta}\right)\bar{L}\left[\frac{\partial}{\partial x^\mu}\left(\frac{\partial x^\mu}{\partial \varepsilon}\right)_{\bar{x}=\text{const}}\right]_{\varepsilon=\text{const}}$$

$$=-L\frac{\partial}{\partial x^\mu}\left(\frac{\partial x^\mu}{\partial \varepsilon}\right).$$

Thus

$$\left(\frac{\partial L}{\partial \varepsilon}\right)_{x=\text{const}}=-\frac{\partial}{\partial x^\mu}\left(\frac{\partial x^\mu}{\partial \varepsilon}L\right),$$

which is just (9.1.10).

Noether's Theorem

We are now in a position to state and prove Noether's theorem. We assume that the Lagrangian transforms as a scalar density under the transformations we have in mind, or at least that (9.1.9) holds:

$$\delta\mathfrak{L}=\partial_\mu \delta B^\mu.$$

We can calculate $\delta\mathfrak{L}$ directly. Using the notation

$$\delta\phi_J(x)=\left(\frac{\partial \phi_J}{\partial \varepsilon}\right)_{x=\text{const}}\delta\varepsilon,$$

we find

$$\delta\mathfrak{L}=\frac{\partial \mathfrak{L}}{\partial \phi_J}\delta\phi_J+\frac{\partial \mathfrak{L}}{\partial(\partial_\mu \phi_J)}\partial_\mu\delta\phi_J+\frac{\partial \mathfrak{L}}{\partial(\partial_\mu\partial_\nu\phi_J)}\partial_\mu\partial_\nu\delta\phi_J+\cdots$$

$$=\left[\frac{\partial \mathfrak{L}}{\partial \phi_J}-\partial_\mu\frac{\partial \mathfrak{L}}{\partial(\partial_\mu\phi_J)}+\partial_\mu\partial_\nu\frac{\partial \mathfrak{L}}{\partial(\partial_\mu\partial_\nu\phi_J)}-\cdots\right]\delta\phi_J$$

$$+\partial_\mu\left(\frac{\partial \mathfrak{L}}{\partial(\partial_\mu\phi_J)}\delta\phi_J\right)+\partial_\mu\left(\frac{\partial \mathfrak{L}}{\partial(\partial_\mu\partial_\nu\phi_J)}\partial_\nu\delta\phi_J\right)-\partial_\nu\left(\partial_\mu\frac{\partial \mathfrak{L}}{\partial(\partial_\mu\partial_\nu\phi_J)}\delta\phi_J\right)+\cdots.$$

The first term is just $[\delta A/\delta\phi_J(x)]\delta\phi_J(x)$ and the other terms are divergences of various quantities. Thus a comparison of these two expressions

*To differentiate the Jacobean here, recall that $\partial \det \mathbf{A}^{-1}/\partial A_{ij}=-(\det\mathbf{A}^{-1})A_{ji}^{-1}$.

for $\delta\mathcal{L}$ gives

$$\partial_\mu \mathcal{J}^\mu \,\delta\varepsilon = -\frac{\delta A}{\delta\phi_J(x)}\delta\phi_J(x), \tag{9.1.13}$$

where the current $\mathcal{J}^\mu$ is

$$\mathcal{J}^\mu \,\delta\varepsilon = -\delta B^\mu + \frac{\partial\mathcal{L}}{\partial(\partial_\mu\phi_J)}\delta\phi_J$$

$$+\frac{\partial\mathcal{L}}{\partial(\partial_\mu\partial_\nu\phi_J)}\partial_\nu\delta\phi_J - \partial_\nu\left(\frac{\partial\mathcal{L}}{\partial(\partial_\nu\partial_\mu\phi_J)}\right)\delta\phi_J + \cdots. \tag{9.1.14}$$

Apparently the current $\mathcal{J}^\mu$ is conserved when the fields satisfy the Euler-Lagrange equations of motion, $\delta A/\delta\phi_J(x)=0$. This is the content of Noether's theorem.

Calculation of J^μ

In most applications of Noether's theorem the Lagrangian transforms as a scalar density under the transformation in question, so that (9.1.10) for δB^μ holds. In addition, the Lagrangian usually does not depend on the second or higher order derivatives of the fields $\phi_J(x)$. Under these assumptions the general formula (9.1.14) reads

$$\mathcal{J}^\mu \,\delta\varepsilon = \mathcal{L}\frac{\partial X^\mu}{\partial\varepsilon}\delta\varepsilon + \frac{\partial\mathcal{L}}{\partial(\partial_\mu\phi_J)}\delta\phi_J. \tag{9.1.15}$$

This expression can be made more explicit by evaluating the variations $\delta\phi_J$ of the fields in terms of the functions $X^\mu(\bar{x};\varepsilon)$ and $\Phi_J(\bar{\phi},x;\varepsilon)$ that appear in the transformation law (9.1.4):

$$\bar{x}^\mu \to x^\mu = X^\mu(\bar{x},\varepsilon),$$

$$\bar{\phi}_J(\bar{x}) \to \phi_J(x) = \Phi_J(\bar{\phi}(\bar{x}),x;\varepsilon).$$

For this purpose we may assume that the Noether current is to be evaluated at $\varepsilon=0$, the point at which the transformation reduces to the identity transformation.* We have

$$\delta\phi_J(x) \equiv \left(\frac{\partial\phi_J}{\partial\varepsilon}\right)_{x=\text{const}}\delta\varepsilon$$

$$=\left[\frac{\partial\Phi_J}{\partial\bar{\phi}_K}\frac{\partial\bar{\phi}_K}{\partial\bar{x}^\nu}\left(\frac{\partial\bar{x}^\nu}{\partial\varepsilon}\right)_{x=\text{const}} + \frac{\partial\Phi_J}{\partial\varepsilon}\right]\delta\varepsilon.$$

*If it were desired to calculate $\mathcal{J}^\mu$ for $\varepsilon=\varepsilon_0\neq 0$, the calculation could be reduced to the simpler $\varepsilon=0$ case by using a new $\varepsilon'=\varepsilon-\varepsilon_0$ in place of ε and choosing the coordinates and fields corresponding to $\varepsilon'=0$ as the "untransformed" coordinates and fields.

Since the transformation is the identity transformation at $\varepsilon=0$, we can substitute

$$\frac{\partial \Phi_J}{\partial \bar{\phi}_K}=\delta_{JK}, \quad \frac{\partial \bar{\phi}_K}{\partial \bar{x}^{\mu}}=\partial_{\mu}\phi_K, \quad \frac{\partial \bar{x}^{\nu}}{\partial \varepsilon}=-\frac{\partial X^{\nu}(\bar{x},\varepsilon)}{\partial \varepsilon}.$$

Then the expression (9.1.15) for $\mathcal{J}^{\mu}$ becomes

$$\mathcal{J}^{\mu}=\mathcal{L}\frac{\partial X^{\mu}}{\partial \varepsilon}+\frac{\partial \mathcal{L}}{\partial(\partial_{\mu}\phi_J)}\left[\frac{\partial \Phi_J}{\partial \varepsilon}-(\partial_{\nu}\phi_J)\frac{\partial X^{\nu}}{\partial \varepsilon}\right]. \tag{9.1.16}$$

We will refer to this formula in future sections.

Two Examples

Suppose that a Lagrangian of the form $\mathcal{L}(\phi_J,\partial_{\mu}\phi_J,x^{\mu})$ does not depend on one of the coordinates x^0, x^1, x^2, or x^3, say x^{λ}. Then $\mathcal{L}$ transforms as a scalar under a translation of the coordinates along the x^{λ}-axis,

$$X^{\mu}(\bar{x},\varepsilon)=\bar{x}^{\mu}+\varepsilon\delta^{\mu}{}_{\lambda}, \tag{9.1.17}$$

$$\Phi_J(\bar{\phi},x,\varepsilon)=\bar{\phi}_J.$$

The corresponding conserved current is easily calculated using (9.1.16):

$$\mathcal{J}^{\mu}=\delta^{\mu}_{\lambda}-(\partial_{\lambda}\phi_J)\frac{\partial \mathcal{L}}{\partial(\partial_{\mu}\phi_J)} \tag{9.1.18}$$

Apparently this is just the momentum current T^{μ}_{λ} that we have used extensively in the preceding chapters.

Consider another simple example, two scalar fields $\phi_1(x)$, $\phi_2(x)$ with the Lagrangian

$$\mathcal{L}=-\tfrac{1}{2}(\partial_{\mu}\phi_J)(\partial^{\mu}\phi_J)-\frac{m^2}{2}\phi_J\phi_J. \tag{9.1.19}$$

The Lagrangian is drawn from quantum field theory, where it is used to describe free charged particles with mass m and spin zero. The Lagrangian is invariant under a transformation that "rotates" the fields, but does nothing to the coordinates:

$$X^{\mu}(\bar{x},\varepsilon)=\bar{x}^{\mu},$$

$$\Phi_J(\bar{\phi},x,\varepsilon)=R_{JK}(\varepsilon)\bar{\phi}_K, \tag{9.1.20}$$

$$R_{JK}(\varepsilon)=\begin{pmatrix}\cos\varepsilon & \sin\varepsilon\\ -\sin\varepsilon & \cos\varepsilon\end{pmatrix}.$$

The corresponding conserved current, calculated from (9.1.16), is

$$\mathcal{J}^{\mu}=\phi_1(\partial^{\mu}\phi_2)-(\partial^{\mu}\phi_1)\phi_2.$$

In this application of Noether's theorem to quantum field theory, $\mathcal{J}^{\mu}$ represents the electromagnetic current associated with the charged particles.

9.2 LORENTZ INVARIANCE AND ANGULAR MOMENTUM

We have just seen that invariances of the action lead to the existence of conserved quantities, and that the conserved quantities derived from invariance under the translations $\bar{x}^{\mu}\to\bar{x}^{\mu}+\varepsilon\delta^{\mu}_{\lambda}$ are the energy and momentum. What are the conserved quantities related to Lorentz invariance?

Lorentz Transformations

It will be helpful for us to explore how finite Lorentz transformations can be built from infinitesimal transformations. We begin with the equation defining a finite Lorentz transformation, $\bar{x}^{\mu}\to x^{\mu}=\Lambda^{\mu}{}_{\nu}\bar{x}^{\nu}$:

$$\Lambda^{\mu}{}_{\alpha}g_{\mu\nu}\Lambda^{\nu}{}_{\beta}=g_{\alpha\beta}. \tag{9.2.1}$$

When written in matrix notation, this is $\mathbf{\Lambda}^T\mathbf{g}\mathbf{\Lambda}=\mathbf{g}$, or

$$\mathbf{g}\mathbf{\Lambda}\mathbf{g}=\mathbf{\Lambda}^{-1T}. \tag{9.2.2}$$

Suppose that $\mathbf{\Lambda}$ has the form*

$$\mathbf{\Lambda}=e^{\mathbf{A}} \tag{9.2.3}$$

with

$$\mathbf{g}\mathbf{A}\mathbf{g}=-\mathbf{A}^T. \tag{9.2.4}$$

Then $\mathbf{\Lambda}$ satisfies (9.2.2):

$$\mathbf{g}e^{\mathbf{A}}\mathbf{g}=e^{\mathbf{gAg}}=e^{-\mathbf{A}^T}=(e^{-\mathbf{A}})^T=(e^{\mathbf{A}})^{-1T}. \tag{9.2.5}$$

*The matrix $e^{\mathbf{A}}$ is defined by its power series expansion $e^{\mathbf{A}}=\mathbf{1}+\sum_{n=1}^{\infty}(1/n!)\mathbf{A}^n$. This expansion can be used to prove the equation (9.2.5).

Thus it is plausible that any Lorentz transformation Λ can be represented as an exponential $\exp \mathbf{A}$ with $\mathbf{A}$ satisfying (9.2.4). (Moreover, it is true—as long as Λ is a proper, orthochronous Lorentz transformation.*) This exponential representation is useful because it shows how a finite Lorentz transformation Λ can be formed as the product of many infinitesimal Lorentz transformations

$$\Lambda = e^{\mathbf{A}} = \left[e^{(1/N)\mathbf{A}}\right]^N = \lim_{N\to\infty}\left[\mathbf{1} + \frac{1}{N}\mathbf{A}\right]^N.$$

In order to study Lorentz transformations, we have only to study the "infinitesimal generators" of Lorentz transformations, that is, the matrices $\mathbf{A}$ that satisfy the simple linear equation (9.2.4). Let us write (9.2.4) as $\mathbf{Ag} = -\mathbf{gA}^T$:

$$A^{\mu}{}_{\alpha}g^{\alpha\nu} = -g^{\mu\alpha}A^{\nu}{}_{\alpha},$$

or

$$A^{\mu\nu} = -A^{\nu\mu}. \tag{9.2.6}$$

There are six linearly independent solutions to (9.2.6):

$$A^{\mu\nu} = \begin{bmatrix} 0 & -1 & 0 & 0 \\ 1 & 0 & 0 & 0 \\ 0 & 0 & 0 & 0 \\ 0 & 0 & 0 & 0 \end{bmatrix}, \begin{bmatrix} 0 & 0 & -1 & 0 \\ 0 & 0 & 0 & 0 \\ 1 & 0 & 0 & 0 \\ 0 & 0 & 0 & 0 \end{bmatrix}, \ldots .$$

We name these six matrices $\mathbf{M}_{01}$, $\mathbf{M}_{02}$, $\mathbf{M}_{03}$, $\mathbf{M}_{12}$, $\mathbf{M}_{13}$, $\mathbf{M}_{23}$. For notational convenience, we also define $\mathbf{M}_{\alpha\beta} = 0$ for $\alpha = \beta$, $\mathbf{M}_{\alpha\beta} = -\mathbf{M}_{\beta\alpha}$ for $\alpha > \beta$. By definition, then, the matrix elements $(M_{\alpha\beta})^{\mu}{}_{\nu}$ of the infinitesimal generator $\mathbf{M}_{\alpha\beta}$ are

$$(M_{\alpha\beta})^{\mu}{}_{\nu} = -g^{\mu}_{\alpha}g_{\beta\nu} + g^{\mu}_{\beta}g_{\alpha\nu}. \tag{9.2.7}$$

*The representation (9.2.3), (9.2.4) can be easily established, as long as the matrix elements of $\Lambda - 1$ are small enough, by defining $\mathbf{A} = \ln \Lambda = -(1-\Lambda) - \frac{1}{2}(1-\Lambda)^2 - \frac{1}{3}(1-\Lambda)^3 - \cdots$. For our purposes this is all that is necessary. However, one can extend the representation to all proper orthochronous Lorentz transformations by an explicit construction. For this purpose it is best to use the correspondence between the proper orthochronous Lorentz group and the group of 2×2 complex matrices with determinant 1, $SL(2,C)$. [See, for example, R. F. Streater and A. S. Wightman, *PCT, Spin and Stalitics, and All That* (Benjamin, New York, 1964).]

Transformations of Vectors and Tensors

We have seen that a finite Lorentz transformation has the form $e^{\mathbf{A}}$ where $\mathbf{A}$ is a linear combination of the six matrices $\mathbf{M}_{\alpha\beta}$:

$$\Lambda = e^{1/2\omega^{\alpha\beta}\mathbf{M}_{\alpha\beta}},$$

$$\omega_{\alpha\beta} = -\omega_{\beta\alpha}. \tag{9.2.8}$$

One can specify the Lorentz transformation by giving the six independent parameters $\omega^{01}, \omega^{02}, \omega^{03}, \omega^{12}, \omega^{13}, \omega^{23}$.

A vector V^μ transforms under the Lorentz transformation specified by $\omega^{\alpha\beta}$ according to $\bar{V}^\mu \to V^\mu$ with

$$V^\mu = \left[e^{1/2\omega^{\alpha\beta}\mathbf{M}^{\alpha\beta}}\right]^\mu{}_\nu \bar{V}^\nu. \tag{9.2.9}$$

The transformation law for a second rank tensor $T^{\mu\nu}$ can be written in the same way if we think of the pair of indices (μ,ν) as one index that takes 16 values. We simply need new 16×16 generating matrices.

$$(\tilde{\mathbf{M}}_{\alpha\beta})^{\mu\nu}{}_{\rho\sigma} = (\mathbf{M}_{\alpha\beta})^\mu{}_\rho g^\nu_\sigma + g^\mu_\rho(\mathbf{M}_{\alpha\beta})^\nu{}_\sigma. \tag{9.2.10}$$

Then

$$T^{\mu\nu} = \Lambda^\mu{}_\rho\Lambda^\nu{}_\sigma\bar{T}^{\rho\sigma} = \left[e^{1/2\omega^{\alpha\beta}\tilde{\mathbf{M}}_{\alpha\beta}}\right]^{\mu\nu}{}_{\rho\sigma}\bar{T}^{\rho\sigma}. \tag{9.2.11}$$

To verify this formula, note that $\mathbf{M}$ has the form $\mathbf{C}+\mathbf{D}$, and that $\exp(\mathbf{C}+\mathbf{D}) = \exp(\mathbf{C})\exp(\mathbf{D})$, since $\mathbf{CD}=\mathbf{DC}$.

It should be apparent that the transformation law for a tensor of any rank can be written in a form like (9.2.11). In the case of a scalar field $T(x)$, the corresponding generating matrices $\tilde{\mathbf{M}}_{\alpha\beta}$ are 1×1 matrices (i.e., numbers) whose value is zero.

There is another class of fields other than tensors that is used in relativistic quantum field theory. These fields are called spinors, and also transform according to a law of the form

$$\bar{\psi}_K(\bar{x}) \to \psi_K(x) = \left[e^{1/2\omega^{\alpha\beta}\tilde{\mathbf{M}}_{\alpha\beta}}\right]_{KL}\psi_L(x),$$

but with another form for the generating matrices $\tilde{\mathbf{M}}_{\alpha\beta}$.

Lorentz Invariance in a General Theory

Suppose that we are dealing with a relativistic field theory involving several fields, $\phi(x)$, $T^{\nu\mu}(x)$, $A^\mu(x)$, etc. In order to have a flexible notation, we give each component of each field a new name, $\phi_J(x)$, $J=1,2,\ldots,N$. Then the transformation law for the fields under Lorentz transformations

$$\bar{x}^{\mu} \rightarrow x^{\mu} = \left[e^{1/2\omega^{\alpha\beta}\mathbf{M}_{\alpha\beta}}\right]^{\mu}{}_{\nu}\bar{x}^{\nu} \tag{9.2.12}$$

has the form

$$\bar{\phi}_J(\bar{x}) \rightarrow \phi_J(x) = \left[e^{1/2\omega^{\alpha\beta}\tilde{\mathbf{M}}_{\alpha\beta}}\right]_{JK}\bar{\phi}_K(\bar{x}). \tag{9.2.13}$$

The generating matrices $\tilde{\mathbf{M}}_{\alpha\beta}$ have a block diagonal form, so that $(\tilde{M}_{\alpha\beta})_{JK} = 0$ unless $\phi_J(x)$ and $\phi_K(x)$ are two components of the same vector or tensor field.

Application of Noether's Theorem

If the Lagrangian is a scalar under Lorentz transformations, then Noether's theorem tells us that there are six conserved quantities $J_{\alpha\beta}$, one for each independent parameter $\omega^{\alpha\beta}$. (We take $J_{\alpha\beta} = -J_{\beta\alpha}$ for convenience.) The corresponding conserved currents $J_{\alpha\beta}{}^{\mu}$ can be calculated by using (9.1.16) and taking ε equal to one of the parameters $\omega_{\alpha\beta} = -\omega_{\beta\alpha}$, with the other parameters equal to zero.

$$\frac{\partial}{\partial\varepsilon}x^{\mu} = (M_{\alpha\beta})^{\mu}{}_{\nu}x^{\nu},$$

$$\frac{\partial}{\partial\varepsilon}\phi_J = (\tilde{M}_{\alpha\beta})_{JK}\phi_K.$$

Thus

$$\begin{aligned} J_{\alpha\beta}{}^{\mu} &= \mathcal{L}(M_{\alpha\beta})^{\mu}{}_{\nu}x^{\nu} \\ &\quad + \frac{\partial\mathcal{L}}{\partial(\partial_{\mu}\phi_J)}\left[(\tilde{M}_{\alpha\beta})_{JK}\phi_K - (\partial_{\nu}\phi_J)(M_{\alpha\beta})^{\nu}{}_{\lambda}x^{\lambda}\right] \\ &= x_{\alpha}\left[g^{\mu}_{\beta}\mathcal{L} - (\partial_{\beta}\phi_J)\frac{\partial\mathcal{L}}{\partial(\partial_{\mu}\phi_J)}\right] \\ &\quad - x_{\beta}\left[g^{\mu}_{\alpha}\mathcal{L} - (\partial_{\alpha}\phi_J)\frac{\partial\mathcal{L}}{\partial(\partial_{\mu}\phi_J)}\right] \\ &\quad + \frac{\partial\mathcal{L}}{\partial(\partial_{\mu}\phi_J)}(\tilde{M}_{\alpha\beta})_{JK}\phi_K \end{aligned}$$

or

$$J_{\alpha\beta}{}^{\mu}=x_\alpha T_\beta{}^\mu - x_\beta T_\alpha{}^\mu + \frac{\partial \mathcal{L}}{\partial(\partial_\mu \phi_J)}(\tilde{M}_{\alpha\beta})_{JK}\phi_K. \tag{9.2.14}$$

The current $J_{\alpha\beta}{}^{\mu}$ is called the angular momentum current, and the conserved tensor $J_{\alpha\beta}=\int d^3\mathbf{x} J_{\alpha\beta}{}^{0}$ is called the angular momentum. We relate $J_{\alpha\beta}$ to the angular momentum $\mathbf{x}\times\mathbf{p}$ of particle mechanics in the next section.

9.3 PHYSICAL INTERPRETATION OF THE ANGULAR MOMENTUM TENSOR

Theories with Scalar Fields

In order to get a feeling for what the conserved quantities $J_{\alpha\beta}$ are, let us look first at theories with only scalar fields. For example, we can think of the theory of elastic materials, in which we deal with three scalar fields $R_a(x)$. Since all of the fields are scalars, the matrix $\tilde{\mathbf{M}}_{\alpha\beta}$ in (9.2.14) is zero and

$$J_{\alpha\beta}{}^{\mu}=x_\alpha T_\beta{}^\mu - x_\beta T_\alpha{}^\mu. \tag{9.3.1}$$

What is the quantity J_{12}? It is conserved because of the invariance of the action under Lorentz transformations

$$\Lambda^\mu{}_\nu=\left[e^{\omega \mathbf{M}_{12}}\right]^\mu{}_\nu. \tag{9.3.2}$$

If we use the definition (9.2.7) of $\mathbf{M}_{12}$ and expand the exponential in its power series, we find

$$\Lambda^\mu{}_\nu=\begin{bmatrix}1 & 0 & 0 & 0\\ 0 & \cos\omega & -\sin\omega & 0\\ 0 & \sin\omega & \cos\omega & 0\\ 0 & 0 & 0 & 1\end{bmatrix}. \tag{9.3.3}$$

Thus conservation of J_{12} arises from invariance of the action under rotations about the x^3-axis. Looking at the expression (9.3.1) for the density $J_{12}{}^0$, we find that it is related to the momentum density T^{k0} by

$$J_{12}{}^0=x^1T^{20}-x^2T^{10}. \tag{9.3.4}$$

This is just the third component of $\mathbf{x}\times\mathbf{p}$, where $p^k=T^{k0}$ is the momentum density.

These results generalize nicely. If we define a three-vector $J_i=\frac{1}{2}\varepsilon_{ijk}J_{jk}$, then

$$\mathbf{J}=\int d\mathbf{x}\,\mathbf{x}\times\mathbf{p} \tag{9.3.5}$$

and conservation of the component $\mathbf{n}\cdot\mathbf{J}$ arises from invariance of the action under rotations about the vector $\mathbf{n}$. If we are describing an elastic material, then each small piece of material contributes to the angular momentum an amount $\mathbf{x}\times\mathbf{p}$ equal to its "orbital" angular momentum. There is no allowance in (9.3.5) for any extra contributions due to spinning motions of the small piece or the spin angular momentum of the electrons or nuclei in the material.

What about the other components of $J_{\alpha\beta}: J_{01}, J_{02}, J_{03}$? The conservation of these components is due to the invariance of the action under pure Lorentz boosts like

$$\Lambda^{\mu}{}_{\nu}=\left[e^{\omega M_{01}}\right]^{\mu}{}_{\nu}=\begin{pmatrix}\cosh\omega & -\sinh\omega & 0 & 0\\ -\sinh\omega & \cosh\omega & 0 & 0\\ 0 & 0 & 1 & 0\\ 0 & 0 & 0 & 1\end{pmatrix}. \tag{9.3.6}$$

Let us write out J_{0k} as given by (9.3.1):

$$J_{0k}=\int d\mathbf{x}(-tT^{k0}+x^kT^{00}). \tag{9.3.7}$$

The quantity $\int d\mathbf{x}\,T^{k0}$ is the total momentum P^k of the system. The quantity $\int d\mathbf{x}\,x^kT^{00}$ is the total energy E of the system times the position $X^k(t)$ of the center of energy. Thus

$$J_{0k}=-tP^k+EX^k(t). \tag{9.3.8}$$

Conservation of J_{0k} means that

$$0=\frac{d}{dt}J_{0k}=-P^k+E\frac{dX^k}{dt}. \tag{9.3.9}$$

Thus the velocity $d\mathbf{X}/dt$ of the center of energy of the system is a constant equal to $\mathbf{P}/E$, just as it is for point particles.

Before we leave the scalar field case, we should recall that the momentum tensor for elastic materials was symmetric. This was no accident. It is a consequence of angular momentum conservation, since

$$0=\partial_\mu J_{\alpha\beta}{}^\mu=\partial_\mu\left[x_\alpha T_\beta{}^\mu - x_\beta T_\alpha{}^\mu\right]$$
$$=T_{\beta\alpha}-T_{\alpha\beta}. \tag{9.3.10}$$

When we deal with theories with vector fields, we will find that the expression (9.3.1) for $J_{\alpha\beta}{}^\mu$ is modified, so that (9.3.10) no longer holds.

Theories with Vector Fields

The physics of angular momentum becomes more interesting in theories with vector, tensor, or spinor fields. In such theories the angular momentum current has the form (9.2.14),

$$J_{\alpha\beta}{}^\mu = x_\alpha T_\beta{}^\mu - x_\beta T_\alpha{}^\mu + S_{\alpha\beta}{}^\mu, \tag{9.3.11}$$

where

$$S_{\alpha\beta}{}^\mu = \frac{\partial \mathcal{L}}{\partial(\partial_\mu \phi_J)}\left[\tilde{M}_{\alpha\beta}\right]_{JK}\phi_K. \tag{9.3.12}$$

We have seen that the terms $x_\alpha T_\beta{}^\mu - x_\beta T_\alpha{}^\mu$ can be interpreted as the orbital angular momentum associated with the momentum current $T^{\mu\nu}$. Thus the remaining term is $S_{\alpha\beta}{}^\mu$ is best described as an "intrinsic" or "spin" angular momentum carried by the tensor fields. The terminology here is meant to suggest the situation in quantum mechanics, where an electron has an orbital angular momentum $\mathbf{x}\times\mathbf{p}$ plus a spin angular momentum $\mathbf{S}$.*

Electrodynamics

The most important classical field theory in which the fields carry spin angular momentum is electrodynamics. We recall the Lagrangian (8.3.4) for the electromagnetic field coupled to charged matter,

$$\mathcal{L} = -\rho U(G_{ab}, R_a) - \tfrac{1}{4}F_{\mu\nu}F^{\mu\nu} + \mathcal{J}_\mu A^\mu. \tag{9.3.13}$$

The spin current $S_{\alpha\beta}{}^\mu$ can be calculated using (9.3.12) and a knowledge of

*This terminology is actually more than just suggestive. In the description of electrons in quantum field theory the operator that measures the electron spin is precisely $\int d\mathbf{x}\, S_{\alpha\beta}{}^0$.

the transformation law for the fields,

$$\begin{pmatrix} \bar{R}_a \\ \bar{A}^\mu \end{pmatrix} \to \begin{pmatrix} R_a \\ A^\mu \end{pmatrix} = e^{1/2\omega^{\alpha\beta}\tilde{\mathbf{M}}_{\alpha\beta}} \begin{pmatrix} \bar{R}_a \\ \bar{A}^\mu \end{pmatrix},$$

$$\tilde{M}_{\alpha\beta} = \begin{pmatrix} 0 & 0 \\ 0 & [\mathbf{M}_{\alpha\beta}]^\mu{}_\nu \end{pmatrix}.$$

We find

$$S_{\alpha\beta}{}^\mu = \frac{\partial \mathcal{L}}{\partial(\partial_\mu A^\lambda)} [M_{\alpha\beta}]^\lambda{}_\sigma A^\sigma$$

$$= -F^\mu{}_\lambda[-g^\lambda_\alpha g_{\beta\sigma} + g^\lambda_\beta g_{\alpha\sigma}]A^\sigma,$$

or

$$S_{\alpha\beta}{}^\mu = F^\mu{}_\alpha A_\beta - F^\mu{}_\beta A_\alpha. \tag{9.3.14}$$

In order to interpret $S_{\alpha\beta}{}^\mu$, let us calculate the spin density

$$s_j = \tfrac{1}{2}\varepsilon_{jkl} S_{kl}{}^0 \tag{9.3.15}$$

for an electromagnetic wave. Using (9.3.14) we find

$$\mathbf{s} = \mathbf{E} \times \mathbf{A}. \tag{9.3.16}$$

To describe a circularly polarized plane wave traveling in the z-direction, we can choose* a potential

$$A^\mu(x) = (0, a\cos[\omega(z-t)], \mp a\sin[\omega(z-t)], 0). \tag{9.3.17}$$

The minus (plus) sign corresponds to a left (right) circularly polarized wave. The corresponding electric field is

$$E_k = \partial^0 A^k - \partial^k A^0 = -\partial_0 A^k. \tag{9.3.18}$$

Thus the spin density carried by this wave is

$$\mathbf{s} = \pm a^2\omega\hat{\mathbf{z}}, \tag{9.3.19}$$

where $\hat{\mathbf{z}}$ is a unit vector pointing in the z-direction.

*The spin density **s** is gauge dependent, so we have made a specific choice of gauge. However, **s** is unaffected by a gauge change of the form $A^\mu(x) \to A^\mu(x) + \varepsilon\partial^\mu \cos[\omega(z-t)+\phi]$.

This result can be understood very simply by thinking of the wave as a beam of photons. Each photon carries an energy $\hbar\omega$. If there are N photons per unit volume, the energy density is $T^{00} = N\hbar\omega$. On the other hand, the energy density for this wave, calculated using (8.3.7), is $T^{00} = a^2\omega^2$. Thus $a^2 = N\hbar/\omega$. This result enables us to write the spin density in terms of the photon density N:

$$\mathbf{s} = \pm N\hbar\hat{\mathbf{z}}. \tag{9.3.20}$$

Apparently each left circularly polarized photon traveling in the z-direction carries spin angular momentum $+\hbar\hat{\mathbf{z}}$; each right polarized photon carries angular momentum $-\hbar\hat{\mathbf{z}}$.

We have given a simple picture of the spin density of the electromagnetic field by using the language of quantum electrodynamics (spinning photons). Nevertheless, this spin density is a feature of classical electrodynamics. Furthermore, the angular momentum of a circularly polarized light beam can be measured in a classical macroscopic experiment. One passes the beam through a "half wave plate," which changes left polarized light into right polarized light. In the process, the plate absorbs $2\hbar$ of angular momentum from each photon that passes through it. If the plate is suspended on a thin thread, it can be observed to rotate.*

9.4 THE SYMMETRIZED MOMENTUM TENSOR

We have seen that the angular momentum tensor in theories with vector (or tensor or spinor) fields contains a spin term $S_{\alpha\beta}{}^{\mu}$. In such theories the tensor $T^{\mu\nu}$, which we will call the "canonical" momentum tensor, is not symmetric.

$$0 = \partial_\mu J_{\alpha\beta}{}^{\mu} = \partial_\mu\left[x_\alpha T_\beta{}^{\mu} - x_\beta T_\alpha{}^{\mu} + S_{\alpha\beta}{}^{\mu}\right],$$

so that

$$T_{\alpha\beta} - T_{\beta\alpha} = \partial_\mu S_{\alpha\beta}{}^{\mu}.$$

In this section we find that it is always possible to define a new, symmetric momentum tensor $\Theta^{\mu\nu}$ which leads to the same total momentum $\int d\mathbf{x}\,\Theta^{\mu 0}$ as the canonical tensor.†

*R. A. Beth, *Phys. Rev.* **50**, 115 (1936); A. H. S. Holbourn, *Nature* **137**, 31 (1936).

†The construction of the symmetrized momentum tensor is due to F. J. Belinfante, *Physica* **7**, 449 (1940), and L. Rosenfeld, *Mem. Acad. R. Belg. Sci.* **18**, No. 6 (1940).

Why is $\Theta^{\mu\nu}$ important? It would seem that there is little basis for preferring one of the two tensors $\Theta^{\mu\nu}$ and $T^{\mu\nu}$ if they lead to the same total momentum and energy. The only difference lies in the different description of where the energy is located. But when we investigate gravity in Chapter 11, the location of energy will be important, since energy is the source of the gravitational field. At this point it is sufficient to say that in the theory of general relativity, the tensor $\Theta^{\mu\nu}$ rather than $T^{\mu\nu}$ is the source of the gravitational field. In electrodynamics the symmetric tensor $\Theta^{\mu\nu}(x)$ has the additional advantage that it is gauge invariant while $T^{\mu\nu}$ is not.

The construction of $\Theta^{\mu\nu}$ begins with the spin tensor $S_{\alpha\beta}{}^{\lambda}$. From its definition (9.3.12) we see that $S_{\alpha\beta}{}^{\lambda}$ is antisymmetric in its first two indices. Thus the tensor

$$G_{\mu\nu\rho}=\tfrac{1}{2}\left[S_{\nu\rho\mu}+S_{\mu\rho\nu}-S_{\mu\nu\rho}\right] \tag{9.4.1}$$

is antisymmetric in its last two indices. We define

$$\Theta^{\mu\nu}=T^{\mu\nu}+\partial_{\rho}G^{\mu\nu\rho}. \tag{9.4.2}$$

Since $G^{\mu\nu\rho}$ is antisymmetric in the indices $\nu\rho$, the conservation of $\theta^{\mu\nu}$ is equivalent to conservation of $T^{\mu\nu}$:

$$\partial_{\nu}\Theta^{\mu\nu}=\partial_{\nu}T^{\mu\nu}+\partial_{\nu}\partial_{\rho}G^{\mu\nu\rho}=\partial_{\nu}T^{\mu\nu}=0. \tag{9.4.3}$$

Using the antisymmetry of $G^{\mu\nu\rho}$ and an integration by parts,* we find that $\Theta^{\mu\nu}$ and $T^{\mu\nu}$ give the same total momentum:

$$\int d\mathbf{x}\,\Theta^{\mu 0}-\int d\mathbf{x}\,T^{\mu 0}=\int d\mathbf{x}\,\partial_{\rho}G^{\mu 0\rho}=\sum_{k=1}^{3}\int d\mathbf{x}\,\partial_{k}G^{\mu 0k}=0. \tag{9.4.4}$$

Finally, we can use conservation of momentum and angular momentum to show that $\Theta^{\mu\nu}$ is symmetric:

$$\Theta^{\mu\nu}-\Theta^{\nu\mu}=T^{\mu\nu}-T^{\nu\mu}-\partial_{\rho}S^{\mu\nu\rho},$$

but

$$\begin{aligned}0&=\partial_{\rho}J^{\nu\mu\rho}\\&=\partial_{\rho}\left[x^{\nu}T^{\mu\rho}-x^{\mu}T^{\nu\rho}+S^{\nu\mu\rho}\right]\\&=T^{\mu\nu}-T^{\nu\mu}-\partial_{\rho}S^{\mu\nu\rho},\end{aligned}$$

*We assume that the fields fall off sufficiently rapidly as $|\mathbf{x}|\to\infty$ so that the surface term in (9.4.4) vanishes.

so that

$$\Theta^{\mu\nu}-\Theta^{\nu\mu}=0. \tag{9.4.5}$$

Angular Momentum

In theories with scalar fields only, the canonical momentum tensor $T^{\mu\nu}$ is symmetric and the "canonical" angular momentum tensor $J^{\alpha\beta\lambda}$ is equal to $x^{\alpha}T^{\beta\lambda}-x^{\beta}T^{\alpha\lambda}$. Now that we have a symmetric momentum tensor $\theta^{\mu\nu}$ at hand for any theory, we are tempted to use it to define an angular momentum tensor

$$J_{\theta}^{\alpha\beta\lambda}=x^{\alpha}\theta^{\beta\lambda}-x^{\beta}\theta^{\alpha\lambda}. \tag{9.4.6}$$

This tensor is distinguished by the lack of a spin term like that occurring in the canonical angular momentum tensor

$$J^{\alpha\beta\lambda}=x^{\alpha}T^{\beta\lambda}-x^{\beta}T^{\alpha\lambda}+S^{\alpha\beta\lambda}.$$

Thus the angular momentum $J_{\theta}{}^{\alpha\beta}=\int d\mathbf{x}J_{\theta}^{\alpha\beta 0}$ contains only "orbital" angular momentum.

Can we get away with this? We first note that $J_{\theta}^{\alpha\beta\lambda}$ is conserved as a direct consequence of the fact that $\theta^{\mu\nu}$ is symmetric and conserved. In order to compare the total angular momenta found by integrating $J_{\theta}^{\alpha\beta\lambda}$ and $J^{\alpha\beta\lambda}$, we can insert the definition (9.4.2) of $\theta^{\mu\nu}$ into (9.4.6):

$$\begin{aligned}J_{\theta}^{\alpha\beta\lambda}&=x^{\alpha}T^{\beta\lambda}-x^{\beta}T^{\alpha\lambda}+x^{\alpha}\partial_{\rho}G^{\beta\lambda\rho}-x^{\beta}\partial_{\rho}G^{\alpha\lambda\rho}\\&=x^{\alpha}T^{\beta\lambda}-x^{\beta}T^{\alpha\lambda}-G^{\beta\lambda\alpha}+G^{\alpha\lambda\beta}\\&\quad+\partial_{\rho}(x^{\alpha}G^{\beta\lambda\rho}-x^{\beta}G^{\alpha\lambda\rho}).\end{aligned}$$

From the definition (9.4.1) of $G_{\mu\nu\rho}$ we can deduce that $G^{\alpha\lambda\beta}-G^{\beta\lambda\alpha}=S^{\alpha\beta\lambda}$, so that

$$J_{\theta}^{\alpha\beta\lambda}=J^{\alpha\beta\lambda}+\partial_{\rho}(x^{\alpha}G^{\beta\lambda\rho}-x^{\beta}G^{\alpha\lambda\rho}). \tag{9.4.7}$$

If we now recall that $G^{\beta\lambda\rho}$ is antisymmetric in its last two indices and use an integration by parts we find

$$\begin{aligned}&\int d\mathbf{x}J_{\theta}{}^{\alpha\beta 0}-\int d\mathbf{x}J^{\alpha\beta 0}\\&=\sum_{\rho=1}^{3}\int d^{3}x\,\partial_{\rho}(x^{\alpha}G^{\beta 0\rho}-x^{\beta}G^{\alpha 0\rho})\\&=0.\end{aligned} \tag{9.4.8}$$

Thus we can indeed use $J_{\theta}^{\alpha\beta\lambda}$ in place of $J^{\alpha\beta\lambda}$ as the angular momentum current, if we so wish.

A Mechanical Analogy

A simple mechanical example may serve to clarify the formal construction just presented. Imagine that on a Sunday morning everybody in London drives his car continuously around his block in a counterclockwise direction, as shown in Figure 9.1. The cars circling block number (N_1, N_2) have a net momentum of zero and a net angular momentum $s(N_1, N_2)$, where $s(\mathbf{N})$ is proportional to the number of cars circling the block.

If $s(\mathbf{N})$ is a slowly varying function of $\mathbf{N}$, it is sensible to give a macroscopic description of this situation. We may say that the macroscopic momentum density T^{k0} is zero and that the macroscopic angular momentum density is $J^{120} = S^{120} = s(\mathbf{N})$.

On the other hand, we can base a macroscopic description on a street by street accounting instead of a block by block accounting. Suppose, as shown in Figure 9.1, that $s(\mathbf{N})$ increases as one moves east. Then each north-south street carries a net flow of cars toward the south. The magnitude of this flow is proportional to $\partial s / \partial N_1$. Thus there is "really" an

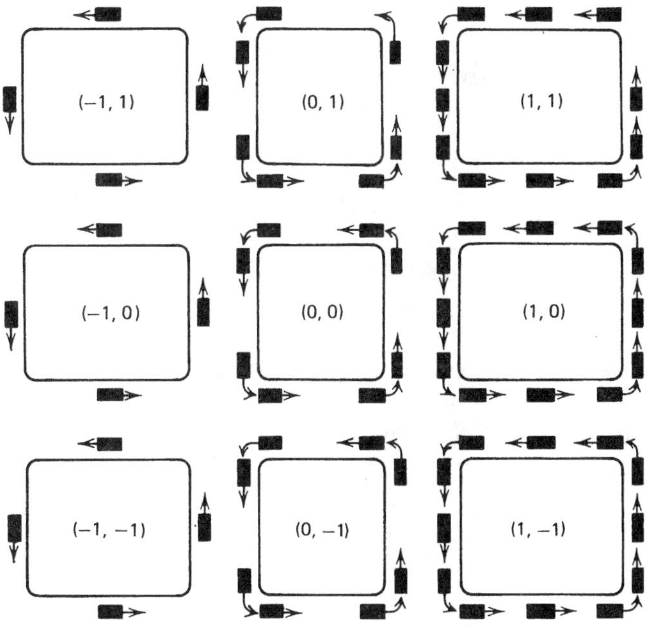

Figure 9.1 Angular momentum of traffic flow in London.

average momentum per block of

$$\theta^{10} = (\text{const})\frac{\partial s}{\partial N_2}, \quad \theta^{20} = -(\text{const})\frac{\partial s}{\partial N_1}.$$

A straightforward calculation shows that the constant appearing here is $\frac{1}{2}$. This result is the same as the formal definition (9.4.2) of θ^{j0} if we set $S^{0k\mu} = 0$:

$$\theta^{j0} = T^{j0} + \tfrac{1}{2}\,\partial_k S^{jk0}. \tag{9.4.10}$$

When the traffic pattern is viewed in this street by street manner, the angular momentum of the traffic flow arises from the net flow of cars counterclockwise around high s areas and clockwise around low s areas:

$$J_\theta^{120} = N^1\Theta^{20} - N^2\Theta^{10}.$$

The reader may ponder whether the $(T^{\mu\nu}, J^{\mu\nu\lambda})$ description or the $(\Theta^{\mu\nu}, J_\theta^{\mu\nu\lambda})$ description is preferable in this situation.

9.5 THE SYMMETRIZED MOMENTUM TENSOR IN ELECTRODYNAMICS

Let us try out the formalism for constructing a symmetric momentum tensor by using it on electrodynamics. We recall from Section 9.3 that the Lagrangian

$$\mathcal{L} = -\rho U(G_{ab}, R_a) - \tfrac{1}{4}F_{\mu\nu}F^{\mu\nu} + \mathcal{J}_\mu A^\mu \tag{9.5.1}$$

for the electromagnetic field interacting with charged matter leads to a spin tensor

$$S^{\alpha\beta\mu} = F^{\mu\alpha}A^\beta - F^{\mu\beta}A^\alpha. \tag{9.5.2}$$

Thus the tensor $G^{\mu\nu\rho}$ defined by (9.4.1) is

$$G^{\mu\nu\rho} = -F^{\nu\rho}A^\mu. \tag{9.5.3}$$

The canonical momentum tensor is given by (8.3.6):

$$T^{\mu\nu} = T_M^{\mu\nu} + (\partial^\mu A_\rho)F^{\nu\rho} - \tfrac{1}{4}g^{\mu\nu}F_{\alpha\beta}F^{\alpha\beta} + A^\mu \mathcal{J}^\nu, \tag{9.5.4}$$

where

$$T_M^{\mu\nu} = \rho U u^\mu u^\nu + 2\rho(\partial^\mu R_a)(\partial^\nu R_b)\frac{\partial U}{\partial G_{ab}}$$

is the momentum tensor of the matter. We form $\theta^{\mu\nu}$ by adding $\partial_\rho G^{\mu\nu\rho}$ to $T^{\mu\nu}$:

$$\Theta^{\mu\nu} = T_M^{\mu\nu} + (\partial^\mu A_\rho - \partial_\rho A^\mu)F^{\nu\rho} - \tfrac{1}{4} g^{\mu\nu} F_{\alpha\beta}F^{\alpha\beta}$$
$$+ A^\mu(\mathcal{J}^\nu - \partial_\rho F^{\nu\rho}).$$

In the second term we recognize $F^{\mu\rho}$, and we note that the last term is zero on account of the equation of motion* for $F^{\nu\rho}$. Thus

$$\Theta^{\mu\nu} = \Theta_M^{\mu\nu} + \Theta_E^{\mu\nu},$$
$$\Theta_M^{\mu\nu} = T_M^{\mu\nu}, \tag{9.5.5}$$
$$\Theta_E^{\mu\nu} = F^\mu{}_\rho F^{\nu\rho} - \tfrac{1}{4} g^{\mu\nu} F_{\alpha\beta}F^{\alpha\beta}.$$

Three nice things have happened here. The new tensor is symmetric, as expected. It is gauge invariant. Finally, it splits into two terms, one involving only the matter fields R_a, the other involving only the electromagnetic field $F^{\mu\nu}$. Thus one may speak of the energy of matter Θ_M^{00} and the energy of the electromagnetic field Θ_E^{00} without needing an interaction energy like $T_I^{00} = A^0 \mathcal{J}^0$.

The electromagnetic part of $\Theta^{\mu\nu}$ is probably already familiar to the reader. The energy density is

$$\Theta_E^{00} = \tfrac{1}{2}(\mathbf{E}^2 + \mathbf{B}^2). \tag{9.5.6}$$

The energy current, usually called Poynting's vector, is

$$\Theta_E^{0k} = (\mathbf{E} \times \mathbf{B})_k. \tag{9.5.7}$$

The momentum density Θ_E^{k0} is also equal to Poynting's vector, since $\Theta^{\mu\nu}$ is symmetric. Finally, the momentum current, often called the Maxwell stress

*When the Lagrangian is a function of the fields and their first derivatives, so is the canonical momentum tensor $T^{\mu\nu}$. However, the symmetrized tensor $\Theta^{\mu\nu}$ can depend also on the second derivatives of the fields, since its definition involves derivatives of the spin current $S^{\alpha\beta\mu}$. Fortunately, the second derivatives of the fields can usually be eliminated from $\Theta^{\mu\nu}$ by using the equations of motion, as was done here.

tensor, is

$$\Theta^{ij} = -E_iE_j - B_iB_j + \tfrac{1}{2}\delta_{ij}(\mathbf{E}^2 + \mathbf{B}^2). \tag{9.5.8}$$

We may recall from (8.3.9) that the rate of momentum transfer from the electromagnetic field to matter is

$$\partial_\nu \Theta_M^{\mu\nu} = F^{\mu\nu} \mathcal{J}_\nu. \tag{9.5.9}$$

(This is just the force law usually thought of as defining $F^{\mu\nu}(x)$.) Since $\partial_\nu(\Theta_M^{\mu\nu} + \Theta_E^{\mu\nu}) = 0$, we can also write

$$-\partial_\nu \Theta_E^{\mu\nu} = F^{\mu\nu} \mathcal{J}_\nu. \tag{9.5.10}$$

It is left as an exercise (Problem 1) to verify (9.5.10) directly from the equations of motion for $F^{\mu\nu}$.

Angular Momentum

If one uses $\Theta_E^{\mu\nu}$ to describe the momentum flow in an electromagnetic field, then one might naturally use $J_\theta^{\alpha\beta\mu} = x^\alpha \theta^{\beta\mu} - x^\beta \theta^{\alpha\mu}$ to describe the flow of angular momentum. Consider the case of a left circularly polarized plane wave propagating in the z-direction, as discussed in Section 9.3. We recall that such a wave carries an angular momentum density

$$J^{120} = S^{120} = +N\hbar,$$

where $N = T^{00}/\hbar\omega$ is the number of photons per unit volume. However, direct calculation of J_θ gives

$$J_\Theta^{120} = x^1\Theta^{20} - x^2\Theta^{10} = 0$$

since the momentum density $\Theta^{j0} = (\mathbf{E} \times \mathbf{B})_j$ is a vector pointing precisely in the z-direction.

What happened to the spin angular momentum of the circularly polarized photons? The example in Section 9.4 of the traffic flow in London should suggest the answer. Near the edge of a beam of light the fields $F^{\mu\nu}$ must behave in a complicated way in order to satisfy Maxwell's equations while changing from a plane wave inside the beam to zero outside the beam. Thus the energy current $\mathbf{E} \times \mathbf{B}$ near the edge need not be precisely in the z-direction. In fact, there must be an energy flow in a counterclockwise direction around the edge of the beam which gives the angular momentum of the beam. Thus by changing descriptions of the momentum current, we have assigned the angular momentum to the edge of the polarized beam of light instead of to the middle.

9.6 A SYMMETRIZED MOMENTUM TENSOR FOR NONRELATIVISTIC SYSTEMS

We have seen that every field theory derived from an action which is invariant under translations and Lorentz transformations possesses a conserved symmetric momentum tensor $\theta^{\mu\nu}$. If the action is invariant under translations and rotations but not Lorentz boosts, a similar but less powerful theorem can be proved.

Translational invariance tells us that there is a conserved momentum current $T^{k\mu}(k=1,2,3;\mu=0,1,2,3)$. If vector fields are involved in the theory, the stress tensor T^{kl} will not be symmetric. However, rotational invariance implies the existence of a conserved angular momentum current $J^{kl\mu}$, which we choose to write in the form

$$J^{kl\mu}=x^kT^{l\mu}-x^lT^{k\mu}+S^{kl\mu}. \tag{9.6.1}$$

With the aid of the spin current $S^{kl\mu}$ we can construct a new momentum current $\theta^{k\mu}$ which gives the same total momentum as $T^{k\mu}$ and for which the stress θ^{kl} is symmetric. To do so, we define*

$$\theta^{k0}=T^{k0}+\tfrac{1}{2}\,\partial_l S^{kl0}, \tag{9.6.2}$$

$$\theta^{kl}=T^{kl}-\tfrac{1}{2}\,\partial_0 S^{kl0}+\tfrac{1}{2}\,\partial_j\left[S^{ljk}+S^{kjl}-S^{klj}\right]. \tag{9.6.3}$$

Note that θ^{k0} is precisely the expression (9.4.10) we obtained for the "real" momentum density in the example of traffic flow in London.

We must show that $\theta^{k\mu}$ has the desired properties. First note that $\theta^{k\mu}$ is conserved, because $T^{k\mu}$ is conserved and $S^{kj\mu}$ is antisymmetric in its first two indices:

$$\partial_\mu\theta^{k\mu}=\partial_\mu T^{k\mu}+\tfrac{1}{2}\,\partial_l\partial_j S^{ljk}=0.$$

Second, the total momenta defined by θ^{k0} and T^{k0} are the same because $(\theta^{k0}-T^{k0})$ is a divergence. Finally, we prove that $\theta^{kl}=\theta^{lk}$ by exploiting the conservation of momentum and angular momentum:

$$0=\partial_\mu J^{lk\mu}=T^{kl}-T^{lk}-\partial_\mu S^{kl\mu}=\theta^{kl}-\theta^{lk}.$$

*This construction is not useful in a Lorentz invariant theory because $\theta^{k\mu}$ is not part of a tensor under Lorentz transformations. Also, nothing is said about the relation of the momentum density θ^{k0} to an energy current θ^{0k}. The present definition of $\theta^{k\mu}$ can be obtained from the Lorentz covariant definition by setting $S^{0k\mu}=0$.

9.7 THE LIMITATIONS OF HAMILTON'S PRINCIPLE

The content of each of the field theories we have discussed can be compactly summarized by giving the physical interpretation of the fields and then simply writing down the Lagrangian. Once the Lagrangian is known, it is easy to derive the equations of motion and the momentum tensor. There are also several technical advantages to basing the theory on a variational principle, as discussed in Section 2.1. These relate to the ease of dealing with constraints, changing variables, finding conserved quantities, and making the transition to quantum mechanics.

Nevertheless, the content of the theory can also be summarized by giving the equations of motion directly. It can therefore be argued* that Hamilton's principle is only an elegant crutch used by theoretical physicists to avoid wrestling directly with the equations of motion or momentum conservation equations in a complicated field theory.

The principle may even be a crutch whose use limits one to walking on well worn paths. This is because the variational approach is restricted to those physical sytems for which a sensible Lagrangian exists. If one wants to formulate new theories within the limits of Hamilton's principle, it is important to have an idea how wide those limits are.

Construction of $A[\phi]$ from $\delta A/\delta\phi(x)$

Can *any* equations of motion be generated from an appropriately chosen action? If this question is put narrowly enough, the answer is definitely "no."

If an action $A[\phi]$ that depends on a set of fields $\phi_K(x)$ is given, then the equations of motion are $\delta A/\delta\phi_K(x)=0$, where $\delta A/\delta\phi_K(x)$ is the variational derivative of $A[\phi]$ with respect to $\phi_K(x)$:

$$\delta A = \int dx \frac{\delta A}{\delta\phi_K(x)}\delta\phi_K(x).$$

Suppose now that $\delta A/\delta\phi_K(x)$ is a given function of the fields and their derivatives. Can we use $\delta A/\delta\phi_K(x)$ to reconstruct $A[\phi]$? If there really is such an $A[\phi]$, we can indeed reconstruct it. We use the same method one

*For instance, the Lagrangian is described as a useful "crutch" by J. D. Bjorken and S. D. Drell, *Relativistic Quantum Fields* (McGraw-Hill, New York, 1965, p. 23), although other quantum field theorists view the role of the Lagrangian as more central. In continuum mechanics, variational principles are not often used; an extreme view is presented by W. Jaunzemis, *Continuum Mechanics* (Macmillan, New York, 1967, p. 398): "The most commonly used method is *Hamilton's principle*, which is often presented as a kind of ultimate law, but would be perhaps better described as a kind of ultimate notation."

would use to reconstruct a function of several variables from its partial derivatives with respect to those variables.*

First choose a reference configuration $\bar{\phi}_K(x)$ of the fields. Then, in order to evaluate $A[\psi]$ for an arbitrary choice of fields $\phi_K(x)=\psi_K(x)$, choose a "path" $\phi_K(x;\lambda)$, $0\leqslant\lambda\leqslant 1$, that begins at the reference point, $\phi_K(x;0)=\bar{\phi}_K(x)$, and leads to the desired final point, $\phi_K(x;1)=\psi_K(x)$. For instance, a straight line path $\phi_K(x,\lambda)=\lambda\psi_K(x)+(1-\lambda)\bar{\phi}_K(x)$ may be convenient. (If an action $A[\phi]$ really exists, the construction will give the same result for any choice of path.) Now consider the action $A[\phi(\lambda)]$ evaluated along the path; its derivative with respect to λ is

$$\frac{d}{d\lambda}A[\phi(\lambda)]=\int dx\left.\frac{\delta A}{\delta\phi_K(x)}\right|_{\phi=\phi(\lambda)}\frac{\partial}{\partial\lambda}\phi_K(x,\lambda).$$

Integration of this equation from $\lambda=0$ to $\lambda=1$ gives

$$A[\psi]=\int dx\int_0^1 d\lambda\left.\frac{\delta A}{\delta\phi_K(x)}\right|_{\phi=\phi(\lambda)}\frac{\partial}{\partial\lambda}\phi_K(x,\lambda)$$
$$+\text{const},\tag{9.7.1}$$

where the integration constant is $A[\bar{\phi}]$. This equation determines $A[\psi]$, up to an arbitrary additive constant.

As an example, let us construct the action which has $\partial_\mu\partial^\mu\phi-\sin\phi=0$ as its Euler-Lagrange equation. It is convenient to choose $\bar{\phi}(x)=0$ as the reference field and to choose a straight line path $\phi(x,\lambda)=\lambda\psi(x)$ from $\phi=\bar{\phi}$ to $\phi=\psi$. Then (9.7.1) reads

$$\begin{aligned}A[\psi]&=\int dx\int_0^1 d\lambda\left[\partial_\mu\partial^\mu\lambda\psi-\sin\lambda\psi\right]\psi+\text{const}\\&=\int dx\left\{\tfrac{1}{2}(\partial_\mu\partial^\mu\psi)\psi+\cos\psi-1\right\}+\text{const}\\&=\int dx\left\{-\tfrac{1}{2}(\partial_\mu\psi)(\partial^\mu\psi)+\cos\psi-1\right\}+\text{const}.\end{aligned}$$

Variation of this action does indeed give $\delta A[\phi]/\delta\phi(x)=\partial_\mu\partial^\mu\phi-\sin\phi$.

*The construction presented here is due to E. P. Hamilton and B. E. Goodwin in R. P. Gilbert and R. G. Newton, eds., *Analytic Methods in Mathematical Physics* (Gordon and Breach, New York, 1970), pp. 461–466, and M. M. Vainberg, *Variational Methods for the Study of Non Linear Operators* (Holden Day, San Francisco, 1964). See also R. W. Atherton and G. M. Homsy, *Stud. Appl. Math.* **54**, 31 (1975).

If the given functions "$\delta A/\delta\phi_K(x)$" are chosen arbitrarily, then the action constructed according to (9.7.1) need not have these functions as its variational derivatives. When this happens, we conclude that there is *no* action which has the given functions as variational derivatives. For example, there is no action such that $\delta A/\delta\phi(x)=\dot{\phi}(x)$: if we choose "$\delta A/\delta\phi$" $=\dot{\phi}$, $\bar{\phi}=0$ and $\phi(x,\lambda)=\lambda\psi(x)$ in (9.7.1) we obtain

$$A[\psi]=\int dx\int_0^1 d\lambda(\lambda\dot{\psi}(x))\psi(x)$$

$$=\tfrac{1}{2}\int dx\,\dot{\psi}(x)\psi(x)=\int dx\frac{\partial}{\partial t}\psi(x)^2$$

$$=0,$$

so $\delta A[\phi]/\delta\phi(x)=0$.

We have seen that it may be impossible to find an action that identically reproduces a given set of equations of motion as its Euler-Lagrange equations. Nevertheless one can sometimes change the form of the equations without changing their content in such a way that the modified equations do follow from an action principle. For instance, one can introduce a potential as in electrodynamics. In the example just given, if one introduces a potential $\omega(x)$ such that $\dot{\omega}(x)=\phi(x)$, then the equation of motion is $\ddot{\omega}(x)=0$ and the action is $A[\omega]=-\frac{1}{2}\int dx\,\dot{\omega}^2$. (By treating $\omega(x)$ instead of $\phi(x)$ as the basic field in this action, one effectively restricts the allowed variations of $\phi(x)$ and thus enlarges the class of solutions to $\delta A=0$.) One can also multiply the equations of motion by an appropriate function of the fields and their derivatives before constructing the action, as we do in the following subsection.

Friction

Hamilton's principle is not very useful for describing systems in which friction, viscosity, or heat transfer is important. In most such cases no Lagrangian is known. In some simple situations we can find a Lagrangian if we try hard enough. The trouble is that the Lagrangian may be so horrible that we would not wish to use it.

As an example, let us consider the damped harmonic oscillator. The equation of motion is simply

$$m\ddot{x}(t)+f\dot{x}(t)+kx(t)=0, \tag{9.7.2}$$

where the dots indicate differentiation with respect to time. This equation is equivalent to the Euler-Lagrange equation for the following Lagrangian (see Problem 3):

$$L(\dot{x},x)=\dot{x}\int_{v_0}^{\dot{x}} dv\frac{H(v,x)}{v^2}, \tag{9.7.3}$$

where v_0 is an arbitrary constant and $H(\dot{x},x)$, which is the conserved "energy," is*

$$\begin{aligned} H(\dot{x},x)= &-\frac{\sqrt{4mk-f^2}\; l\dot{x}}{2\sqrt{(1/l^2)x^2+(f/kl^2)x\dot{x}+(m/kl^2)\dot{x}^2}} \\ &\times\cos\left\{\frac{\sqrt{4mk-f^2}}{2f}\ln\left[\frac{1}{l^2}x^2+\frac{f}{kl^2}x\dot{x}+\frac{m}{kl^2}\dot{x}^2\right]\right\} \\ &+\frac{2klx+fl\dot{x}}{2\sqrt{(1/l^2)x^2+(f/kl^2)x\dot{x}+(m/kl^2)\dot{x}^2}} \\ &\times\sin\left\{\frac{\sqrt{4mk-f^2}}{2f}\ln\left[\frac{1}{l^2}x^2+\frac{f}{kl^2}x\dot{x}+\frac{m}{kl^2}\dot{x}^2\right]\right\}, \end{aligned} \tag{9.7.4}$$

where l is an arbitrary constant. The Euler-Lagrange equation of motion derived from the Lagrangian (9.7.3) is

$$\frac{1}{m\dot{x}}\frac{\partial H(\dot{x},x)}{\partial\dot{x}}(m\ddot{x}+f\dot{x}+kx)=0, \tag{9.7.5}$$

which is equivalent to the simple equation (9.7.2).†

In this case a Lagrangian exists. However, if we had been trying to generalize the harmonic oscillator Lagrangian

$$L=\tfrac{1}{2}m\dot{x}^2-\tfrac{1}{2}kx^2$$

*The function $H(\dot{x},x)$ labels the orbits of the oscillator in $(\dot{x},x)$ space. The orbit that begins at $\dot{x}=0$, $x=a$ at $t=0$ is labeled by $H=kl^2\sin([4mk-f^2]^{1/2}f^{-1}\ln[a/l])$.

†The function $\dot{x}^{-1}\partial H(\dot{x},x)/\partial\dot{x}$ vanishes only along certain exceptional lines in $(\dot{x},x)$ space. These lines are in fact solutions of (9.7.2) with initial conditions $\dot{x}(0)=0$, $x(0)=l\exp[(2N+1)\pi f/(4mk-f^2)^{1/2}]$.

to include frictional effects, it would have been a long time before we hit upon the Lagrangian (9.7.3). Furthermore, the form of the Lagrangian is so complex that it gives us neither physical insight into the damped oscillator nor any calculational advantages.

Magnetic Poles

Even though Hamilton's principle is not of much use in formulating field theories in which "friction" plays a role, we may still hope that it is generally useful for describing physical theories in which friction can be neglected. It is instructive to test this hypothesis on a modified theory of electrodynamics in which both electric charge and magnetic charge exist.

In order to motivate the proposed generalization of electrodynamics, let us consider the electromagnetic field $F^{\mu\nu}$ and the "dual" field

$$\tilde{F}^{\mu\nu} = \tfrac{1}{2}\varepsilon^{\mu\nu\rho\sigma}F_{\rho\sigma}. \tag{9.7.6}$$

The components of the dual field may be obtained simply by replacing **E** by **B** and **B** by $-$**E** in $F^{\mu\nu}$:

$$\tilde{F}^{\mu\nu} = \begin{bmatrix} 0 & B_1 & B_2 & B_3 \\ -B_1 & 0 & -E_3 & E_2 \\ -B_2 & E_3 & 0 & -E_1 \\ -B_3 & -E_2 & E_1 & 0 \end{bmatrix}. \tag{9.7.7}$$

Notice that Maxwell's equations,

$$\partial_\nu F^{\mu\nu} = \mathcal{J}^\mu,$$
$$\partial_\nu \tilde{F}^{\mu\nu} = 0,$$

exhibit an asymmetry in the roles of $F^{\mu\nu}$ and $\tilde{F}^{\mu\nu}$. This asymmetry can be eliminated by inserting a "magnetic current" $\tilde{\mathcal{J}}^\mu$ on the right hand side of the equation for $\tilde{F}^{\mu\nu}$:

$$\partial_\nu F^{\mu\nu} = \mathcal{J}^\mu,$$
$$\partial_\nu \tilde{F}^{\mu\nu} = \tilde{\mathcal{J}}^\mu. \tag{9.7.8}$$

Since $\tilde{F}^{\mu\nu}$ is antisymmetric, the magnetic current $\tilde{\mathcal{J}}^\mu$ must be conserved. We may imagine that it is produced by a new kind of particle that carries magnetic charge; these hypothetical particles are often called magnetic monopoles. If, for instance, some magnetic monopoles were fixed on a piece of matter, the magnetic current would be

$$\tilde{\mathcal{J}}^{\mu} = \tilde{q}(R_a)J^{\mu}, \tag{9.7.9}$$

Where $\tilde{q}(R_a)$ is the number of monopoles per atom of material times the strength of one monopole, and J^{μ} is the familiar matter current.

Let us look now at the force on matter produced by the interaction of the electromagnetic field and the currents $\mathcal{J}^{\mu}$ and $\tilde{\mathcal{J}}^{\mu}$. In ordinary electrodynamics this force is given by equation (8.3.9):

$$\partial_\nu \theta_M^{\mu\nu} = F^{\mu\nu}\mathcal{J}_\nu,$$

where $\theta_M^{\mu\nu}$ is the momentum tensor of matter. In order to generalize this law so that symmetry between $F^{\mu\nu}$ and $\tilde{F}^{\mu\nu}$ is maintained, we adopt

$$\partial_\nu \theta_M{}^{\mu\nu} = F^{\mu\nu}\mathcal{J}_\nu + \tilde{F}^{\mu\nu}\tilde{\mathcal{J}}_\nu. \tag{9.7.10}$$

The equations of motion (9.7.8) for $F^{\mu\nu}$ together with the force law (9.7.10) constitute an apparently complete theory of the interaction of $F^{\mu\nu}$ with electrically and magnetically charged matter. The system of equations tells us how the motion of the charged matter affects the electromagnetic field and how the field affects the motion of the matter.

Do these proposed interactions conserve momentum and energy? To answer this question, we need to see if we can find a momentum tensor for the electromagnetic field such that

$$\partial_\nu(\Theta_M^{\mu\nu} + \Theta_E^{\mu\nu}) = 0$$

that is,

$$-\partial_\nu \Theta_E^{\mu\nu} = F^{\mu\nu}\mathcal{J}_\nu + \tilde{F}^{\mu\nu}\tilde{\mathcal{J}}_\nu. \tag{9.7.11}$$

Fortunately the usual tensor

$$\Theta_E^{\mu\nu} = F^{\mu\alpha}F^{\nu}{}_{\alpha} - \tfrac{1}{4}g^{\mu\nu}F^{\alpha\beta}F_{\alpha\beta} \tag{9.7.12}$$

is admirably suited to this purpose, since it is already invariant under the substitution $F^{\mu\nu} \to \tilde{F}^{\mu\nu}$. [This is best seen by examining (9.5.6), (9.5.7), and (9.5.8), which exhibit the components of $\Theta_E^{\mu\nu}$ in terms of **E** and **B**.] It is a straightforward exercise to differentiate (9.7.12) and use the modified Maxwell equations (9.7.8) to verify the momentum conservation law (9.7.11). (See Problem 4.) We may also note that since $\theta_E{}^{\mu\nu}$ is symmetric, angular momentum is conserved.

Now all we need is a Lagrangian. But despite the fact that the theory seems sensible, is Lorentz covariant, and incorporates conservation of momentum, energy, and angular momentum, no formulation using Hamilton's principle is known.* The most obvious scheme would be to introduce two potentials, $A^\mu(x)$ and $\tilde{A}^\mu(x)$, related to $F^{\mu\nu}$ by

$$F^{\mu\nu} = (\partial^\mu A^\nu - \partial^\nu A^\mu) - \tfrac{1}{2}\varepsilon^{\mu\nu\rho\sigma}(\partial_\rho \tilde{A}_\sigma - \partial_\sigma \tilde{A}_\rho).$$

Then the equations of motion for the two potentials are

$$\partial_\nu(\partial^\mu A^\nu - \partial^\nu A^\mu) = \mathcal{J}^\mu,$$

$$\partial_\nu(\partial^\mu \tilde{A}^\nu - \partial^\nu \tilde{A}^\mu) = \tilde{\mathcal{J}}^\mu.$$

These equations of motion can be reproduced by choosing the Lagrangian

$$\mathcal{L} = \mathcal{L}_M - \tfrac{1}{4}F_{\mu\nu}F^{\mu\nu} + \mathcal{J}_\mu A^\mu - \tilde{\mathcal{J}}_\mu \tilde{A}^\mu,$$

where $\mathcal{L}_M$ is the Lagrangian for matter alone. However, the force on matter arising from this Lagrangian is not given by (9.7.10) but rather by

$$\partial_\nu \Theta_M^{\mu\nu} = (\partial^\mu A^\nu - \partial^\nu A^\mu)\mathcal{J}_\nu$$
$$- (\partial^\mu \tilde{A}^\nu - \partial^\nu \tilde{A}^\mu)\tilde{\mathcal{J}}_\nu.$$

Thus this particular scheme fails. The reader is invited to make up an alternative scheme and try it out.

We may tentatively conclude from this discussion that not all theoretically sensible systems of differential equations can be derived from a variational principle. However, this example need cause no despair about the applicability of Hamilton's principle to physics, for the hypothetical magnetic monopoles seem not to exist in the real world.†

*See, for example, F. Rohrlich, *Phys. Rev.* **150**, 1104 (1966). However, monopoles do emerge as particular solutions in some more complicated field theories based on Hamilton's principle; see G. 't Hooft, *Nucl. Phys.* **B79**, 276 (1974).

†P. Price, E. Shirk, W. Osborne, and L. Pinski, *Phys. Rev. Lett.* **35**, 487 (1975), have reported finding the track of a monopole in a cosmic ray experiment. However, this result appears to be in conflict with the negative results of several monopole searches. See, for example, P. Eberhard, R. Ross, L. Alvarez, and R. Watt, *Phys. Rev. D* **4**, 3260 (1971); H. Kolm, F. Villa, and A. Odian, *Phys. Rev. D* **4** 1285 (1971); R. Fleischer, H. Hart, I. Jacobs, P. Price, W. Schwarz, and F. Aumento, *Phys. Rev.* **184**, 1393 (1969); R. Fleischer, P. Price, and R. Woods, *Phys. Rev.* **184**, 1398 (1969); and R. Fleischer, H. Hart, I. Jacobs, P. Price, W. Schwarz, and R. Woods, *J. Appl. Phys.* **41**, 958 (1970).

PROBLEMS

1. Use Maxwell's equations to verify directly that $\partial_\nu \theta_E^{\mu\nu} = -F^{\mu\nu} \mathcal{J}_\nu$, where $\theta_E^{\mu\nu}$ is the symmetrized electromagnetic momentum tensor $\theta_E^{\mu\nu} = F^\mu{}_\rho F^{\nu\rho} - \frac{1}{4} g^{\mu\nu} F_{\alpha\beta} F^{\alpha\beta}$.
2. Let $H(\dot{x},x)$ be the conserved energy arising from a Lagrangian $L(\dot{x},x)$: $H = \dot{x}\, \partial L/\partial \dot{x} - L$. Show that L is given in terms of H by (9.7.3), except that in general a function of the form $\dot{x} f(x)$ can be added to L.
3. Verify (9.7.5). *Hint*: You must show that $\mathfrak{D} H \equiv 0$ where $\mathfrak{D} = (f\dot{x} + kx)\, \partial/\partial \dot{x} - m\dot{x}\, \partial/\partial x$; notice that $\mathfrak{D}[x^2/l^2 + fx\dot{x}/kl^2 + m\dot{x}^2/kl^2]$ is very simple.
4. Use the modified Maxwell's equations (9.7.8) including magnetic monopole sources to verify the momentum conservation law (9.7.11).
5. The action for an infinite homogeneous nonrelativistic Hooke's law elastic solid is $\int dx\, \mathcal{L}$ with

$$\mathcal{L} = \tfrac{1}{2} m J^0 \mathbf{v}^2 - \frac{J^0}{2n} C_{abcd} \tfrac{1}{2}\left[\delta_{ab} - (\partial_k R_a)(\partial_k R_b)\right]$$
$$\times \tfrac{1}{2}\left[\delta_{cd} - (\partial_l R_c)(\partial_l R_d)\right]$$

[see (7.1.8) and (6.1.3)]. Find the conserved quantities F_b that result from the invariance of the action under the transformations $R_a \to R_a + \varepsilon \delta_{ab}$.

One can write an approximate Lagrangian $\mathcal{L}'$ appropriate for the description of small vibrations by writing $R_a = x_a + \phi_a(x)$ and expanding $\mathcal{L}$ to second order in ϕ:

$$\mathcal{L}' = \tfrac{1}{2} mn \dot{\phi}_i \dot{\phi}_j - \tfrac{1}{2} C_{ijkl} (\partial_i \phi_j)(\partial_k \phi_l).$$

This Lagrangian $\mathcal{L}'$ suggests a conserved "momentum" $P_i' = -\int d\mathbf{x}\, mn (\partial_i \phi_j) \dot{\phi}_j$. Show that P_i' is a linear combination of the actual momentum $\int d\mathbf{x}\, mJ^0 v_i$ and the conserved quantity F_i discussed above.
6. Calculate the energy density and spin angular momentum density in an elliptically polarized light wave and interpret the result in terms of left and right polarized photons.

CHAPTER 10

Electromagnetically Polarized Materials

Many of the most interesting phenomena observed in bulk matter are associated with the partial alignment of the electric or magnetic dipole moments of the atomic constituents of the matter. In this chapter we discuss the macroscopic theory of these polarization phenomena (with the omission of ferromagnetism).

The chapter begins with a definition of the macroscopic dipole moment tensor $M^{\mu\nu}$ in terms of the microscopic currents in the material. This construction leads to the macroscopic Maxwell's equation $\partial_\nu F^{\mu\nu} = \mathcal{J}^\mu + \partial_\nu M^{\mu\nu}$; the addition of an appropriate term to the Lagrangian for matter and electromagnetic fields reproduces this equation as the equation of motion for $F^{\mu\nu}$. To complete the Lagrangian, we add a term involving $M^{\mu\nu}$ and the matter fields; this gives the equation of motion for $M^{\mu\nu}$, with a number of parameters free to be determined from experiment.

With a Lagrangian in hand, we examine the mechanical effects of polarization by computing the energy momentum tensor $\theta^{\mu\nu}$. We then discuss some special cases in which the interplay between electrodynamics and mechanics is particularly interesting: a light beam in a fluid, piezoelectricity, and a perfectly conducting fluid.

10.1 SOURCES OF THE ELECTROMAGNETIC FIELD IN MATTER

The Average Field

When, in the context of continuum mechanics, we speak of the electromagnetic field $F_{\mu\nu}(x)$ inside a material body, we have in mind a smoothly varying field. In a classical mechanical model for the atomic structure of matter, we may define this macroscopic field $F_{\mu\nu}(y)$ to be an average of the "real" microscopic field $f_{\mu\nu}(x)$ over a small region in space-time near the observation point y^μ. To be definite, let us define

$$F_{\mu\nu}(y)=\int d^4x\,\Delta(y-x)f_{\mu\nu}(x), \tag{10.1.1}$$

with

$$\Delta(y-x)=\pi^{-2}\lambda^{-3}\bar{\lambda}^{-1}\exp\left(-\bar{\lambda}^{-2}(y^0-x^0)^2-\lambda^{-2}(\mathbf{y}-\mathbf{x})^2\right). \tag{10.1.2}$$

This smoothing function $\Delta(y-x)$ has been normalized so that

$$\int d^4x\,\Delta(x)=1.$$

The averaging distance λ should be small compared with any relevant macroscopic distances, but still large enough so that the microscopic fluctuations in $F_{\mu\nu}(x)$ are smoothed out—for instance, $\lambda \sim 10^{-5}$ cm. The corresponding averaging time should be short compared to any relevant macroscopic times but long compared to atomic time scales—for instance $\bar{\lambda}\sim 10^{-10}$ sec.*

The macroscopic field so defined will be useful provided that the relevant macroscepic distances are much larger than atomic distances. Only then can we choose an averaging distance λ which is small enough to be neglected in macroscopic problems, but still large when compared to atomic distances. Indeed, this requirement is essential to the usefulness of all continuum mechanics.

Assuming that the continuum requirement is met, the macroscopic field $F_{\mu\nu}(y)$ calculated using (10.1.1) will not depend on λ or $\bar{\lambda}$ or on the details of the shape of $\Delta(y-x)$. The situation is indicated schematically in Figure 10.1. The graph shows the Fourier transform

$$\tilde{e}(k)=\int_{-\infty}^{\infty} dx e^{ikx}e(x)$$

of the electric field inside a crystal in a one dimensional world. The wiggles on the left represent an externally applied field or the field produced by macroscopic inhomogeneities in the material. It is these wiggles that command our interest. The wiggles on the right, which are compressed into small regions near the inverse of the lattice distance and its harmonics, represent the interatomic fields. The macroscopic field is

$$E(y)=\int dx\,\Delta(y-x)e(x),$$

*The average over time is probably superfluous once the average over space has been carried out; it is included for formal convenience.

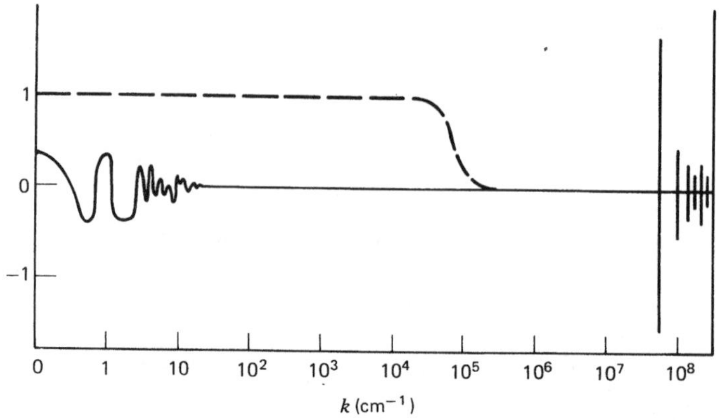

Figure 10.1 Idealized Fourier transform, $\tilde{e}(k)$, of the electric field in a one dimensional crystal (solid lines). The dashed line is the Fourier transform of tha averaging function $\Delta(x)$. (The scale for the graph of $\tilde{e}(k)$ is arbitrary and, furthermore, changes as k increases in order that the spikes on the right hand side will fit into the figure.)

or, Fourier transformed,

$$\tilde{E}(k)=\tilde{\Delta}(k)\tilde{e}(k).$$

Here $\tilde{\Delta}(k)$ is the Fourier transform of $\Delta(x)$; it is shown as a dashed line in Figure 10.1. We see that $\tilde{E}(k)$ contains only the interesting wiggles in this idealized example.

Furthermore, it does not make any difference what averaging function we pick, as long as $\Delta(x)$ is sharply peaked enough so that $\tilde{\Delta}(k)\cong 1$ for small k, but smooth enough so that $\tilde{\Delta}(k)\cong 0$ for large k. This is important because an observer who used another reference frame to construct the averaging function according to (10.1.2) would be using a different function.

Multipole Expansion of the Average Current

In the same way that $F_{\mu\nu}$ is the average of the microscopic field $f_{\mu\nu}$, so the source $\mathfrak{J}^{\mu}=\partial_{\nu}F^{\mu\nu}$ of $F^{\mu\nu}$ is the average of the microscopic current j^{μ}:

$$\mathfrak{J}^{\mu}(y)=\int d^4x\,\Delta(y-x)j^{\mu}(x). \qquad (10.1.3)$$

We seek an approximate expression for the average current; this expression will be obtained by systematically expanding $\mathfrak{J}^{\mu}$ in powers of d/λ, where d is the size of an atom and λ is the averaging distance or averaging time, then keeping only the first two terms.* We will find that these first two terms have the form

$$\bar{\mathcal{J}}^{\mu}(y)=\mathcal{J}^{\mu}(y)+\partial_{\nu}M^{\mu\nu}(y)+O\left(\frac{d^2}{\lambda^2}\right), \tag{10.1.4}$$

where $\mathcal{J}^{\mu}(y)$ is the "macroscopic current" and $M^{\mu\nu}$ is an antisymmetric tensor called the "macroscopic dipole moment tensor".†

We derive (10.1.4) within the context of a classical model for the microscopic current in a material. We assume that the material consists of a large number of atoms (or molecules) which in turn are made of several charged constituents—electrons and nuclei. Let us call the contribution to the microscopic current from the Nth atom $j_N^{\mu}(x)$, so that

$$j^{\mu}(x)=\sum_N j_N^{\mu}(x). \tag{10.1.5}$$

For each atom, we select some representative point and use the position $\mathbf{X}_N(t)$ of this point to stand for the position of the atom. For instance, $\mathbf{X}_N$ may be the position of the nucleus of the atom.‡ We assume that the constituents of the atom are bound together and thus remain within a small distance $d\sim 10^{-8}$ cm of the representative position $\mathbf{X}_N$.§ Thus

$$j_N^{\mu}(t,\mathbf{x})=0 \qquad \text{when} \quad |\mathbf{x}-\mathbf{X}_N(t)|>d. \tag{10.1.6}$$

We can make use of the smallness of d compared to the averaging distance λ in (10.1.3) by expanding $\Delta(y-x)$ about $\mathbf{x}=\mathbf{X}_N(t)$. In order to have a compact notation and to facilitate our later analysis of the Lorentz invariance of $\mathcal{J}^{\mu}$ and $M^{\mu\nu}$, we denote $X_N^0(t)=t$. Then the first two terms in the expression for $\bar{\mathcal{J}}^{\mu}$ are

$$\bar{\mathcal{J}}^{\mu}(y)=\sum_N\int d^4x\left\{\Delta(y-X_N(t))\,j_N^{\mu}(x)\right.$$
$$\left.-(x^{\nu}-X_N^{\nu}(t))\frac{\partial}{\partial y^{\nu}}\Delta(y-X_N(t))\,j_N^{\mu}(x)+O\left(\frac{d^2}{\lambda^2}\right)\right\}. \tag{10.1.7}$$

*The covariant multipole analysis presented here is a modified version of that given by S. R. de Groot and L. G. Suttorp, *Foundations of Electrodynamics* (North-Holland, Amsterdam, 1972).

†We call $\mathcal{J}^{\mu}(y)$ the "macroscopic" current at some risk of confusion with the average current $\bar{\mathcal{J}}^{\mu}(y)$. Some authors call $\mathcal{J}^{\mu}(y)$ the "true" current and $\partial_{\nu}M^{\mu\nu}(y)$ the "polarization current."

‡In the case of a solid, one can imagine that $\mathbf{X}_N(t)$ is the "average" position of the Nth atom, defined so that the path $(t,\mathbf{X}_N(t))$ in space-time is a line of constant material coordinates $R_a(t,\mathbf{x})$.

§Therefore conduction electrons, if any, must be treated as separate "atoms."

Since $|\mathbf{x}-\mathbf{X}_N(t)|<d$ in (10.1.7) while $|\partial\Delta/\partial y^\nu|\cong\Delta/\lambda$, each term in this expansion is smaller than the preceding term by a factor d/λ.

Next we add to (10.1.7) two identities that result from current conservation* (we denote dX^μ/dt by $\dot{X}^\mu$):

$$0=\sum_N\int d^4x\,\Delta(y-X_N(t))[x^\mu-X_N^\mu(t)]\partial_\lambda j_N^\lambda(x)$$

$$=\sum_N\int d^4x\Big\{\Delta(y-X_N(t))[-j_N^\mu+\dot{X}_N^\mu j_N^0]$$

$$+\frac{\partial}{\partial y^\nu}\Delta(y-X_N(t))\dot{X}_N^\nu[x^\mu-X_N^\mu]j_N^0\Big\} \tag{10.1.8}$$

and

$$0=-\tfrac{1}{2}\sum_N\int d^4x[x^\nu-X_N^\nu(t)]\frac{\partial}{\partial y^\nu}\Delta(y-X_N(t))[x^\mu-X_N^\mu(t)]\partial_\lambda j_N^\lambda(x)$$

$$=\tfrac{1}{2}\sum_N\int d^4x\frac{\partial}{\partial y^\nu}\Delta(y-X_N)([x^\mu-X_N^\mu][j_N^\nu-\dot{X}_N^\nu j_N^0]$$

$$+[x^\nu-X_N^\nu][j_N^\mu-\dot{X}_N^\mu j_N^0])$$

$$+0(d^2/\lambda^2). \tag{10.1.9}$$

The sum of (10.1.7), (10.1.8), and (10.1.9) is

$$\bar{\mathcal{J}}^\mu(y)=\mathcal{J}^\mu(y)+\frac{\partial}{\partial y^\nu}M^{\mu\nu}(y)+0\left(\frac{d^2}{\lambda^2}\right), \tag{10.1.10}$$

where

$$\mathcal{J}^\mu(y)=\sum_N\int d^4x\,\Delta(y-X_N(t))\dot{X}_N^\mu(t)j_N^0(x) \tag{10.1.11}$$

*In order to generate the entire multipole expansion, we must add to (10.1.7) an infinite sequence of identities, $0=z_m$, similar to (10.1.8) and (10.1.9) but with m factors of $(x^\nu-X^\nu)(-\partial/\partial y^\nu)$ and a factor $1/(m+1)!$ See de Groot and Suttorp, *op. cit.*

and

$$M^{\mu\nu}(y) = \sum_N \tfrac{1}{2} \int d^4x\, \Delta(y - X_N(t)) \{ [x^\mu - X_N^\mu(t)][j_N^\nu(x) + \dot{X}_N^\nu(t) j_N^0(x)]$$

$$- [x^\nu - X_N^\nu(t)][j_N^\mu(x) + \dot{X}_N^\mu(t) j_N^0(x)] \}. \tag{10.1.12}$$

These are the desired first two terms in the multipole expansion of the average current. We discuss the leading term first.

Monopole Term

The "macroscopic current" $\mathcal{J}^\mu(y)$ can be rewritten by recognizing that the space integral, $\int d\mathbf{x} j_N^0(\mathbf{x})$ is the charge of the Nth atom, Q_N:

$$\mathcal{J}^\mu(y) = \sum_N \int dt\, \Delta(y - X_N(t)) \dot{X}^\mu(t) Q_N.$$

Then we can note that $\dot{X}_N^\mu(t)\,dt$ is the line element dX_N^μ along the path of the Nth atom in space time:

$$\mathcal{J}^\mu(y) = \sum_N \int dX_N^\mu \Delta(y - X_N) Q_N. \tag{10.1.13}$$

This expression makes the Lorentz covariance of $\mathcal{J}^\mu(y)$ manifest.

The current $\mathcal{J}^\mu(y)$ is the average current we would obtain if we were to assume that each atom acts like a point charge which follows the path $\mathbf{x} = \mathbf{X}_N(t)$. This is the current we had in mind in Section 8.3 when we wrote $\mathcal{J}^\mu(x) = q(R_a(x)) J^\mu(x)$ to describe the current produced by atoms fixed in place in the matter.

As we have seen, the macroscopic current $\mathcal{J}^\mu(y)$ gives the dominant contribution to the average current $\bar{\mathcal{J}}^\mu(y)$. The remaining contributions will be smaller by a factor d/λ unless there is some cancellation in $\mathcal{J}^\mu(y)$.

As a matter of fact, there normally is a remarkable cancellation in $\mathcal{J}^\mu(y)$. A typical macroscopic piece of matter contains many thousands of coulombs of positive charge which is almost precisely balanced by an equal amount of negative charge. In nonconducting materials, this cancellation occurs at the molecular level, so that almost all of the Q_N are zero. Because of this cancellation in the lowest order approximation to $\bar{\mathcal{J}}^\mu(y)$, it is possible to observe the effects of the next term in the expansion.

Dipole Term

The macroscopic dipole moment tensor, (10.1.12), can be rewritten in a way that facilitates interpretation:

$$M^{\mu\nu}(y)=\sum_N \int dt\,\Delta\big(y^0-t,\mathbf{y}-\mathbf{X}_N(t)\big)m_N^{\mu\nu}(t), \qquad (10.1.14)$$

where

$$m_N^{\mu\nu}(t)=\tfrac{1}{2}\int d\mathbf{x}\Big\{\big[x^\mu-X_N^\mu(t)\big]\big[j_N^\nu(x)+\dot{X}_N^\nu(t)j_N^0(x)\big]$$
$$-\big[x^\nu-X_N^\nu(t)\big]\big[j_N^\mu(x)+\dot{X}_N^\mu(t)j_N^0(x)\big]\Big\}. \qquad (10.1.15)$$

The components of $m_N^{\mu\nu}$ are familiar objects, at least for atoms at rest. If $\dot{\mathbf{X}}=0$ we have

$$m_N^{i0}(t)=\int d\mathbf{x}(x^i-X_N^i)j_N^0$$

$$m_N^{ij}(t)=\tfrac{1}{2}\int d\mathbf{x}\{(x^i-X_N^i)j_N^j-(x^j-X_N^j)j_N^i\}.$$

We recognize m^{i0} as the ith component of the electric dipole moment of the atom and $\frac{1}{2}\epsilon_{ijk}m^{jk}$ as the ith component of its magnetic dipole moment. Thus $M^{\mu\nu}$ is the average dipole moment per unit volume. In the case of a moving atom we can still interpret $M^{\mu\nu}(y)$ as a dipole moment per unit volume, provided that we use (10.1.15) as the definition of the electric and magnetic atomic dipole moments when $\dot{\mathbf{X}}\neq 0$.

The electric dipole moment per unit volume M^{i0} is often denoted P_i and called the polarization vector. Similarly the magnetic dipole moment per unit volume $\frac{1}{2}\epsilon_{ijk}M^{jk}$ is often denoted M_i and called the magnetization vector.

Frame Dependence of $M^{\mu\nu}$

Equation 10.1.12 can be viewed as defining a tensor $M^{\mu\nu}(y)$ in a preferred reference frame. Let T_α be a vector whose components in this preferred frame are $T_\alpha=(1,0,0,0)$; thus T_α specifies the time axis of this frame. Does the choice of this timelike vector T_α really affect $M^{\mu\nu}(y)$?

To find out, let us first choose a fixed parameterization for the path of the Nth atom in space-time; instead of $X_N^\mu(t)$ we write $X_N^\mu(\tau_N)$, where τ_N

can be any* convenient parameter that labels points along the path and is chosen independently of T_α. We write $\dot{X}_N^\mu(\tau)$ for $dX^\mu/d\tau$. The parameterization τ_N can be extended to provide an atomic "time function" $\tau_N(x^\mu)$ for the whole atom with the aid of the vector T_α. We let $\tau_N(x^\mu)$ be a function that is constant along the planes of constant coordinate time $T_\alpha x^\alpha$ in the preferred reference frame and equals the parameter τ_N when x^μ lies on the path of the atom:

$$T_\alpha\left[x^\alpha - X_N^\alpha\left(\tau_N(x)\right)\right]=0. \tag{10.1.16}$$

In the definition (10.1.12) of $M^{\mu\nu}(y)$ we use τ_N as the curve parameter and we use the identity $dX_N^\mu/dt\, g_\lambda^0 = \dot{X}_N^\mu(\tau)(\partial_\lambda \tau_N(x))$; this gives

$$\begin{aligned} M^{\mu\nu}(y) = {} & \sum_N \tfrac{1}{2}\int d^4x\, \Delta(y - X_N(\tau_N(x))) \\ & \times\left\{\left[x^\mu - X_N^\mu(\tau_N(x))\right]\left[j_N^\nu(x) + \dot{X}_N^\nu(\tau_N(x))(\partial_\lambda \tau_N(x)) j_N^\lambda(x)\right]\right. \\ & \left. -(\mu\leftrightarrow\nu)\right\}. \end{aligned} \tag{10.1.17}$$

We are now in a position to calculate the change $\delta M^{\mu\nu}(y)$ produced by a small change δT_α in the time axis of the preferred reference frame. According to (10.1.16) the corresponding change in $\tau_N(x)$ is†

$$\delta\tau_N(x) = \frac{\delta T_\alpha\left(x^\alpha - X_N^\alpha(\tau_N)\right)}{T_\beta \dot{X}_N^\beta(\tau_N)}. \tag{10.1.18}$$

The variation in $M^{\mu\nu}(y)$ is

$$\begin{aligned} \delta M^{\mu\nu}(y) = {} & \sum_N \tfrac{1}{2}\int d^4x \\ & \times\left\{\left(-\frac{\partial\Delta}{\partial y^\gamma}\dot{X}_N^\gamma\left[x^\mu - X_N^\mu\right] - \Delta\dot{X}_N^\mu\right)\left(j_N^\nu + \dot{X}_N^\nu(\partial_\lambda\tau_N) j_N^\lambda\right)\delta\tau_N\right. \\ & \left. + \Delta\left[x^\mu - X_N^\mu\right] j_N^\lambda \partial_\lambda\left(\dot{X}_N^\nu \delta\tau_N\right) - (\mu\leftrightarrow\nu)\right\}. \end{aligned}$$

*It will be apparent that the choice of curve parameters does not affect $M^{\mu\nu}$.

†The averaging function $\Delta(y-x)$ is also frame dependent. However as we argued at the beginning of the chapter, $M^{\mu\nu}$ does not really depend on Δ, up to corrections of order (averaging distance)/(macroscopic distance).

The $\partial/\partial x^\lambda$ in the second term can be integrated by parts to give (using $\partial_\lambda j_N^\lambda = 0$),

$$\Delta[x^\mu - X_N^\mu] j_N^\lambda \partial_\lambda(\dot{X}_N^\nu \delta\tau_N)$$

$$\to + \frac{\partial\Delta}{\partial y^\gamma} \dot{X}^\gamma (\partial_\lambda \tau_N)[x^\mu - X_N^\mu] j_N^\lambda \dot{X}_N^\nu \delta\tau_N$$

$$+ \Delta \dot{X}_N^\mu (\partial_\lambda \tau_N) j_N^\lambda \dot{X}_N^\nu \delta\tau_N - \Delta j_N^\mu \dot{X}_N^\nu \delta\tau_N.$$

After some cancellation and the insertion of (10.1.18) for $\delta\tau$, we find

$$\delta M^{\mu\nu}(y) = -\frac{\partial}{\partial y^\gamma} \sum_N \frac{1}{2} \int d^4x \left\{ \Delta(y - X_N(\tau_N(x))) \dot{X}_N^\gamma \right.$$
$$\times [x^\mu - X_N^\mu] j_N^\nu [x^\alpha - X_N^\alpha] \frac{\delta T_\alpha}{T_\beta \dot{X}_N^\beta}$$
$$\left. - (\mu \leftrightarrow \nu) \right\}. \qquad (10.1.19)$$

We see that $\delta M^{\mu\nu}(y)/\delta T_\alpha$ is smaller than $M^{\mu\nu}(y)$ by a factor d/λ because of the extra factor $[x^\alpha - X_N^\alpha](\partial/\partial y^\gamma)$ in $\delta M^{\mu\nu}$. Thus $M^{\mu\nu}(y)$ does not "really" depend on the choice of the preferred frame specified by T_α: the change in $M^{\mu\nu}$ caused by changing T_α is no larger than the terms in the multipole expansion that we are already ignoring.

10.2 THE ELECTROMAGNETIC FIELD-DIPOLE MOMENT INTERACTION

The dipole moment tensor $M^{\mu\nu}(x)$ introduced in the preceding section enters Maxwell's equations in the form

$$\partial_\nu F^{\mu\nu} = \mathcal{J}^\mu + \partial_\nu M^{\mu\nu}. \qquad (10.2.1)$$

This equation of motion for A^μ can be obtained from Hamilton's principle by adding an interaction term $\frac{1}{2} M^{\mu\nu} F_{\mu\nu}$ to the Lagrangian for the electromagnetic field:

$$\mathcal{L}_E + \mathcal{L}_{E\mathcal{J}} + \mathcal{L}_{EP} = -\tfrac{1}{4} F^{\mu\nu} F_{\mu\nu} - \mathcal{J}_\mu A^\mu + \tfrac{1}{2} M^{\mu\nu} F_{\mu\nu}. \qquad (10.2.2)$$

When evaluated in the local rest frame of the material, this added interaction term reads

$$\tfrac{1}{2}M^{\mu\nu}F_{\mu\nu} = \mathbf{P}\cdot\mathbf{E} + \mathbf{M}\cdot\mathbf{B}. \tag{10.2.3}$$

The reader may check that variation of the action defined by (10.2.2) with respect to the potential $A_\mu(x)$ does indeed give Maxwell's equations.

Of course, we need to include in the Lagrangian the term $\mathcal{L}_M = -\rho U$ describing the polarizable matter. We also need a new term $\mathcal{L}_{PM}$ to describe how the polarization tensor interacts with the matter. This is the topic of the next section.

10.3 THE MATTER-DIPOLE MOMENT INTERACTION

There are many possibilities for the interaction of the dipole moment field $M^{\mu\nu}(x)$ with the matter fields $R_a(x)$. In this and the following sections we consider some of the most interesting and important types of interaction.*

The simplest interaction is very simple indeed. We suppose that it costs some action to polarize the material. In analogy with the potential energy term in the Lagrangian for a harmonic oscillator, $\mathcal{L} = \frac{1}{2}m\dot{x}(t)^2 - \frac{1}{2}kx(t)^2$, we take this action to be proportional to the squares of the electric and magnetic polarizations:

$$\mathcal{L}_{PM} = -\tfrac{1}{2}\frac{1}{\kappa}\mathbf{P}(x)^2 - \tfrac{1}{2}\frac{1}{\chi}\mathbf{M}(x)^2. \tag{10.3.1}$$

The "spring constants" $1/\kappa$ and $1/\chi$ are to be characteristics of the material.

The interaction Lagrangian (10.3.1) is written as it appears in a local rest frame of the material. The corresponding expression valid in any reference frame is

$$\begin{aligned}\mathcal{L}_{PM} = &-\frac{1}{2\kappa}u_\mu M^{\lambda\mu}u_\nu M_\lambda{}^\nu \\ &-\frac{1}{4\chi}M^{\mu\nu}(g_{\mu\rho}+u_\mu u_\rho)(g_{\nu\sigma}+u_\nu u_\sigma)M^{\rho\sigma},\end{aligned} \tag{10.3.2}$$

where u^μ is the four-velocity of the matter. When we use $\mathcal{L}_{PM}$ as the

*The action used here was given (for a fluid, and with somewhat different variables) by H. G. Schöpf, *Ann. Phys.* **9**, 301 (1962). See also R. A. Grot, *J. Math. Phys.* **11**, 109 (1970); G. A. Maugin and A. C. Eringen, *J. Math. Phys.* **13**, 1777 (1972); G. A. Maugin, *Ann. Inst. Henri Poincaré* **16**, 133 (1972).

interaction between $M^{\mu\nu}(x)$ and $R_a(x)$, the complete Lagrangian for matter, polarization, and electromagnetic field is

$$\begin{aligned} \mathcal{L} = & -\tfrac{1}{4}F_{\mu\nu}F^{\mu\nu} + \mathcal{J}_\mu A^\mu + \tfrac{1}{2}M_{\mu\nu}F^{\mu\nu} \\ & + \mathcal{L}_{PM} - \rho U(G_{ab}, R_a), \end{aligned} \tag{10.3.3}$$

where $\mathcal{L}_M = -\rho U(G_{ab}, R_a)$ is the Lagrangian for the material in the absence of polarization.

We have seen that variation of the action defined by (10.3.3) with respect to $A^\mu(x)$ gives the correct Maxwell equation (10.2.1). Variation with respect to $M_{\mu\nu}(x)$ gives*

$$\begin{aligned} \mathbf{P}(x) &= \kappa \mathbf{E}(x), \\ \mathbf{M}(x) &= \chi \mathbf{B}(x) \end{aligned} \tag{10.3.4}$$

in a local rest frame of the material. The covariant form of this equation is

$$\begin{aligned} 0 &= \frac{\partial \mathcal{L}}{\partial M_{\mu\nu}} - \frac{\partial \mathcal{L}}{\partial M_{\nu\mu}} \\ &= F^{\mu\nu} - \frac{1}{\kappa}(M^{\mu\alpha}u_\alpha u^\nu - M^{\nu\alpha}u_\alpha u^\mu) \\ &\quad - \frac{1}{\chi}(g^{\mu\rho} + u^\mu u^\rho)(g^{\nu\sigma} + u^\nu u^\sigma)M_{\rho\sigma}. \end{aligned} \tag{10.3.5}$$

The linear relations between **P** and **E** and between **M** and **B** indicated by (10.3.4) are observed in a wide variety of materials. The dimensionles constant κ is called the electric susceptibility of the material. It is normally positive and ranges between 10^{-3} and 10^{+2} in common materials. The magnetic susceptibility[†,‡] χ may be very small and positive (paramagnetism) or very small and negative (diamagnetism). We will not discuss another important class of materials called ferromagnetic, in which **M** is large but is not uniquely determined by **B**. Some typical values of κ and χ are shown in Table 10.1.

*When we form the variation δA of the action corresponding to a variation $\delta M_{\mu\nu}$ of $M_{\mu\nu}$, we constrain $M_{\mu\nu}$ to remain antisymmetric. Thus only antisymmetric variations $\delta M_{\mu\nu}$ are allowed, and Hamilton's principle implies that the antisymmetric part of $\partial \mathcal{L} / \partial M_{\mu\nu}$ must be zero.

†Actually, it is $\chi' = \chi/(1-\chi)$ that is called the magnetic susceptibility; χ' is defined by $\mathbf{M} = \chi'(\mathbf{B} - \mathbf{M}) = \chi'\mathbf{H}$. For paramagnetic and diamagnetic materials, χ is so small that $\chi \cong \chi'$.

‡The reader should be warned that authors who use "unrationalized" systems of electrical units, such as the Gaussian CGS system, define $\kappa_U = \kappa/4\pi$ and $\chi_U = \chi/4\pi$. Authors who use "rationalized" units, such as the standard MKS system, use the susceptibilities presented here.

Table 10.1 Electric and Magnetic Susceptibilities of Three Common Substances (at 20°C and Atmospheric Pressure)[a]

Substance	κ	χ
O_2 gas	5.2×10^{-4}	1.8×10^{-6}
H_2O	79	-9.1×10^{-6}
NaC	5.1	-1.4×10^{-5}

[a]*Source*: *Handbook of Chemistry and Physics*, 43rd edition (Chemical Rubber Publishing Co., Cleveland, 1961).

Alternative Choice Of Polarization Fields

We have been thinking of the components of $M^{\mu\nu}$ as the fundamental fields describing the polarization of the material. But since the polarization is really a property of matter, it is somewhat more natural to use fields that describe the polarization and magnetization in the material coordinate system. We define

$$\mathcal{P}_a(x) = \frac{1}{\rho}\frac{\partial x^i}{\partial R_a} P_i(x), \tag{10.3.6}$$

$$\mathcal{M}_a(x) = \frac{1}{\rho}\frac{\partial x^i}{\partial R_a} M_i(x)$$

in a local rest frame of the material.* The inverse transformation in the rest frame is

$$P_i = \rho\frac{\partial R_a}{\partial x^i}\mathcal{P}_a,$$

$$M_i = \rho\frac{\partial R_a}{\partial x^i}\mathcal{M}_a, \tag{10.3.7}$$

or, in any frame,

$$M^{\mu\nu} = (\partial^\mu R_a)\mathcal{P}_a J^\nu - (\partial^\nu R_a)\mathcal{P}_a J^\mu$$
$$- J_\lambda \epsilon^{\lambda\mu\nu\sigma}(\partial_\sigma R_a)\mathcal{M}_a. \tag{10.3.8}$$

When we write the Lagrangian (10.3.3) in terms of $\mathcal{P}_a(x)$ and $\mathcal{M}_a(x)$ as

*Readers familiar with Riemannian geometry will recognize $\mathcal{P}_a$ and $\mathcal{M}_a$ as the covariant components in the material coordinate system of the average electric and magnetic dipole moments per atom.

the independent fields we obtain

$$\mathcal{L} = -\tfrac{1}{4}F_{\mu\nu}F^{\mu\nu} + \mathcal{J}_\mu A^\mu + \tfrac{1}{2}F_{\mu\nu}M^{\mu\nu}$$
$$-\tfrac{1}{2}\frac{\rho^2}{\kappa}\mathcal{P}_a G_{ab}\mathcal{P}_b - \tfrac{1}{2}\frac{\rho^2}{\chi}\mathcal{M}_a G_{ab}\mathcal{M}_b$$
$$-\rho U(\mathbf{G},\mathbf{R}). \tag{10.3.9}$$

The equations of motion are unchanged except for the substitution (10.3.7), since we have only made a change of variables in the Lagrangian. The advantage of the present formulation is not that it contains new physics, but that it suggests a generalization of the Lagrangian that does contain new physics.

The Electric Susceptibility Matrix

We chose for the electric polarization part of the Lagrangian (10.3.9) the term

$$\mathcal{L}_{\mathcal{P}M} = -\frac{1}{2}\frac{\rho^2}{\kappa}\mathcal{P}_a G_{ab}\mathcal{P}_b. \tag{10.3.10}$$

An important generalization suggests itself: there is no sacred principle that prevents us from replacing the matrix G_{ab} by some other symmetric matrix. Let us, therefore, examine the implications of replacing (10.3.10) with

$$\mathcal{L}_{\mathcal{P}M} = -\tfrac{1}{2}\rho n\,\mathcal{P}_a \kappa_{ab}^{-1}\mathcal{P}_b, \tag{10.3.11}$$

where κ_{ab} is a symmetric matrix, κ_{ab}^{-1} is its inverse. Recall that $n(R_a)$ is the number of atoms per unit $d\mathbf{R}$.*

The corresponding equation of motion for $\mathcal{P}_a(x)$ is

$$n\mathcal{P}_a(x) = \kappa_{ab}\mathcal{E}_b(x), \tag{10.3.12}$$

where

$$\mathcal{E}_b(x) = -(\partial_\mu R_b)u_\nu F^{\nu\mu}(x). \tag{10.3.13}$$

In the special case of a stationary undeformed piece of material with

*The choice of factors, ρn, in (10.3.11) is made for later convenience. The factor of ρ has been included so that $-\mathcal{L}/\rho$, the interaction energy per atom, is simple. The factor n is included so that κ_{ab} is dimensionless.

$R_a(t,\mathbf{x})=x^a$, this is

$$P_i=\kappa_{ij}E_j. \tag{10.3.14}$$

The linear relationship between **P** and **E** has been retained; but the susceptibility κ has now been replaced by a susceptibility matrix, so that **P** does not necessarily point in the direction of **E**. In the general case of a deformed material the relation between **P** and **E** in a local rest frame of the material is similar to (10.3.14):

$$P_i=\left[\frac{\rho}{n}(\partial_i R_a)(\partial_j R_b)\kappa_{ab}\right]E_j. \tag{10.3.15}$$

The equations of motion for $F_{\mu\nu}$ and $\mathcal{P}_a$ are not affected if we allow κ_{ab} to be a function of the material coordinates $R_a(x)$, reflecting an inhomogeneity in the material. The susceptibility can also depend on the derivatives $\partial_\mu R_a$, although, for reasons of Lorentz invariance, κ_{ab} can depend on these derivatives only through their scalar products $(\partial_\mu R_a)(\partial^\mu R_b)=G_{ab}$.

The matrix κ_{ab} may, in principle, be an arbitrary function of the deformation matrix G_{ab}. In many cases, however, this messy situation can be considerably simplified. In a solid dielectric one can often assume that the strain $s_{ab}=\frac{1}{2}(\delta_{ab}-G_{ab})$ is small. Then it is sensible to replace κ_{ab} by the first two terms in its Taylor expression in powers of s_{ab}:

$$\kappa_{ab}(\mathbf{s})=\alpha_{ab}+\beta_{abcd}s_{cd}. \tag{10.3.16}$$

Here α_{ab} is a symmetric matrix, and β_{abcd} is symmetric in the indices ab and in the indices cd. Thus α_{ab} has 6 independent components and β_{abcd} has 36 independent components.

If the material is crystalline, there are symmetry restrictions on the components of α_{ab} and β_{abcd} appropriate to the crystal class, just as there are restrictions on the elastic constants C_{abcd} discussed in Section 6.1.

The requirement that a material be isotropic is very restrictive indeed: invariance of $\mathcal{L}$ under rotations of the material coordinates implies that

$$\boldsymbol{\kappa}(\mathbf{G})=\mathbf{A}^T\boldsymbol{\kappa}(\mathbf{A}\mathbf{G}\mathbf{A}^T)\mathbf{A} \tag{10.3.17}$$

for any rotation matrix A_{ab}. This implies

$$\alpha_{ab}=\alpha\delta_{ab},$$

$$\beta_{abcd}=\beta_1\delta_{ab}\delta_{cd}+\beta_2\left(\delta_{ac}\delta_{bd}+\delta_{ad}\delta_{bc}-\tfrac{2}{3}\delta_{ab}\delta_{cd}\right). \tag{10.3.18}$$

The constant α is the inverse of the electric susceptibility of the unstrained material; β_1 and β_2 tell how the susceptibility matrix changes under compression and shear.

Fluid dielectrics are particularly simple. The Lagrangian for a fluid should be invariant under any change of material coordinates $R_a \to \bar{R}_a$ with $\det[\partial \bar{R}_a / \partial R_b] = 1$. (That is, if you stir a beaker of fluid while keeping the density constant, its physical properties are unchanged.) Thus (10.3.17) must hold for any matrix **A** with $\det \mathbf{A} = 1$. This implies that $\kappa(\mathbf{G})_{ab}$ has the form (see Problem 1)

$$\kappa(\mathbf{G})_{ab} = \kappa(\mathcal{V}) G_{ab}^{-1}. \tag{10.3.19}$$

This is precisely the form (10.3.10) with which we began our discussion of the interaction of the polarization tensor and matter. We recall that with this form for κ_{ab}, the polarization **P** in the fluid rest frame is always in the direction of the electric field, even when the fluid is deformed.

The Magnetic Susceptibility Matrix

It is apparent that the magnetization term in the Lagrangian should also be generalized in the case of a solid material. Instead of

$$\mathcal{L}_{\mathfrak{M}M} = -\frac{1}{2} \frac{\rho^2}{\chi} \mathfrak{M}_a G_{ab} \mathfrak{M}_b, \tag{10.3.20}$$

we should choose

$$\mathcal{L}_{\mathfrak{M}M} = -\tfrac{1}{2} \rho n \mathfrak{M}_a \chi_{ab}^{-1} \mathfrak{M}_b, \tag{10.3.21}$$

where χ_{ab} is a symmetric matrix.

The equation of motion for $\mathfrak{M}_a$ corresponding to this interaction term in $\mathcal{L}$ is

$$n \mathfrak{M}_a = \chi_{ab} \mathcal{B}_b, \tag{10.3.22}$$

where

$$\mathcal{B}_b = -\tfrac{1}{2} u_\lambda F_{\alpha\beta} \epsilon^{\lambda\alpha\beta\sigma} (\partial_\sigma R_b). \tag{10.3.23}$$

In a local rest frame of the material this is

$$M_i = \left[\frac{\rho}{n} (\partial_i R_a)(\partial_j R_b) \chi_{ab} \right] B_j. \tag{10.3.24}$$

Thus we retain the linear relation between **M** and **B**, but do not require that **M** be parallel to **B**.

The magnetic susceptibility matrix χ_{ab} can depend on $R_a(x)$ and $G_{ab} = (\partial_\mu R_a)(\partial^\mu R_b)$, just as the electric susceptibility matrix κ_{ab} could depend on R_a and G_{ab}. Also, the form of χ_{ab} is restricted by symmetry considerations. In the case of a fluid, these symmetry considerations lead us back to our starting point by requiring $\mathcal{L}_{\mathfrak{M}M}$ have the form (10.3.20) with χ a function of the volume per atom $\mathcal{V}$.

10.4 THE MOMENTUM TENSOR

How does a piece of dielectric material bend when it is placed between capacitor plates? We can answer this and similar questions if we have equations of motion for the matter fields $R_a(x)$ of a polarizable material.

Since by now we have a complete Lagrangian for polarizable materials, these equations of motion are completely determined; we need only compute them. What has happened is that we have chosen the terms in the Lagrangian in order to describe how the electromagnetic fields $F_{\mu\nu}(x)$, $\mathcal{P}_a(x)$ and $\mathfrak{M}_a(x)$ depend on each other and on the motion of the material. This choice determines how the motion of the material depends on the electromagnetic fields.

The most illuminating procedure for determining the motion of the matter is to use the momentum conservation equations $\partial_\nu \theta^{\mu\nu} = 0$ as equations of motion for $R_a(x)$. (Equivalently, one might use the Euler-Lagrange equations for $R_a(x)$ or the momentum conservation equations $\partial_\nu T^{\mu\nu} = 0$ involving the canonical, unsymmetrized momentum tensor $T^{\mu\nu}$.) Therefore we construct the symmetrized momentum tensor $\theta^{\mu\nu}(x)$ corresponding to our Lagragian.

Since the complete Lagrangian for matter and fields has grown to be long and complicated, if not chaotic, it will pay us to adopt a systematic approach to the calculation* of $\theta^{\mu\nu}$. The Lagrangian consists of seven terms:

$$\mathcal{L} = \mathcal{L}_E + \mathcal{L}_{E\mathcal{J}} + \mathcal{L}_{E\mathcal{P}} + \mathcal{L}_{E\mathfrak{M}} + \mathcal{L}_{\mathcal{P}M} + \mathcal{L}_{\mathfrak{M}M} + \mathcal{L}_M,$$

$$\mathcal{L}_E = -\tfrac{1}{4} F^{\mu\nu} F_{\mu\nu},$$

$$\mathcal{L}_{E\mathcal{J}} = + \mathcal{J}_\mu A^\mu,$$

*There is an easier way to do this calculation than the one to be described here. One differentiates each term with respect to $g_{\mu\nu}$, using a method to be explained in the next chapter. I suggest that readers wishing to do the calculating themselves do it by this easier method.

$$\begin{aligned}
\mathcal{L}_{E\mathcal{P}} + \mathcal{L}_{E\mathfrak{M}} &= \tfrac{1}{2}F^{\mu\nu}M_{\mu\nu} \\
&= \tfrac{1}{2}F^{\mu\nu}\left[(\partial_\mu R_a)\mathcal{P}_a J_\nu - (\partial_\nu R_a)\mathcal{P}_a J_\mu\right] \\
&\quad - \tfrac{1}{2}J_\lambda \epsilon^{\lambda\alpha\beta\sigma}F_{\alpha\beta}(\partial_\sigma R_a)\mathfrak{M}_a, \qquad (10.4.1)
\end{aligned}$$

$$\mathcal{L}_{\mathcal{P}M} = -\tfrac{1}{2}\rho n \mathcal{P}_a \kappa_{ab}^{-1}\mathcal{P}_b,$$

$$\mathcal{L}_{\mathfrak{M}M} = -\tfrac{1}{2}\rho n \mathfrak{M}_a \chi_{ab}^{-1}\mathfrak{M}_b,$$

$$\mathcal{L}_M = -\rho U(G_{ab}, R_a),$$

where κ_{ab} and χ_{ab} are functions of G_{ab} and R_a. Each term contributes a term to the canonical momentum tensor

$$T_\mu{}^\nu = g_\mu^\nu \mathcal{L} - \sum_{\substack{\text{fields}\\ \phi_A}} (\partial_\mu \phi_A)\frac{\partial \mathcal{L}}{\partial(\partial_\nu \phi_A)}. \qquad (10.4.2)$$

In order to form $\theta^{\mu\nu}$, we have to add to $T^{\mu\nu}$ the tensor $\partial_\rho G^{\mu\nu\rho}$ where $G^{\mu\nu\rho} = \tfrac{1}{2}[S^{\nu\rho\mu} + S^{\mu\rho\nu} - S^{\mu\nu\rho}]$ and $S^{\mu\nu\rho}$ is the spin angular momentum tensor

$$\begin{aligned}
S_{\mu\nu}{}^\rho &= \frac{\partial \mathcal{L}}{\partial(\partial_\rho A^\lambda)}\left[-g_\mu^\lambda g_{\nu\sigma} + g_\nu^\lambda g_{\mu\sigma}\right]A^\sigma \\
&= (F^\rho{}_\mu - M^\rho{}_\mu)A_\nu - (F^\rho{}_\nu - M^\rho{}_\nu)A_\mu. \qquad (10.4.3)
\end{aligned}$$

Thus, making use of Maxwell's equations (10.2.1),

$$\begin{aligned}
\partial_\rho G^{\mu\nu\rho} &= -\partial_\rho\left[A^\mu(F^{\nu\rho} - M^{\nu\rho})\right] \\
&= -(\partial_\rho A^\mu)F^{\nu\rho} + (\partial_\rho A^\mu)M^{\nu\rho} - A^\mu \mathcal{J}^\nu. \qquad (10.4.4)
\end{aligned}$$

We associate the term $-(\partial_\rho A^\mu)F^{\nu\rho}$ with the term in $T^{\mu\nu}$ arising from $\mathcal{L}_E$. Similarly, we associate the term $-A^\mu \mathcal{J}^\nu$ with $\mathcal{L}_{E\mathcal{J}}$, and we associate the electric polarization and magnetic polarization parts of $(\partial_\rho A^\mu)M^{\nu\rho}$ with $\mathcal{L}_{E\mathcal{P}}$ and $\mathcal{L}_{E\mathfrak{M}}$ respectively. In this way we arrive at a momentum tensor $\theta^{\mu\nu}$ divided into seven terms,

$$\theta^{\mu\nu} = \theta_E^{\mu\nu} + \theta_{E\mathcal{J}}^{\mu\nu} + \theta_{E\mathcal{P}}^{\mu\nu} + \theta_{E\mathfrak{M}}^{\mu\nu} + \theta_{\mathcal{P}M}^{\mu\nu} + \theta_{\mathfrak{M}M}^{\mu\nu} + \theta_M^{\mu\nu}, \qquad (10.4.5)$$

corresponding to the seven terms in the Lagrangian.

The actual calculation of the various contributions to $\theta^{\mu\nu}$ is an instructive exercise, but it is somewhat tedious. It is therefore left as an exercise

(Problem 4) in Chapter 12. The results are presented in Tables 10.2 and 10.3. Table 10.2 contains the covariant forms of the terms in $\theta^{\mu\nu}$ as they emerge from the calculation described above. Table 10.3 contains the components θ^{00}, θ^{i0}, and θ^{ij} evaluated in a local rest frame of the material, after some simplification has been achieved by making use of the equations of motion for $\mathcal{P}_a$ and $\mathcal{M}_a$.

Table 10.2 Symmetrized Momentum Tensor in a Linearly Polarizable Material

$$\theta_E^{\mu\nu} = F^{\mu\lambda}F^{\nu}{}_{\lambda} - \tfrac{1}{4}g^{\mu\nu}F^{\alpha\beta}F_{\alpha\beta}$$

$$\theta_{E\mathcal{J}}^{\mu\nu} = 0$$

$$\theta_{E\mathcal{P}}^{\mu\nu} = J_\alpha F^{\alpha\mu}(\partial^\nu R_a)\mathcal{P}_a + J_\alpha F^{\alpha\nu}(\partial^\mu R_a)\mathcal{P}_a$$

$$\theta_{E\mathcal{M}}^{\mu\nu} = -\tfrac{1}{2}J^\mu F_{\alpha\beta}\epsilon^{\nu\alpha\beta\gamma}(\partial_\gamma R_a)\mathcal{M}_a - \tfrac{1}{2}J^\nu F_{\alpha\beta}\epsilon^{\mu\alpha\beta\gamma}(\partial_\gamma R_a)\mathcal{M}_a + \tfrac{1}{2}g^{\mu\nu}J_\lambda F_{\alpha\beta}\epsilon^{\lambda\alpha\beta\gamma}(\partial_\gamma R_a)\mathcal{M}_a$$

$$\theta_{\mathcal{P}M}^{\mu\nu} = \tfrac{1}{2}u^\mu u^\nu \rho n\mathcal{P}_a\kappa_{ab}^{-1}\mathcal{P}_b + \rho n(\partial^\mu R_a)(\partial^\nu R_b)\frac{\partial\kappa_{cd}^{-1}}{\partial G_{ab}}\mathcal{P}_c\mathcal{P}_d$$

$$\theta_{\mathcal{M}M}^{\mu\nu} = \tfrac{1}{2}u^\mu u^\nu \rho n\mathcal{M}_a\chi_{ab}^{-1}\mathcal{M}_b + \rho n(\partial^\mu R_a)(\partial^\nu R_b)\frac{\partial\chi_{cd}^{-1}}{\partial G_{ab}}\mathcal{M}_c\mathcal{M}_d$$

$$\theta_M^{\mu\nu} = \rho U u^\mu u^\nu + 2\rho(\partial^\mu R_a)(\partial^\nu R_b)\frac{\partial U}{\partial G_{ab}}$$

Table 10.3 Energy Density, Momentum Density, and Stress Tensor in a Local Rest Frame of a Linearly Polarizable Material

	θ^{00}	$\theta^{i0}=\theta^{0i}$	θ^{ij}
$\mathcal{L}_E$	$\frac{1}{2}(\mathbf{E}^2+\mathbf{B}^2)$	$(\mathbf{E}\times\mathbf{B})_i$	$-E_iE_j - B_iB_j + \frac{1}{2}\delta_{ij}(\mathbf{E}^2+\mathbf{B}^2)$
$\mathcal{L}_{E\mathcal{J}}$	0	0	0
$\mathcal{L}_{E\mathcal{P}}$	0	0	$-P_iE_j - E_iP_j$
$\mathcal{L}_{E\mathcal{M}}$	$-\mathbf{M}\cdot\mathbf{B}$	$-(\mathbf{E}\times\mathbf{M})_i$	$-\delta_{ij}\mathbf{M}\cdot\mathbf{B}$
$\mathcal{L}_{\mathcal{P}M}$	$\frac{1}{2}\mathbf{P}\cdot\mathbf{E}$	0	$\rho n(\partial_iR_a)(\partial_jR_b)\frac{\partial\kappa_{cd}^{-1}}{\partial G_{ab}}\mathcal{P}_c\mathcal{P}_d$
$\mathcal{L}_{\mathcal{M}M}$	$\frac{1}{2}\mathbf{M}\cdot\mathbf{B}$	0	$\rho n(\partial_iR_a)(\partial_jR_b)\frac{\chi_{cd}^{-1}}{G_{ab}}\mathcal{M}_c\mathcal{M}_d$
$\mathcal{L}_M$	ρU	0	$2\rho(\partial_iR_a)(\partial_jR_b)\frac{\partial U}{\partial G_{ab}}$

A glance at Table 10.3 shows that the forms of θ^{00} and θ^{0i} in a material rest frame are simple and independent of the type of material. The "electromagnetic energy density" $\theta^{00}-\theta^{00}_{M}$ is*

$$\theta^{00}-\theta^{00}_{M}=\tfrac{1}{2}\left[\mathbf{D}\cdot\mathbf{E}+\mathbf{H}\cdot\mathbf{B}\right], \tag{10.4.6}$$

where we have defined the conventional fields **D** and **H**:

$$\begin{aligned}\mathbf{D}&=\mathbf{E}+\mathbf{P},\\ \mathbf{H}&=\mathbf{B}-\mathbf{M}.\end{aligned} \tag{10.4.7}$$

The energy current θ^{0i} and the momentum density θ^{i0} are given by the familiar Poynting's vector

$$\theta^{0i}=(\mathbf{E}\times\mathbf{H})_{i}. \tag{10.4.8}$$

The electromagnetic contribution to the stress tensor θ^{ij} in the material is somewhat more complicated. It is a quadratic function of the fields **E**, **B**, **P**, **M** and also involves the derivatives of the susceptibility matrices κ_{ab}, χ_{ab} with respect to the material deformation matrix G_{ab}.

Let us consider, for example, a solid material with susceptibility matrices

$$\begin{aligned}\kappa_{ab}&=\alpha^{(E)}_{ab}+\beta^{(E)}_{abcd}s_{cd},\\ \chi_{ab}&=\alpha^{(M)}_{ab}+\beta^{(M)}_{abcd}s_{cd},\end{aligned} \tag{10.4.9}$$

where terms of higher order in the strain s_{ab} have been omitted as in (10.3.16). With this choice of κ_{ab}, the contribution to the stress tensor from the polarization-matter interaction is (see Problem 2)

$$\theta^{ij}_{\mathcal{P}M}=\tfrac{1}{2}\,\bar{\beta}^{(E)}_{klij}E_kE_l, \tag{10.4.10}$$

where we have defined

$$\bar{\beta}^{(E)}_{klij}=\frac{\rho}{n}(\partial_kR_a)(\partial_lR_b)(\partial_iR_c)(\partial_jR_d)\,\beta^{(E)}_{abcd}. \tag{10.4.11}$$

*We are considering dielectric materials that are linear in the following sense. Let a sample of the material be clamped into place so that its volume and shape cannot change, and thermally insulated so that its entropy cannot change. If it is then exposed to an electric field, the polarization will be a linear function of the electric field. Some authors consider, for instance, dielectrics that are linear at constant shape and temperature. They obtain (10.4.6) as the electromagnetic contribution to the Helmholtz free energy density, $h=\theta^{00}-T\rho s$, where T is the temperature and s is the entropy per atom. These distinctions are theoretically interesting, but not of much experimental relevance, because a material that is precisely linear under one set of conditions will be almost linear under the other.

Notice that $\theta^{ij}_{\mathcal{P}M}$ vanishes if the susceptibility is a constant matrix $\kappa_{ab} = \alpha^{(E)}_{ab}$. Similarly,

$$\theta^{ij}_{\mathfrak{M}M} = \tfrac{1}{2}\bar{\beta}^{(M)}_{klij} B_k B_l \tag{10.4.12}$$

with $\bar{\beta}^{(M)}$ defined analogously to $\bar{\beta}^{(E)}$. Thus the complete electromagnetic part of the stress tensor in the solid is

$$\begin{aligned}\theta^{ij} - \theta^{ij}_{M} = &- E_i E_j - P_i E_j - E_i P_j + \tfrac{1}{2}\delta_{ij}\mathbf{E}^2 \\ &- B_i B_j - \delta_{ij}\mathbf{M}\cdot\mathbf{B} + \tfrac{1}{2}\delta_{ij}\mathbf{B}^2 \\ &+ \tfrac{1}{2}\bar{\beta}^{(E)}_{klij} E_k E_l + \tfrac{1}{2}\bar{\beta}^{(M)}_{klij} B_k B_l.\end{aligned} \tag{10.4.13}$$

Note the asymmetry between **E** and **B**.

As another example let us consider a polarizable fluid. As we found in the preceding section, the susceptibility matrices for a fluid have the form

$$\begin{aligned}\kappa^{-1}_{ab} &= \frac{1}{\kappa(\mathcal{V})}\frac{\rho}{n}G_{ab},\\ \chi^{-1}_{ab} &= \frac{1}{\chi(\mathcal{V})}\frac{\rho}{n}G_{ab}.\end{aligned} \tag{10.4.14}$$

We can calculate the corresponding contributions to the stress tensor, $\theta^{ij}_{\mathcal{P}\mathrm{M}}$ and $\theta^{ij}_{\mathfrak{M}\mathrm{M}}$, from the general formulas given in Table 10.2 if we recall that $\rho/n = [\det\mathbf{G}]^{1/2}$ and $\partial(\det\mathbf{G})/\partial G_{ab} = \det\mathbf{G}\, G^{-1}_{ab}$. For $\theta^{ij}_{\mathcal{P}\mathrm{M}}$ we find

$$\begin{aligned}\theta^{ij}_{\mathcal{P}M} = &\frac{\rho^2}{\kappa}(\partial_i R_a)\mathcal{P}_a(\partial_j R_b)\mathcal{P}_b \\ &+ \tfrac{1}{2}\delta_{ij}\frac{\rho^2}{\kappa}\left[1 + \mathcal{V}\frac{d\ln\kappa}{d\mathcal{V}}\right]\mathcal{P}_c G_{cd}\mathcal{P}_d.\end{aligned}$$

Making use of the definition (10.3.7) of $\mathcal{P}_a$, we can write this as

$$\theta^{ij}_{\mathcal{P}M} = \frac{1}{\kappa}P_i P_j + \tfrac{1}{2}\delta_{ij}\frac{1}{\kappa}\left[1 + \mathcal{V}\frac{d\ln\kappa}{d\mathcal{V}}\right]\mathbf{P}^2. \tag{10.4.15}$$

Similarly,

$$\theta^{ij}_{\mathfrak{M}M} = \frac{1}{\chi}M_i M_j + \tfrac{1}{2}\delta_{ij}\frac{1}{\chi}\left[1 + \mathcal{V}\frac{d\ln\chi}{d\mathcal{V}}\right]\mathbf{M}^2. \tag{10.4.16}$$

Finally, we combine these terms with the other contributions to θ^{ij} listed in

Table 10.3, using the equations of motion $\mathbf{P}=\kappa\mathbf{E}$ and $\mathbf{M}=\chi\mathbf{B}$. This gives for the electromagnetic part of the stress tensor in the rest frame of a fluid

$$\begin{aligned}\theta^{ij}-\theta^{ij}_M = &-\tfrac{1}{2}(E_iD_j+D_iE_j)+\tfrac{1}{2}\delta_{ij}\mathbf{E}\cdot\mathbf{D}\\ &-\tfrac{1}{2}(B_iH_j+H_iB_j)+\tfrac{1}{2}\delta_{ij}\mathbf{B}\cdot\mathbf{H}\\ &+\tfrac{1}{2}\delta_{ij}\mathcal{V}\frac{d\ln\kappa}{d\mathcal{V}}\mathbf{P}\cdot\mathbf{E}+\tfrac{1}{2}\delta_{ij}\mathcal{V}\frac{d\ln\chi}{d\mathcal{V}}\mathbf{M}\cdot\mathbf{B}.\end{aligned} \tag{10.4.17}$$

If we wish, we can write this in a more conventional form using the constants ϵ and μ defined by $\mathbf{D}=\epsilon\mathbf{E}$ and $\mathbf{B}=\mu\mathbf{H}$ (that is, $\epsilon=1+\kappa$ and $1/\mu=1-\chi$):

$$\begin{aligned}\theta^{ij}-\theta^{ij}_M = &-(\epsilon E_iE_j+\mu H_iH_j)+\tfrac{1}{2}\delta_{ij}(\epsilon\mathbf{E}^2+\mu\mathbf{H}^2)\\ &+\tfrac{1}{2}\delta_{ij}\left\{\mathcal{V}\frac{d\epsilon}{d\mathcal{V}}\mathbf{E}^2+\mathcal{V}\frac{d\mu}{d\mathcal{V}}\mathbf{H}^2\right\}.\end{aligned} \tag{10.4.18}$$

Results of Other Methods

There has been some controversy over the years about the correct form of the electromagnetic momentum density in a material medium. The identification of $\mathbf{E}\times\mathbf{H}$ as the momentum density is due to Abraham; the chief competing value, $\mathbf{D}\times\mathbf{B}$, was proposed by Minkowski.* The dispute is, to some extent, semantic, since one might divide the total momentum tensor $\theta^{\mu\nu}$ into "electromagnetic" and "material" parts in different ways. It is entirely semantic when one does not specify both the material and electromagnetic parts of $\theta^{\mu\nu}$. In the division we have adopted, the material part $\theta^{\mu\nu}_M$ is a function only of the matter fields $R_a(x)$ and is independent of $F_{\mu\nu}$, $\mathcal{P}_a$ and $\mathcal{M}_a$. Thus, in particular, $\theta^{i0}_M=0$ in a local rest frame of the material.

In view of this history of controversy, it is noteworthy that one can drive the form of $\theta^{\mu\nu}$ from assumptions other than Hamilton's principle. Grot and Eringen† have done so in an analysis of the relativistic mechanics and

*M. Abraham and R. Becker, *Theorie der Elektrizität*, Vol. 2, 6th ed. (Leipzig, 1933). H. Minkowski, *Gött. Nachr.*, 53 (1908); *Math. Ann.* **68**, 472 (1910). Many of the arguments that have been made on this question are summarized in C. Møller, *The Theory of Relativity*, Chapter 7 (Oxford, London, 1952), and in S. R. deGroot and L. G. Suttorp *Foundations of Electrodynamics* (North-Holland, Amsterdam, 1972).

†R. A. Grot and A. C. Eringen, *Int. J. Eng. Sci.* **4**, 611 (1966); see also P. Penfield and H. A. Haus in *Recent Advances in Engineering Science*, edited by A. C. Eringen (Gordon and Breach, New York, 1966).

thermodynamics of electromagnetic interactions with matter. They take as postulates the conservation of momentum and angular momentum, Maxwell's equations, the second law of thermodynamics, and a plausible assumption—derived from microscopic considerations—about electromagnetic forces on matter. From this, with the aid of a few technical assumptions, they are able to derive a "most general" form for $\theta^{\mu\nu}$. They find $\theta^{i0}=(\mathbf{E}\times\mathbf{H})_i$. Their form for $\theta^{\mu\nu}$, when specialized to linear media with no dissipative effects, is that given in Table 10.2.

10.5 USE OF THE MOMENTUM TENSOR

Knowledge of the energy density θ^{00} alone is sufficient to solve easily several simple problems often found on university examinations in physics, if nowhere else. For instance, one can find the total force on a piece of dielectric placed in an external static electric field by calculating the change in energy caused by making a rigid infinitesimal displacement $\delta\mathbf{x}$ of the whole body. This change in energy is equal to the work, $-\mathbf{F}\cdot\delta\mathbf{x}$, done in moving the body against the electrical force $\mathbf{F}$.

In nonstatic problems and in problems in which one wants to know how the electromagnetic force is distributed within a body, it is easiest to use the stress tensor θ^{jk}. By definition, the electromagnetic force per unit volume on matter is $f^j=\partial_\mu\theta_M^{j\mu}=-\partial_\mu[\theta^{j\mu}-\theta_M^{j\mu}]$. We can calculate f^j if we know the fields $\mathbf{E}$, $\mathbf{B}$, $\mathbf{P}$, $\mathbf{M}$.

To see how this works, we analyze a simple problem in some detail. Consider a laser beam directed downward at a glass of water. What happens to the water? A sensible first reaction is that all of those photons hitting the water probably push the surface of the water down a little. We find that this guess is correct except for the sign of the effect.

The Fields

To state our problem more precisely, we suppose that a liquid with electricity susceptibility $\kappa(\mathcal{V})$ and magnetic susceptibility $\chi(\mathcal{V})$ is initially at rest in the region $z<0$, with vacuum in the region $z>0$. A beam of electromagnetic radiation with frequency ω shines onto the surface of the liquid. (In most real liquids, χ is negligible and one finds $\mathbf{P}=\kappa\mathbf{E}$; but for high frequencies ω, the effective κ depends on ω. In order to keep within the bounds of the theory developed in this chapter, we must suppose that ω is sufficiently small so that we can ignore this dispersion in κ.)

We suppose that the radiation within the incident beam is described by fields

$$\mathbf{E} = E_I \cos(\omega[t+z])\hat{\mathbf{x}},$$
$$\mathbf{H} = -E_I \cos(\omega[t+z])\hat{\mathbf{y}}, \tag{10.5.1}$$

where $\hat{\mathbf{x}}$, $\hat{\mathbf{y}}$, $\hat{\mathbf{z}}$ are unit vectors pointing in the directions of the coordinate axes. When the radiation strikes the liquid surface, part is reflected:

$$\mathbf{E} = -E_R \cos(\omega[t-z])\hat{\mathbf{x}},$$
$$\mathbf{H} = -E_R \cos(\omega[t-z])\hat{\mathbf{y}}. \tag{10.5.2}$$

The rest is transmitted into the liquid:

$$\mathbf{E} = E_T \cos\left(\omega\left[t+\sqrt{\epsilon\mu}\, z\right]\right)\hat{\mathbf{x}},$$
$$\mathbf{H} = -\sqrt{\frac{\epsilon}{\mu}}\; E_T \cos\left(\omega\left[t+\sqrt{\epsilon\mu}\, z\right]\right)\hat{\mathbf{y}}. \tag{10.5.3}$$

The quantities E_R and E_T are determined by the boundary conditions that E_x and H_y be continuous at $z=0$:

$$E_R = \frac{\sqrt{\epsilon/\mu}\, -1}{\sqrt{\epsilon/\mu}\, +1} E_I,$$
$$E_T = \frac{2}{\sqrt{\epsilon/\mu}\, +1} E_I. \tag{10.5.4}$$

Energy Conservation

We have been supposing that the liquid is stationary—at least when the beam is first switched on. If so, the electromagnetic forces acting on the liquid are not doing any work. Therefore, the electromagnetic part of the energy current should be conserved. In particular, the z-component of the energy current $\mathbf{E} \times \mathbf{H}$ should be continuous across the surface $z=0$. We can check this:

$$\mathbf{E} \times \mathbf{H} = (E_R{}^2 - E_I{}^2)\cos^2(\omega t)\hat{\mathbf{z}} \qquad z>0$$
$$= -\sqrt{\epsilon/\mu}\; E_T{}^2 \cos^2(\omega t)\hat{\mathbf{z}} \qquad z<0.$$

Using (10.5.4), we find that $E_I^2 = E_R^2 + \sqrt{\epsilon/\mu}\ E_T^2$, so $\mathbf{E} \times \mathbf{H}$ is indeed continuous.

Forces Inside the Liquid

What forces are exerted on the liquid? The electromagnetic force per unit volume is

$$f^j = -\partial_0\left(\theta^{j0} - \theta_M^{j0}\right) - \partial_k\left(\theta^{jk} - \theta_M^{jk}\right). \tag{10.5.5}$$

We begin by looking at part of the stress tensor (10.4.18):

$$\bar{\theta}^{jk} \equiv -\left[\epsilon E_j E_k + \mu H_j H_k\right] + \tfrac{1}{2}\delta_{jk}\left[\epsilon \mathbf{E}^2 + \mu \mathbf{H}^2\right]. \tag{10.5.6}$$

The complete electromagnetic stress tensor is $\bar{\theta}^{jk}$ plus the so called "electrostriction" and "magnetostriction" terms

$$\tfrac{1}{2}\,\delta_{jk}\left\{ \mathcal{V}\frac{d\epsilon}{d\mathcal{V}}\mathbf{E}^2 + \mathcal{V}\frac{d\mu}{d\mathcal{V}}\mathbf{H}^2 \right\}. \tag{10.5.7}$$

We can calculate* the divergence of $\bar{\theta}^{jk}$ by using Maxwell's equations (10.2.1):

$$\begin{aligned}
\partial_k \bar{\theta}^{jk} = &-\epsilon E_k \partial_k E_j - E_j \partial_k(\epsilon E_k) \\
&-\mu H_k \partial_k H_j - H_j \partial_k(\mu H_k) \\
&+\epsilon E_k \partial_j E_k + \tfrac{1}{2}\mathbf{E}^2 \partial_j \epsilon \\
&+\mu H_k \partial_j H_k + \tfrac{1}{2}\mathbf{H}^2 \partial_j \mu.
\end{aligned} \tag{10.5.8}$$

The second term is $-E_j\partial_k D_k = -E_j \mathcal{J}^0$. The fourth term is $-H_j \partial_k B_k = 0$. The sum of the first and fifth terms is $D_k(\partial_j E_k - \partial_k E_j) = -(\mathbf{D} \times \partial_0 \mathbf{B})_j$. The sum of the third and seventh terms is $B_k(\partial_j H_k - \partial_k H_j) = -(\boldsymbol{\mathcal{J}} \times \mathbf{B})_j - (\partial_0 \mathbf{D} \times \mathbf{B})_j$. Thus

$$\begin{aligned}
-\partial_k \bar{\theta}^{jk} = &\mathcal{J}^0 E_j + (\boldsymbol{\mathcal{J}} \times \mathbf{B})_j - \tfrac{1}{2}\left[(\partial_j \epsilon)\mathbf{E}^2 + (\partial_j \mu)\mathbf{H}^2\right] \\
&+ \frac{\partial}{\partial t}(\mathbf{D} \times \mathbf{B}).
\end{aligned} \tag{10.5.9}$$

Notice that in a homogeneous stationary fluid with no macroscopic

*This calculation follows that of Panofsky and Phillips, *Classical Electricity and Magnetism* (Addison-Wesley, Reading, Mass., 1962).

current $\mathcal{J}^{\mu}(x)$, the case we are considering, (10.5.9) has the form of a conservation law:

$$\partial_0 \bar{\theta}^{j0} + \partial_k \bar{\theta}^{jk} = 0,$$

where $\bar{\theta}^{j0} \equiv (\mathbf{D} \times \mathbf{B})_j = \epsilon\mu(\mathbf{E} \times \mathbf{H})_j$. We might have expected such a conservation law. The Lagrangian whose variation gives the equations of motion for $F_{\mu\nu}$, $\mathcal{P}_a$, and $\mathfrak{M}_a$ when the fluid is constrained to be motionless is given by (10.4.1) without the terms $\mathcal{L}_{E\mathcal{J}}$ and $\mathcal{L}_M$ and with $R_a(x)$ held fixed, $R_a(x) = x^a$. Since the corresponding action is invariant under translations $\mathbf{x} \to \mathbf{x} + \mathbf{a}$, there must be a conserved current. This current is in fact $\bar{\theta}^{i\mu}$ (see Problem 3).

Now we can write down the force per unit volume on the fluid, using the result (10.5.9) with $\mathcal{J}^{\mu} = 0$ and $\nabla \epsilon = \nabla \mu = 0$ and $(\theta^{j0} - \theta^{j0}_M) = (\mathbf{E} \times \mathbf{H})_j$:

$$f^j = -\partial_0 \left[\theta^{j0} - \theta_M{}^{j0} \right] - \partial_k \bar{\theta}^{jk} - \tfrac{1}{2} \partial_j \left[\mathcal{V} \frac{d\epsilon}{d\mathcal{V}} E^2 + \mathcal{V} \frac{d\mu}{d\mathcal{V}} \mathbf{H}^2 \right],$$

or

$$\mathbf{f} = (\epsilon\mu - 1) \frac{\partial}{\partial t} (\mathbf{E} \times \mathbf{H}) - \tfrac{1}{2} \nabla \left[\mathcal{V} \frac{d\epsilon}{d\mathcal{V}} \mathbf{E}^2 + \mathcal{V} \frac{d\mu}{d\mathcal{V}} \mathbf{H}^2 \right]. \tag{10.5.10}$$

The first term in $\mathbf{f}$ can be neglected if ω is reasonably large because its time average is zero. It does, however, contribute a net impulse to the liquid at the leading edge of the beam when the beam is first turned on. This same impulse is subtracted from the liquid at the trailing edge of the beam when the beam is turned off.

The second term arises from the electrostriction and magnetostriction terms (10.5.7) in θ^{ij}. Its time average is not zero, and provides an inward force on the liquid at the outside edges of the beam if $d\epsilon/d\mathcal{V}$ and $d\mu/d\mathcal{V}$ are negative. This force will compress the liquid inside the beam until the liquid's pressure $P(\mathcal{V}) = -dU/d\mathcal{V}$ becomes

$$P = -\frac{1}{2} \left\langle \mathcal{V} \frac{d\epsilon}{d\mathcal{V}} \mathbf{E}^2 + \mathcal{V} \frac{d\mu}{d\mathcal{V}} \mathbf{H}^2 \right\rangle_{\text{time average}}. \tag{10.5.11}$$

When P has risen to this value, the time average of the net force (electromagnetic plus mechanical) on each drop of liquid is zero:*

$$\langle f^j_{\rm net}\rangle = \partial_0\langle \theta^{j0}_M\rangle$$
$$= -\partial_i\langle \delta_{ji}\tfrac{1}{2}\left[\mathcal{V}\frac{d\epsilon}{d\mathcal{V}}\mathbf{E}^2 + \mathcal{V}\frac{d\mu}{d\mathcal{V}}\mathbf{H}^2\right] + \delta_{ji}P\rangle$$
$$= 0.$$

Momentum Conservation at the Surface

What happens at the surface of the liquid? The time averaged force per unit area exerted on the surface is

$$\langle f^j\rangle = \langle \theta^{j3}_{\rm liq} - \theta^{j3}_{\rm vac}\rangle. \tag{10.5.12}$$

Making use of the general expression (10.4.18) for $\theta^{jk} - \theta^{jk}_M$, the relation $\Theta^{ij}_M = \delta_{ij}P$, and the expressions (10.5.1), (10.5.2), and (10.5.3) for the fields, we find

$$\theta^{j3}_{\rm liq} = \theta^{j3}_{\rm vac} = 0 \qquad j \neq 3, \tag{10.5.13}$$

$$\langle \Theta^{33}_{\rm vac}\rangle = \langle \tfrac{1}{2}(\mathbf{E}^2 + \mathbf{H}^2)\rangle = \frac{\epsilon + \mu}{\left[\sqrt{\epsilon} + \sqrt{\mu}\,\right]^2} E_I^2. \tag{10.5.14}$$

Also,

$$\langle \Theta^{33}_{\rm liq}\rangle = \langle \tfrac{1}{2}(\epsilon\mathbf{E}^2 + \mu\mathbf{H}^2) + \tfrac{1}{2}\left(\mathcal{V}\frac{d\epsilon}{d\mathcal{V}}\mathbf{E}^2 + \mathcal{V}\frac{d\mu}{d\mathcal{V}}\mathbf{H}^2\right) + P\rangle.$$

As we have just seen, the fluid inside the beam will be compressed until its pressure P is enough to cancel the electrostriction and magnetostriction pressures $\frac{1}{2}[\mathcal{V}(d\epsilon/d\mathcal{V})\mathbf{E}^2 + \mathcal{V}(d\mu/d\mathcal{V})\mathbf{H}^2]$. If the fluid is nearly incompressible, the time it takes for this to happen will be small. After this readjustment has taken place we will have

$$\langle \theta^{33}{}_{\rm liq}\rangle = \langle \tfrac{1}{2}(\epsilon\mathbf{E}^2 + \mu\mathbf{H}^2)\rangle$$
$$= \frac{2\epsilon\mu}{\left[\sqrt{\epsilon} + \sqrt{\mu}\,\right]^2} E_I{}^2. \tag{10.5.15}$$

*Once the liquid inside the beam has been compressed the term $-\frac{1}{2}[(\partial_j\epsilon)\mathbf{E}^2 + (\partial_j\mu)\mathbf{H}^2]$ in the force will no longer be zero. However this term is of order E_I^4 and can be neglected in comparison to the forces of order E_I^2 being considered.

Thus the force per unit area on the surface is

$$\langle \mathbf{f} \rangle = \frac{2\epsilon\mu - \epsilon - \mu}{\left[\sqrt{\epsilon} + \sqrt{\mu}\,\right]^2} E_I^{\,2} \hat{\mathbf{z}}. \tag{10.5.16}$$

In the usual case $\epsilon > 1$, $\mu \cong 1$ this is

$$\langle f^3 \rangle \cong \frac{\epsilon - 1}{\left[\sqrt{\epsilon} + 1\right]^2} E_I^{\,2} > 0.$$

That is, the radiation sucks the liquid upward.

The argument given above is not quite correct, since the upward force will not act on the surface layer only. The liquid near the surface will be dilated, causing a drop in pressure. The resulting pressure gradient will redistribute the upward surface force into the liquid. The motion of the fluid that results from this unbalanced upward force can be expected to be quite complicated.

10.6 PIEZOELECTRIC CRYSTALS

The Lagrangian (10.4.1) for a polarizable material suggests a generalization that, it turns out, is found in nature and is of some practical importance. The materials described by this generalization are called piezoelectric because they become electrically polarized if you squeeze them. The most common example is quartz.

The term $\mathcal{L}_{\mathcal{P}M}$ in (10.4.1) which describes the interaction between the polarization and the matter fields is

$$\mathcal{L}_{\mathcal{P}M} = -\tfrac{1}{2}\rho n \mathcal{P}_a \kappa_{aa'}^{-1} \mathcal{P}_{a'}. \tag{10.6.1}$$

This term is analogous to the potential energy term in the Lagrangian for a mass on a spring that is stretched by an amount $\mathcal{P}$. The equilibrium point of the spring is $\mathcal{P}_a = 0$.

One can generalize (10.6.1) by letting the equilibrium point be a nonzero function of the strain matrix s_{ab}:

$$n\mathcal{P}_a(\text{equilibrium}) = \gamma_a + \gamma_{abc} s_{bc} + \gamma_{abcde} s_{bc} s_{de} + \cdots.$$

We will choose to examine materials with $\gamma_a = 0$ but $\gamma_{abc} \neq 0$, and we will neglect the terms of higher order in the strain.* Thus we take

$$\mathcal{L}_{\mathcal{P}M} = -\tfrac{1}{2}\frac{\rho}{n}(n\mathcal{P}_a - \gamma_{abc}s_{bc})\kappa_{aa'}^{-1}(n\mathcal{P}_{a'} - \gamma_{a'b'c'}s_{b'c'}). \tag{10.6.2}$$

The piezoelectric coefficients γ_{abc} are characteristic of the material. Since s_{bc} is symmetric we may take $\gamma_{abc} = \gamma_{acb}$. Thus there are $3\times 6 = 18$ independent coefficients. Symmetry considerations may provide relations among the coefficients. In particular, any isotropic material must have $\gamma_{abc} = 0$ since the only array that satisfies

$$R_{aa'}R_{bb'}R_{cc'}\gamma_{a'b'c'} = \gamma_{abc} \tag{10.6.3}$$

for all rotation matrices $R_{aa'}$ is $\gamma_{abc} \propto \epsilon_{abc}$, but ϵ_{abc} fails to meet the symmetry requirement $\gamma_{abc} = \gamma_{acb}$. Similarly, if the crystal looks the same when the material coordinates are inverted, $R_a \to -R_a$, then (10.6.3) must be satisfied when $R_{aa'} = -\delta_{aa'}$. Hence $\gamma_{abc} = 0$.

An important crystal for which γ_{abc} does not vanish is quartz. The crystal symmetries of quartz allow two independent nonzero components of γ_{abc}. They are†

$$\gamma_{111} = -\gamma_{122} = -\gamma_{212} = -\gamma_{221} = 0.17 \quad \mathrm{C/m^2}, \tag{10.6.4}$$

$$\gamma_{123} = \gamma_{132} = -\gamma_{231} = -\gamma_{213} \cong 0.04 \quad \mathrm{C/m^2}.$$

All of the other components of γ_{abc} are zero. The approximate magnitude of these numbers can be made more meaningful by writing them in units appropriate for atomic physics. For instance

$$\gamma_{111} = 0.17\ \mathrm{C/m^2} = 0.011\frac{e}{(10^{-8}\,\mathrm{cm})^2}.$$

When we substitute the new interaction term (10.6.2) into the Lagrangian (10.4.1) we obtain a new equation of motion for $\mathcal{P}_a$:

$$\begin{aligned} n\mathcal{P}_a &= \gamma_{abc}s_{bc} + \kappa_{ab}\mathcal{E}_b, \\ \mathcal{E}_b &= -(\partial_\mu R_b)u_\nu F^{\nu\mu}. \end{aligned} \tag{10.6.5}$$

*There are a few materials with $\gamma_a \neq 0$, called electrets. They are permanently polarized even in the absence of an electric field or strain.

†From W. G. Cady, *Piezolectricity* (McGraw-Hill, New York, 1946), p. 187.

We see that piezoelectrics can be polarized in the usual way by applying an electric field, and that they can also be polarized by squeezing (so $s_{bc} \neq 0$). This property makes them useful for turning mechanical signals into electrical signals.

What is the momentum tensor for a piezoelectric crystal? Since we have altered only one term in the Lagrangian (10.4.1), we have only to calculate the corresponding revised term in $\theta^{\mu\nu}$. The result of a straightforward computation is

$$\theta^{\mu\nu}_{\mathcal{P}M} = \frac{1}{2}\frac{\rho}{n}(n\mathcal{P}_a - \gamma_{abc}s_{bc})\kappa^{-1}_{aa'}(n\mathcal{P}_{a'} - \gamma_{a'b'c'}s_{b'c'})u^\mu u^\nu$$

$$+\frac{\rho}{n}(\partial^\mu R_b)(\partial^\nu R_c)\gamma_{abc}\kappa^{-1}_{aa'}(n\mathcal{P}_{a'} - \gamma_{a'b'c'}s_{b'c'})$$

$$-\frac{\rho}{n}(\partial^\mu R_e)(\partial^\nu R_f)(n\mathcal{P}_a - \gamma_{abc}s_{bc})\kappa^{-1}_{ad}\frac{\partial\kappa_{dd'}}{\partial G_{ef}}\kappa^{-1}_{d'a'}(n\mathcal{P}_{a'} - \gamma_{a'b'c'}s_{b'c'}). \qquad (10.6.6)$$

The result looks a bit simpler if we eliminate $\mathcal{P}_a$ in favor of $\mathcal{E}_a$ by using (10.6.5):

$$\theta^{\mu\nu}_{\mathcal{P}M} = \frac{1}{2}\frac{\rho}{n}\mathcal{E}_a\kappa_{ab}\mathcal{E}_b u^\mu u^\nu$$

$$+\frac{\rho}{n}(\partial^\mu R_a)(\partial^\nu R_b)\left\{\gamma_{cab}\mathcal{E}_c - \mathcal{E}_c\frac{\partial\kappa_{cd}}{\partial G_{ab}}\mathcal{E}_d\right\}. \qquad (10.6.7)$$

This expression replaces the expression for $\theta^{\mu\nu}_{\mathcal{P}M}$ listed in Tables 10.2 and 10.3, which was just (10.6.7) with $\gamma_{abc} = 0$. The contribution to the energy density in a local rest frame of the material is the same quadratic function of the electric field as it was for nonpiezoelectric materials.

$$\theta^{00}_{\mathcal{P}M} = \frac{1}{2}\frac{\rho}{n}\mathcal{E}_a\kappa_{ab}\mathcal{E}_b. \qquad (10.6.8)$$

We cannot, however, write this as $\Theta^{00}_{\mathcal{P}M} = \frac{1}{2}\mathbf{P}\cdot\mathbf{E}$ as we did in Table 10.2, since the relation between $\mathbf{P}$ and $\mathbf{E}$ has been altered.

The contributions to the momentum density and the energy flow are zero, just as in the case $\gamma_{abc} = 0$. However, there is an important change in the stress tensor in a local rest frame:

$$\theta^{ij}_{\mathcal{P}M} = \frac{\rho}{n}(\partial_i R_a)(\partial_j R_b)\left\{\gamma_{cab}\mathcal{E}_c - \mathcal{E}_c\frac{\partial\kappa_{cd}}{\partial G_{ab}}\mathcal{E}_d\right\}.$$

We see that when a piezoelectric material is subjected to an electric field, a stress linear in the electric field is produced. This stress will cause the material to deform. Thus piezoelectrics are also useful for turning electrical signals into mechanical signals.

10.7 PERFECTLY CONDUCTING FLUIDS

Consider a fluid that is a perfect* conductor of electricity. The Lagrangian is

$$\mathcal{L} = -\tfrac{1}{4}F_{\mu\nu}F^{\mu\nu} - \rho U(\mathcal{V}, R_a), \tag{10.7.1}$$

with the subsidiary condition that only potentials $A_\mu(x)$ are considered which satisfy

$$u_\mu F^{\mu\nu}(x) = 0. \tag{10.7.2}$$

That is, the electric field measured in a local rest frame of the fluid must vanish.

In Section 5.4 we learned how to reformulate such variational problems with constraint conditions by using Lagrange multipliers. In the present case we would use Lagrange multiplier fields $\mathcal{P}_a(x)$ and write:

$$\begin{aligned}\mathcal{L} = &-\tfrac{1}{4}F_{\mu\nu}F^{\mu\nu} + \tfrac{1}{2}F^{\mu\nu}\left[(\partial_\mu R_a)\mathcal{P}_a J_\nu - (\partial_\nu R_a)\mathcal{P}_a J_\mu\right] \\ &-\rho U(\mathcal{V}, R_a)\end{aligned} \tag{10.7.3}$$

with no subsidiary condition. We would obtain (10.7.2) as the Euler-Lagrange equations $0 = \delta A/\delta \mathcal{P}_a(x)$. The reader will recognize that this is the Lagrangian for a fluid with an infinite electric polarizability κ and no magnetic polarization. We simply set $\kappa = \infty$, $\mathfrak{M}_a(x) = 0$ in (10.3.9) to obtain (10.7.3).

It is not, however, necessary to resort to Lagrange multipliers, for we can solve the constraint condition (10.7.2) exactly and eliminate $F_{\mu\nu}(x)$ in favor of $R_a(x)$, $\partial_\mu R_a(x)$ and some functions $b_c(R_a)$ which specify the initial values of $F_{\mu\nu}$. In a sense that we will make precise, the magnetic field is "frozen into" the fluid as the fluid moves. Roughly speaking, we will find that Maxwell's equation $\partial_0 \mathbf{B} = -\nabla \times \mathbf{E}$ and the condition $\mathbf{E} = 0$ imply that $\mathbf{B}$ does not change as seen from the fluid.

*The conductivity σ of a fluid can be treated as infinite if $\sigma \gg t/L^2$, where L is a typical length for the system being considered and t is a typical time. (In MKSQ units, $\sigma \gg t/\mu_0 L^2$). This condition is not ordinarily satisfied in laboratory size experiments, but is well satisfied in astrophysical applications such as study of magnetic phenomena in the sun.

The Magnetic Field

Let us define* the "magnetic field as seen from the fluid":

$$b_a(x) = -\frac{1}{2}\frac{n}{\rho}\epsilon^{\alpha\beta\gamma\delta}u_\alpha F_{\beta\gamma}(\partial_\delta R_a). \tag{10.7.4}$$

In a local rest frame of the fluid this is

$$b_a(x) = \frac{n}{\rho}\frac{\partial R_a}{\partial x^j}B_j(x).$$

The inverse relation is, in a local rest frame,

$$\tfrac{1}{2}\epsilon_{jkl}F_{kl} \equiv B_j(x) = \tfrac{1}{2}\epsilon_{jkl}\epsilon_{abc}\frac{\partial R_b}{\partial x^k}\frac{\partial R_c}{\partial x^l}b_a(x).$$

Since $F_{0l} = F_{l0} = 0$ in a rest frame of the fluid, we obtain

$$F_{\mu\nu} = \epsilon_{abc}(\partial_\mu R_a)(\partial_\nu R_b)b_c. \tag{10.7.5}$$

This is a covariant equation, so it holds in any reference frame. It is the inverse of (10.7.4).

Now let us examine the implications for $b_a(x)$ of the homogeneous Maxwell equations (8.1.4). We have

$$\begin{aligned} 0 &= \epsilon^{\alpha\beta\mu\nu}\partial_\beta F_{\mu\nu} \\ &= \epsilon^{\alpha\beta\mu\nu}\epsilon_{abc}(\partial_\mu R_a)(\partial_\nu R_b)(\partial_\beta b_c). \end{aligned} \tag{10.7.6}$$

Consider the $\alpha \equiv j = 1,2,3$ components of this equation in a local rest frame of the fluid. Since $\partial_0 R_a = \partial_0 R_b = 0$ in this frame, only $\beta = 0$ contributes:

$$\begin{aligned} 0 &= -\epsilon_{jkl}\epsilon_{abc}(\partial_k R_a)(\partial_l R_b)(\partial_0 b_c) \\ &= -2\frac{\rho}{n}\frac{\partial x^j}{\partial R_c}(\partial_0 b_c). \end{aligned}$$

Thus

$$\frac{\partial}{\partial t}b_c(x) = 0$$

*In (10.3.23) we used a similar symbol, $\mathfrak{B}_a(x)$, to denote the quantity given by (10.7.4) without the factor n/ρ. This factor is important for our present purposes.

or, covariantly,

$$u^{\alpha}\partial_{\alpha}b_a(x)=0 \tag{10.7.7}$$

Equation 10.7.7 says that $b_a(x)$ is constant along the path through space-time of each droplet of fluid. Therefore b_a depends only on the labels $R_b(x)$ of the droplet:

$$b_a=b_a(R_b(x)). \tag{10.7.8}$$

This is a precise way of stating that the magnetic field is frozen into the fluid.

We return now to (10.7.6) in a rest frame of the fluid and consider the $\alpha=0$ component. Using (10.7.8) and the chain rule we find

$$0=\epsilon_{jkl}\epsilon_{abc}(\partial_k R_a)(\partial_l R_b)(\partial_j R_d)\frac{\partial b_c}{\partial R_d}$$

$$=2\frac{\rho}{n}\frac{\partial x^j}{\partial R_a}\frac{\partial R_d}{\partial x^j}\frac{\partial b_c}{\partial R_d}$$

or

$$0=\frac{\partial b_a}{\partial R_a}. \tag{10.7.9}$$

This is the analoque of the Maxwell equation $\nabla\cdot\mathbf{B}=0$.

The electromagnetic current that flows in the fluid in such a way as to maintain the condition $u_{\mu}F^{\mu\nu}=0$ can be calculated from the inhomogeneous Maxwell equation $\partial_{\nu}F^{\mu\nu}=\mathcal{J}^{\mu}$.

Frozen-in Lines of Force

We can make the statement about the frozen-in magnetic field more graphic by using the familiar concept of magnetic field lines. Imagine drawing a two parameter family of lines,

$$R_a=R_a(\lambda;\alpha_1,\alpha_2),$$

through the fluid. Here λ is the curve parameter of each line and the parameters α_1, α_2 label the lines. The lines are drawn so that the field $b_a(\mathbf{R})$ at each point $\mathbf{R}$ is parallel to the tangent vector of the line passing through that point:

$$b_a(\mathbf{R})=f(\mathbf{R})\frac{\partial R_a}{\partial\lambda}. \tag{10.7.10}$$

The magnitude of the field is equal to the density of lines—counting one line per unit $d\alpha_1\,d\alpha_2$. To state this condition precisely, we construct the surface $\lambda=$constant in $(\lambda,\alpha_1,\alpha_2)$ space and its image in the fluid. The normal vector to the surface in the fluid is

$$dS_a=\epsilon_{abc}\frac{\partial R_b}{\partial\alpha_1}\frac{\partial R_c}{\partial\alpha_2}d\alpha_1\,d\alpha_2.$$

We demand that

$$b_a\,dS_a=d\alpha_1\,d\alpha_2$$

or

$$b_a\epsilon_{abc}\frac{\partial R_b}{\partial\alpha_1}\frac{\partial R_c}{\partial\alpha_2}=1. \qquad (10.7.11)$$

The condition $\partial b_a/\partial R_a=0$ ensures that the lines can be drawn in the fashion described: no lines end in the fluid.

Equation 10.7.11 gives the normalization constant $f(\mathbf{R})$ in (10.7.10):

$$f(\mathbf{R})\epsilon_{abc}\frac{\partial R_a}{\partial\lambda}\frac{\partial R_b}{\partial\alpha_1}\frac{\partial R_c}{\partial\alpha_2}=1$$

or

$$f(\mathbf{R})\frac{\partial\,(R_1,R_2,R_3)}{\partial\,(\lambda,\alpha_1,\alpha_2)}=1.$$

Thus the relation between $b_a(\mathbf{R})$ and the field lines $R_a(\lambda,\alpha_1,\alpha_2)$ is

$$b_a(\mathbf{R})=\frac{\partial\,(\lambda,\alpha_1,\alpha_2)}{\partial\,(R_1,R_2,R_3)}\frac{\partial R_a}{\partial\lambda}. \qquad (10.7.12)$$

At any instant of time, t, the field lines in the fluid may be considered as lines $x^j(R_a(\lambda,\alpha_1,\alpha_2),t)$ in space. A candidate $B_j(\mathbf{x},t)$ for the magnetic field can be constructed from these field lines according to the standard prescription:

$$B_j=\frac{\partial\,(\lambda,\alpha_1,\alpha_2)}{\partial\,(x_1,x_2,x_3)}\frac{\partial x^j}{\partial\lambda}. \qquad (10.7.13)$$

Using the relation (10.7.12) we find

$$B_j = \frac{\partial(\lambda,\alpha_1,\alpha_2)}{\partial(R_1,R_2,R_3)} \frac{\partial(R_1,R_2,R_3)}{\partial(x_1,x_2,x_3)} \frac{\partial x^j}{\partial R_a} \frac{\partial R_a}{\partial\lambda}$$

$$= \frac{\partial(R_1,R_2,R_3)}{\partial(x_1,x_2,x_3)} \frac{\partial x^j}{\partial R_a} b_a$$

$$= \tfrac{1}{2}\epsilon_{jkl}\epsilon_{abc}(\partial_k R_b)(\partial_l R_c)b_a.$$

Compare this result with (10.7.5) relating $F_{\mu\nu}$ and b_a. We see that the candidate field $B_j(x,t)$ constructed from the field lines is in fact the magnetic field $B_j = \frac{1}{2}\epsilon_{jkl}F_{kl}$. The lines $R_a(\lambda,\alpha_1,\alpha_2)$ which are swept along by the motion of the fluid are precisely the magnetic field lines at all times and in any reference frame.

In a reference frame in which the fluid is moving with velocity $\mathbf{v}$, the electric field is

$$\mathbf{E} = -\mathbf{v}\times\mathbf{B} \tag{10.7.14}$$

(see Problem 4).

The Momentum Tensor

Now that we have the solution (10.7.5) of the constraint condition for $F_{\mu\nu}(x)$, the Lagrangian (10.7.1) reads

$$\mathcal{L} = -\tfrac{1}{4}\epsilon_{abc}(\partial_\mu R_a)(\partial_\nu R_b)b_c(R_a)\epsilon_{a'b'c'}(\partial^\mu R_{a'})(\partial^\nu R_{b'})b_{c'}(R_a)$$
$$-\rho U(\mathcal{V},R_a). \tag{10.7.15}$$

As we see, the only fields left are $R_a(x)$. Thus the conducting fluid under consideration falls under the class of materials we considered in general terms in Chapter 5. Its mechanical properties differ from those of an ordinary fluid because of the presence of the first term in (10.7.15).

The momentum tensor

$$\theta^{\mu\nu} = g^{\mu\nu}\mathcal{L} - \partial^\mu R_a \frac{\partial\mathcal{L}}{\partial(\partial_\nu R_a)}$$

is easy to calculate. One finds

$$\theta^{\mu\nu} = F^{\mu\lambda}F^{\nu}{}_{\lambda} - \tfrac{1}{4}g^{\mu\nu}F^{\alpha\beta}F_{\alpha\beta}$$
$$+\rho U u^\mu u^\nu + P(g^{\mu\nu} + u^\mu u^\nu), \tag{10.7.16}$$

where $F_{\mu\nu}$ is the electromagnetic field given by (10.7.5),

$$F_{\mu\nu}=\epsilon_{abc}(\partial_\mu R_a)(\partial_\nu R_b)b_c, \tag{10.7.17}$$

and $P(\mathcal{V})$ is the pressure the fluid would have in the absense of a frozen-in field,

$$P=-\frac{\partial U}{\partial \mathcal{V}}. \tag{10.7.18}$$

This is formally the same momentum tensor one would find for a nonconducting fluid in an electromagnetic field, except that now the field is, so to speak, attached to the fluid.

The stress tensor in a local rest frame of the fluid is

$$T^{ij}=-B_iB_j+\delta_{ij}\left(\tfrac{1}{2}\mathbf{B}^2+P(\mathcal{V})\right). \tag{10.7.19}$$

If we choose the z direction along the field lines, this is

$$T^{ij}=\begin{bmatrix} P & 0 & 0 \\ 0 & P & 0 \\ 0 & 0 & P \end{bmatrix}+\begin{bmatrix} \frac{1}{2}B^2 & 0 & 0 \\ 0 & \frac{1}{2}B^2 & 0 \\ 0 & 0 & -\frac{1}{2}B^2 \end{bmatrix}.$$

First, there is the ordinary pressure $P(\mathcal{V})$. Then there is a "transverse pressure" $\frac{1}{2}B^2$—as if the magnetic field lines were repelling one another. Finally, there is a tension per unit area of $\frac{1}{2}B^2$ in the direction of the field lines—as if the lines were stretched rubber bands.

Sound Waves

A homogeneous conducting fluid that is at rest in a uniform magnetic field is in mechanical equilibrium, since $\partial_k\theta^{lk}=0$. Small disturbances away from this equilibrium propagate in an interesting kind of wave, which we can investigate by setting

$$R_a(x)=x^a+\phi_a(x)$$

and writing the equations of motion to lowest order in the small displacement field $\phi_a(x)$.

To first order in ϕ, the four-velocity and invariant density are

$$u^0\cong 1$$

$$u^i\cong-\dot{\phi}_i$$

$$\rho\cong n+n(\partial_i\phi_i).$$

The electromagnetic field is

$$F_{0i} \cong \epsilon_{ijk} b_j \dot{\phi}_k,$$

$$F_{ij} \cong \epsilon_{ijk} b_k + (\partial_i \phi_k)\epsilon_{kjl} b_l - (\partial_j \phi_k)\epsilon_{kil} b_l.$$

Finally, we let U_0, P_0, and $\mathcal{V}_0$ be the undisturbed values of U, P, and $\mathcal{V}$. When we use these results in the formula (10.7.16) for $\theta^{\mu\nu}$ we obtain, to first order in ϕ, some cumbersome expressions:

$$\theta^{i0} = -\left(nU_0 + P_0 + \mathbf{b}^2\right)\dot{\phi}_i + \dot{\phi}_k b_k b_i, \tag{10.7.20}$$

$$\begin{aligned}\theta^{ij} = {} & \delta_{ij}\left(P_0 + \tfrac{1}{2}\mathbf{b}^2\right) - b_i b_j - \delta_{ij}\mathcal{V}_0\left(\frac{dP}{d\mathcal{V}}\right)_{\mathcal{V}=\mathcal{V}_0} \partial_k \phi_k \\ & - (\partial_i \phi_k) b_k b_j - (\partial_j \phi_k) b_k b_i + (\partial_i \phi_j + \partial_j \phi_i)\mathbf{b}^2 \\ & + \epsilon_{ikl}\epsilon_{jmn}(\partial_k \phi_m + \partial_m \phi_k) b_l b_n + \delta_{ij}\left[(\partial_k \phi_l) b_k b_l - (\partial_k \phi_k)\mathbf{b}^2\right]. \end{aligned} \tag{10.7.21}$$

Fortunately, the equations of motion,

$$\partial_0 \theta^{i0} + \partial_j \theta^{ij} = 0,$$

are somewhat less complicated:

$$\begin{aligned}\left(nU_0 + P_0 + \mathbf{b}^2\right)\ddot{\phi}_i - b_i b_k \ddot{\phi}_k & \\ = {} & -\mathcal{V}_0 \frac{dP}{d\mathcal{V}}(\partial_i \partial_k \phi_k) - b_i b_j (\partial_k \partial_k \phi_j) \\ & + \mathbf{b}^2(\partial_k \partial_k \phi_i) \\ & + \epsilon_{ikl}\epsilon_{jmn} b_l b_n (\partial_j \partial_k \phi_m). \end{aligned} \tag{10.7.22}$$

We can find out what kind of waves are predicted by this equation of motion by setting $\phi_i(x) = \phi_i \exp(i\mathbf{k}\cdot\mathbf{x} - i\omega t)$. We choose coordinates so that $\mathbf{b}$ lies in the z-direction and k is in the $y-z$ plane. Then the equation of motion reads

$$\begin{aligned} \left(nU_0 + P_0 + \mathbf{b}^2\right)\omega^2 \phi_1 &= \mathbf{b}^2 k_3^2 \phi_1, \\ \left(nU_0 + P_0 + \mathbf{b}^2\right)\omega^2 \phi_2 &= -\mathcal{V}_0 \frac{dP}{d\mathcal{V}} k_2 (k_2 \phi_2 + k_3 \phi_3) \\ &\quad + \mathbf{b}^2 \mathbf{k}^2 \phi_2, \\ (nU_0 + P_0)\omega^2 \phi_3 &= -\mathcal{V}_0 \frac{dP}{d\mathcal{V}} k_3 (k_2 \phi_2 + k_3 \phi_3). \end{aligned} \tag{10.7.23}$$

This is a matrix equation of the form $A\omega^2\boldsymbol{\phi}=B(\mathbf{k})\boldsymbol{\phi}$. For each $\mathbf{k}$ there will be three allowed frequencies ω^2, with corresponding polarization vectors $\boldsymbol{\phi}$.

One of the three solution is easy to find, since ϕ_1 occurs in the first of equations (10.7.23) but not in the other two:

$$\boldsymbol{\phi}_{\mathrm{I}}=(\phi_1,0,0),$$

$$\omega_{\mathrm{I}}{}^2=\frac{(\mathbf{k}\cdot\mathbf{b})^2}{nU_0+P_0+\mathbf{b}^2}. \qquad (10.7.24)$$

This wave, in which the direction of oscillation is orthogonal to both $\mathbf{b}$ and $\mathbf{k}$, has a phase velocity

$$|\mathbf{v}_{\mathrm{I}}|=\frac{\omega_{\mathrm{I}}}{|\mathbf{k}|}=\frac{|\mathbf{b}|\cos\theta}{\sqrt{nU0+P_0+\mathbf{b}^2}},$$

where θ is the angle between $\mathbf{b}$ and $\mathbf{k}$. If a wave packet is made up of a superposition of these waves with similar wave vector $\mathbf{k}$, its group velocity is

$$v_{\mathrm{I}}^{(G)}=\frac{\partial\omega_{\mathrm{I}}(\mathbf{k})}{\partial k_i}=\pm\frac{b_i}{\sqrt{nU_0+P_0+\mathbf{b}^2}}.$$

Note that such a wave packet is always propagated along the field lines.

The other two modes of propagation can be found by solving the second two of equations (10.7.23). If we make the simplifying approximation that the magnetic field energy $\frac{1}{2}\mathbf{b}^2$ and the pressure P_0 can be neglected in comparison to the rest energy U_0 of the fluid, we obtain (see Problem 5):

$$\omega_{\mathrm{II,III}}^2=\frac{\mathbf{k}^2}{2}\left[-\frac{\mathcal{V}}{nU_0}\frac{dP}{d\mathcal{V}}+\frac{\mathbf{b}^2}{nU_0}\pm\sqrt{\left[-\frac{\mathcal{V}}{nU_0}\frac{dP}{d\mathcal{V}}+\frac{\mathbf{b}^2}{nU_0}\right]^2+4\frac{\mathcal{V}}{nU_0}\frac{dP}{d\mathcal{V}}\frac{\mathbf{b}^2}{nU_0}\cos^2\theta}\right]$$

For $\mathbf{b}^2\ll-\mathcal{V}\,dP/d\mathcal{V}$, this is approximately

$$\omega_{\mathrm{II}}{}^2\cong-\frac{\mathbf{k}^2\mathcal{V}}{nU_0}\frac{dP}{d\mathcal{V}},$$

$$\omega_{\mathrm{III}}^2\cong\frac{(\mathbf{b}\cdot\mathbf{k})^2}{nU_0}. \qquad (10.7.26)$$

The approximate solution ω_{II} is the same as for an ordinary sound wave in a fluid with no magnetic field; the corresponding polarization ϕ_{II} is approximately in the direction of **k**. The solution ω_{III} is approximately the same as ω_{I}. The corresponding polarization vector ϕ_{III} is orthogonal to ϕ_{II}, since the ϕ's are eigenvectors of a Hermitian matrix. Thus ϕ_{III} is approximately orthogonal to **k** and lies in the **k**−**b** plane.

PROBLEMS

1. Suppose that $\mathbf{F(g)}$ is a matrix function of a matrix $\mathbf{G}$ with the property that $\mathbf{F(AGA}^T) = \mathbf{AF(G)A}^T$ for any matrix $\mathbf{A}$ with $\det \mathbf{A} = 1$. Show that $\mathbf{F(G)}$ has the form $\mathbf{F(G)} = f(\det \mathbf{G})\,\mathbf{G}$.
2. Substitute the form (10.4.9) for κ_{ab} into the expression given in Table 10.2 for the contribution to θ^{ij} from $\mathcal{L}_{\mathcal{P}M}$ and verify (10.4.10). [Recall that $\partial\kappa_{cd}^{-1}/\partial\kappa_{ef} = \kappa_{ce}^{-1}\kappa_{fd}^{-1}$ and use the equation of motion (10.3.12) of $\mathcal{P}_a$.]
3. Consider the Lagrangian for the electromagnetic field interacting with polarizable matter that is held fixed in space:

$$\mathcal{L} = \tfrac{1}{2}(\mathbf{E}^2 - \mathbf{B}^2) + \mathbf{P}\cdot\mathbf{E} + \mathbf{M}\cdot\mathbf{B} - \tfrac{1}{2}P_i\kappa_{ij}^{-1}P_j - \tfrac{1}{2}M_i\chi_{ij}^{-1}M_j.$$

   Show that the conserved "momentum" current arises from the invariance of $\int dx\,\mathcal{L}$ under translations is $\bar{\theta}^{i0} = (\mathbf{D}\times\mathbf{B})_i$, $\bar{\theta}^{ij} = -\epsilon E_i E_j - \mu H_i H_j + \tfrac{1}{2}\delta_{ij}[\epsilon E^2 + \mu H^2]$.
4. Show that the electric field $F^{0i} = -\epsilon_{abc}(\partial_0 R_a)(\partial_i R_b)b_c$ in a perfectly conducting fluid is related to the magnetic field $F^{ij} = \epsilon_{abc}(\partial_i R_a)(\partial_j R_b)b_c$ by $\mathbf{E} = -\mathbf{v}\times\mathbf{B}$, where $\mathbf{v}$ is the velocity of the fluid.
5. Use (10.7.23) to calculate the allowed frequencies of a magnetohydrodynamic waves with wave vector **k**.
6. A light beam carries an energy flux S in a fluid dielectric. The beam is reflected from a mirror suspended in the fluid and oriented perpendicularly to the light beam. Show that a force per unit area $P_{\text{rad}} = \sqrt{\epsilon\mu}\,2S$ is transmitted to the mirror. (This experiment has been done: R. V. Jones and J. C. S. Richards, *Proc. Roy. Soc. London* **A221**, 480 (1954).)
7. If you squeeze a quartz crystal with your fingers, about how much potential difference can you create across the crystal? If you clamp the crystal into place so that its shape cannot change and apply an electric field of 10 V/cm across the crystal, about how much force is exerted on the clamp? Assume that the crystal is roughly in the shape of a cube with sides 1 cm long.

8. Show that the "transverse pressure" $\frac{1}{2}B^2$ discussed in Section 10.7 can be roughly understood as the result of a repulsive force between pairs of magnetic field lines which falls off like the inverse cube of the distance between the lines.

9. Write an alternative form of the definition (10.1.17) of $M^{\mu\nu}(y)$ by inserting a factor of $1=\dot{X}_N^\lambda(\partial_\lambda \tau_N)$ next to $j_N^\nu(x)$. Use this form to show that $M^{\mu\nu}(y)$ does not depend at all on the timelike vector T_α if the atoms move with constant velocities and the currents $j_N^\mu(x)$ are independent of time in the rest frames of their respective atoms.

CHAPTER 11

Gravity

In this chapter we describe how the gravitational field is coupled to the fields that have appeared previously. The interaction looks simple when written in its most compact form. However, a certain amount of mathematical apparatus is required to build the theory. We set up this apparatus and state the most important theorems, but we do not have enough space to delve deeply into the structure of the mathematics or the reasons why this structure is thought to correspond to the physical world. A reader who is not already acquainted with general relativity is therefore advised to consult a textbook* on the subject if he wants more than a superficial introduction to gravity.

11.1 GENERAL COVARIANCE

We have ensured in previous chapters that equations of physics valid in a coordinate system x^μ are written in such a way that their form is unchanged when they are rewritten using new coordinates $\bar{x}^\mu$ which are related to x^μ by a Lorentz transformation $\bar{x}^\mu = \Lambda^\mu{}_\nu x^\nu$. We now extend this principle by demanding that the equations be left invariant under any coordinate transformation $\bar{x}^\mu = F^\mu(x)$.†

Vector Fields

A few simple extensions of the previous formalism give us some convenient mathematical building blocks with which generally covariant equations can be constructed. We first define a "contravariant vector field" $\boldsymbol{\phi}$. In each coordinate system, such a field $\boldsymbol{\phi}$ assigns to the point with coordinates x^ν four numbers, $\phi^\mu(x^0, x^1, x^2, x^3)$, which are called the components of $\boldsymbol{\phi}$

*See, for example, S. Weinberg, *Gravitation and Cosmology* (Wiley, New York, 1972), or C. Misner, K. Thorne, and J. Wheeler, *Gravitation* (Freeman, San Francisco, 1973).

†We demand only that the functions $F^\mu(x)$ be differentiable a suitable number of times and that the inverse transformation exist and be suitably differentiable.

in that coordinate system. In another coordinate system, the values $\bar{\phi}^\mu(\bar{x}^0,\bar{x}^1,\bar{x}^2,\bar{x}^3)$ are assigned to the same physical point, $\bar{x}^\mu = F^\mu(x)$. We require that these two sets of values be related by the transformation law:

$$\bar{\phi}^\mu(\bar{x}) = \frac{\partial \bar{x}^\mu}{\partial x^\nu}\phi^\nu(x). \tag{11.1.1}$$

In the case of the Lorentz transformation, $\bar{x}^\mu = \Lambda^\mu{}_\nu x^\nu$, this transformation law reduces to its previous form, $\bar{\phi}^\mu = \Lambda^\mu{}_\nu \phi^\nu$.

Scalar fields and tensor fields of any rank are defined analogously. For instance,

$$\bar{\phi}(\bar{x}) = \phi(x),$$

$$\bar{\phi}^{\mu\nu}(\bar{x}) = \frac{\partial \bar{x}^\mu}{\partial x^\alpha}\frac{\partial \bar{x}^\nu}{\partial x^\beta}\phi^{\alpha\beta}(x).$$

Notice that we can define a vector or tensor field, but that we do not try to define "tensors" that are not associated with points x^ν in space-time, since we would not know what matrix to use in place of $M(x)^\mu{}_\nu = \partial \bar{x}^\mu / \partial x^\nu$ in the transformation law.

We can similarly define a "covariant vector field" $\boldsymbol{\phi}$ by demanding that its components ϕ_μ transform according to

$$\bar{\phi}_\mu(\bar{x}) = \frac{\partial x^\nu}{\partial \bar{x}^\mu}\phi_\nu(x). \tag{11.1.2}$$

We can also define tensor fields with mixed covariant and contravariant components. For example,

$$\bar{\phi}^\mu_{\nu\rho}(\bar{x}) = \frac{\partial \bar{x}^\mu}{\partial x^\alpha}\frac{\partial x^\beta}{\partial \bar{x}^\nu}\frac{\partial x^\gamma}{\partial \bar{x}^\rho}\phi^\alpha_{\beta\gamma}(x). \tag{11.1.3}$$

If $\phi^\mu{}_\nu$ is a second rank tensor field, the field $\psi(x) = \phi^\mu{}_\mu(x)$, formed by contracting the indices, is a scalar field:

$$\bar{\psi}(\bar{x}) = \frac{\partial \bar{x}^\mu}{\partial x^\alpha}\frac{\partial x^\beta}{\partial \bar{x}^\mu}\phi^\alpha_\beta(x) = \psi(x).$$

Similar examples abound. For instance, from a contravariant vector field ϕ^μ and a covariant tensor field $G_{\mu\nu}$ one can form a new covariant vector field

$$\psi_\mu = G_{\mu\nu}\phi^\nu.$$

The Metric Tensor

In special relativity we measured the invariant "magnitude" of a vector field $A^{\mu}(x)$ by writing $(\mathbf{A})^2 = A^{\mu} g_{\mu\nu} A^{\nu}$. We can continue to do so provided that we make $g_{\mu\nu}$ into a tensor field $g_{\mu\nu}(x)$.

We can take the metric tensor to have components

$$g_{\mu\nu}(x) = \eta_{\mu\nu} \equiv \begin{pmatrix} -1 & 0 & 0 & 0 \\ 0 & +1 & 0 & 0 \\ 0 & 0 & +1 & 0 \\ 0 & 0 & 0 & +1 \end{pmatrix}$$

in an inertial coordinate system, provided that we neglect gravity. In any other coordinate system the components $\bar{g}_{\mu\nu}$ of the metric tensor can then be found by using the tensor transformation law. The special metric $\eta_{\mu\nu}$ is called the Minkowski metric.

A region of space-time in which one can make $g_{\mu\nu}(x) \equiv \eta_{\mu\nu}$ is called "flat." We also consider "curved" space-time, in which the components of the metric tensor are not identically equal to $\eta_{\mu\nu}$ in any coordinate system. We still require that $g_{\mu\nu}(x)$ be a symmetric tensor. We also ask that at each point x, $g_{\mu\nu}(x)$ have one negative eigenvalue and three positive eigenvalues. Then if we use $dS^2 = g_{\mu\nu}(x)\, dx^{\mu}\, dx^{\mu}$ to measure the length of the interval between x^{μ} and $x^{\mu} + dx^{\mu}$, we find that there are one "timelike" direction and three "spacelike" directions in space-time at the point x.

The claim of general relativity is that the effects of gravity can be described by letting space-time be curved. The role of the Newtonian gravitational potential $\phi(t, \mathbf{x})$ is taken over by the metric tensor $g_{\mu\nu}(x)$. Accordingly, $g_{\mu\nu}(x)$ will be a field that appears in an action principle along with the matter (and electromagnetic) fields. Variation of this action with respect to the matter fields will give the equations of motion for matter in the presence of gravity; these equations will in fact be just the old equations made generally covariant. Variations of the action with respect to $g_{\mu\nu}(x)$ will give the equation of motion for $g_{\mu\nu}$. We will construct this Lagrangian for matter and metric in Section 11.2, after we finish producing the necessary building blocks.

Raising and Lowering Indices

The metric tensor can be used to form covariant vectors $g_{\mu\nu}\phi^{\nu}$ from contravariant vectors ϕ^{ν}. In keeping with our earlier notation, we use the same symbol ϕ both for ϕ^{ν} and $\phi_{\mu} \equiv g_{\mu\nu}\phi^{\nu}$, and regard both forms as simply different ways of expressing the same physical vector. The contravariant form of ϕ can be recovered from the covariant form using

$$\phi^{\mu}=g^{\mu\nu}\phi_{\nu}, \tag{11.1.4}$$

where $g^{\mu\nu}(x)$ is the inverse matrix to $g_{\mu\nu}(x)$. It is a simple exercise to prove that $g^{\mu\nu}(x)$ is indeed a contravariant tensor (see Problem 1). Notice that the notation $g^{\mu\nu}$ is consistent with the "lowering indices" rule: $g_{\alpha\mu}g_{\beta\nu}g^{\mu\nu} = g_{\alpha\beta}$. The metric tensor with one lower and one upper index is the Kroneker delta: $g^{\alpha}{}_{\beta}\equiv g^{\alpha\gamma}g_{\gamma\beta}=\delta^{\alpha}_{\beta}$.

Locally Inertial Coordinates

In a curved space the metric tensor is not identically equal to the Minkowski metric $\eta_{\mu\nu}$ in any coordinate system. One can, however, choose a coordinate system in which $g_{\mu\nu}=\eta_{\mu\nu}$ and $\partial g_{\mu\nu}/\partial x_{\rho}=0$ at a given point.

To prove this assertion, let the point in question be $x^{\mu}=0$. Choose new coordinates $\bar{x}^{\mu}$ defined by

$$x^{\alpha}=A^{\alpha}{}_{\mu}\bar{x}^{\mu}-\tfrac{1}{4}A^{\alpha}{}_{\lambda}\eta^{\lambda\sigma}b_{\sigma\mu\gamma}\bar{x}^{\mu}\bar{x}^{\gamma},$$

where $A^{\alpha}{}_{\mu}$ and $b_{\sigma\mu\gamma}$ are constants to be determined and $b_{\sigma\mu\gamma}=b_{\sigma\gamma\mu}$. The new metric tensor is

$$\bar{g}_{\mu\nu}(\bar{x})=\left[A^{\alpha}{}_{\mu}-\tfrac{1}{2}A^{\alpha}{}_{\lambda}\eta^{\lambda\sigma}b_{\sigma\mu\gamma}\bar{x}^{\gamma}\right]g_{\alpha\beta}(x)\left[A^{\beta}{}_{\nu}-\tfrac{1}{2}A^{\beta}{}_{\kappa}\eta^{\kappa\tau}b_{\tau\nu\delta}\bar{x}^{\delta}\right].$$

At $x^{\mu}=0$ this is

$$\bar{g}_{\mu\nu}(0)=A^{\alpha}{}_{\mu}g_{\alpha\beta}(0)A^{\beta}{}_{\nu}.$$

Since, by assumption, $g_{\alpha\beta}(0)$ has one positive and three negative eigenvalues, it is possible to find a matrix $A^{\alpha}{}_{\mu}$ such that $\bar{g}_{\mu\nu}(0)=\eta_{\alpha\beta}$. Having made this choice, we find for $\partial\bar{g}_{\mu\nu}/\partial\bar{x}^{\rho}$ at $x^{\mu}=0$.

$$\frac{\partial\bar{g}_{\mu\nu}}{\partial\bar{x}^{\rho}}=A^{\alpha}{}_{\mu}\frac{\partial g_{\alpha\beta}}{\partial\bar{x}^{\rho}}A^{\beta}{}_{\nu}-\tfrac{1}{2}\left[b_{\mu\nu\rho}+b_{\nu\mu\rho}\right].$$

We choose

$$b_{\mu\nu\rho}=A^{\alpha}{}_{\mu}\frac{\partial g_{\alpha\beta}}{\partial\bar{x}^{\rho}}A^{\beta}{}_{\nu}+A^{\alpha}{}_{\mu}\frac{\partial g_{\alpha\beta}}{\partial\bar{x}^{\nu}}A^{\beta}{}_{\rho}-A^{\alpha}{}_{\nu}\frac{\partial g_{\alpha\beta}}{\partial\bar{x}^{\mu}}A^{\beta}{}_{\rho}$$

Then $\partial\bar{g}_{\mu\nu}/\partial\bar{x}^{\rho}=0$ at $x=0$, as desired.

This is the best we can do: the second derivatives of $g_{\mu\nu}$ cannot all be eliminated by adjusting the coordinate system.

Some Special Tensors

We have made extensive use in the preceding chapters of the symbol $\varepsilon^{\mu\nu\rho\sigma}$ which is defined by the properties that it is completely antisymmetric and that $\varepsilon^{0123}=1$. Although $\varepsilon^{\mu\nu\rho\sigma}$ is a tensor as far as Lorentz transformations are concerned, it is not quite a tensor under general coordinate transformations. If we apply the tensor transformation law to $\varepsilon^{\mu\nu\rho\sigma}$ we find

$$\frac{\partial \bar{x}^{\mu}}{\partial x^{\alpha}}\frac{\partial \bar{x}^{\nu}}{\partial x^{\beta}}\frac{\partial \bar{x}^{\rho}}{\partial x^{\gamma}}\frac{\partial \bar{x}^{\sigma}}{\partial x^{\delta}}\varepsilon^{\alpha\beta\gamma\delta}=\det\frac{\partial \bar{x}}{\partial x}\varepsilon^{\mu\nu\rho\sigma}. \qquad (11.1.6)$$

Except for the determinant of $\partial\bar{x}^{\mu}/\partial x^{\alpha}$ on the right hand side of (11.1.6), $\varepsilon^{\mu\nu\rho\sigma}$ would be a tensor.

We can make an honest tensor out of $\varepsilon^{\mu\nu\rho\sigma}$ by dividing it by

$$\sqrt{g(x)}\equiv\sqrt{-\det g_{\mu\nu}(x)}\ .$$

The transformation law for $\sqrt{g}$ is

$$\sqrt{\bar{g}}=\left[-\det\left(\frac{\partial x^{\alpha}}{\partial \bar{x}^{\mu}}g_{\alpha\beta}\frac{\partial x^{\beta}}{\partial \bar{x}^{\nu}}\right)\right]^{1/2}=\left[-\det\left(\frac{\partial x}{\partial \bar{x}}\right)^{2}\det g_{\alpha\beta}\right]^{1/2}$$

$$=\det\left(\frac{\partial x}{\partial \bar{x}}\right)\sqrt{g}\ . \qquad (11.1.7)$$

Thus the quantity

$$e^{\mu\nu\rho\sigma}(x)=\frac{1}{\sqrt{g(x)}}\varepsilon^{\mu\nu\rho\sigma} \qquad (11.1.8)$$

is a tensor:

$$\frac{1}{\sqrt{\bar{g}}}\varepsilon^{\mu\nu\rho\sigma}=\frac{\det(\partial\bar{x}/\partial x)}{\sqrt{g}}\varepsilon^{\mu\nu\rho\sigma}$$

$$=\frac{\partial \bar{x}^{\mu}}{\partial x^{\alpha}}\frac{\partial \bar{x}^{\nu}}{\partial x^{\beta}}\frac{\partial \bar{x}^{\rho}}{\partial x^{\gamma}}\frac{\partial \bar{x}^{\sigma}}{\partial x^{\delta}}\frac{1}{\sqrt{g}}\varepsilon^{\alpha\beta\gamma\delta}.$$

A quantity like $\varepsilon^{\mu\nu\rho\sigma}$ which equals $\sqrt{g(x)}$ times a tensor is called a "tensor density."

We have also made extensive use in the preceding chapters of the fact that the differential d^4x is a scalar under Lorentz transformations. That is, the integral $\int d^4x\,\phi(x)$ of a scalar field is Lorentz invariant. To write an integral of a scalar field that is invariant under general coordinate transformation we must insert a factor $[g(x)]^{1/2}$:

$$I=\int d^4x\sqrt{g(x)}\;\phi(x). \tag{11.1.9}$$

Then the integral calculated in the $\bar{x}$-system is

$$\bar{I}=\int d^4\bar{x}\sqrt{\bar{g}(\bar{x})}\;\phi(\bar{x})=\int d^4x\,\det\left(\frac{\partial\bar{x}}{\partial x}\right)\det\left(\frac{\partial x}{\partial\bar{x}}\right)\sqrt{g(x)}\;\phi(x)$$

$$=\int d^4x\sqrt{g(x)}\;\phi(x)=I.$$

A field that equals $\sqrt{g(x)}$ times a scalar field is called a scalar density. For instance, the Lagrangian $\mathfrak{L}(x)$ in general relativity is a scalar density, so that the action $\int dx\,\mathfrak{L}(x)$ is invariant.

Derivatives

Let $\phi(x)$ be a scalar field, and consider its gradient. In keeping with the customary notation in general relativity, differentiation is indicated by a comma in this and the next chapters:

$$\phi(x)_{,\mu}\equiv\frac{\partial}{\partial x^\mu}\phi(x). \tag{11.1.10}$$

The gradient $\phi_{,\mu}$ is a covariant vector field:

$$\bar{\phi}(\bar{x})_{,\mu}=\frac{\partial}{\partial\bar{x}^\mu}\bar{\phi}(\bar{x})=\frac{\partial x^\nu}{\partial\bar{x}^\mu}\frac{\partial}{\partial x^\nu}\phi(x)=\frac{\partial x^\nu}{\partial\bar{x}^\mu}\phi(x)_{,\nu}.$$

Now consider the derivative of a vector field

$$A_{\mu,\nu}=\frac{\partial}{\partial x^\nu}A_\mu.$$

Unfortunately, $A_{\mu,\nu}$ is *not* a tensor:

$$\bar{A}_{\mu,\nu}=\frac{\partial}{\partial\bar{x}^\nu}\left(\frac{\partial x^\alpha}{\partial\bar{x}^\mu}A_\alpha\right)$$

$$=\frac{\partial x^\alpha}{\partial\bar{x}^\mu}\frac{\partial x^\beta}{\partial\bar{x}^\nu}A_{\alpha,\beta}+\frac{\partial^2x^\alpha}{\partial\bar{x}^\nu\partial\bar{x}^\mu}A_\alpha.$$

The term proportional to $\partial^2 x^\alpha / \partial\bar{x}^\nu \partial\bar{x}^\mu$ spoils the transformation law. We would like to have an object that *is* a tensor and reduces to $A_{\mu,\nu}$, at least in an inertial coordinate system in a gravity free space. The notation for this "covariant derivative" is $A_{\mu;\nu}$ with a semicolon instead of a comma. Such an object is

$$A_{\mu;\nu} = A_{\mu,\nu} - \Gamma^\lambda_{\mu\nu} A_\lambda \tag{11.1.11}$$

where the "connection coefficients" $\Gamma^\lambda_{\mu\nu}$ are a particular combination of first derivatives of the metric tensor:

$$\Gamma^\lambda_{\mu\nu} = \tfrac{1}{2} g^{\lambda\sigma}\left[g_{\sigma\mu,\nu} + g_{\sigma\nu,\mu} - g_{\mu\nu,\sigma}\right]. \tag{11.1.12}$$

It is apparent that $A_{\mu;\nu}$ and $A_{\mu,\nu}$ are indeed equal in an inertial coordinate system, since then $g_{\alpha\beta,\gamma} = 0$.

To see that $A_{\mu;\nu}$ is a tensor, we must find the transformation law for the connection coefficients. The transformed $\Gamma^\lambda_{\mu\nu}$ symbol is

$$\bar{\Gamma}^\lambda_{\mu\nu} = \tfrac{1}{2} \bar{g}^{\lambda\sigma}\left[\bar{g}_{\sigma\mu,\nu} + \bar{g}_{\sigma\nu,\mu} - \bar{g}_{\mu\nu,\sigma}\right],$$

where

$$\bar{g}^{\lambda\sigma}\bar{g}_{\sigma\mu,\nu} = \frac{\partial\bar{x}^\lambda}{\partial x^\delta}\frac{\partial\bar{x}^\sigma}{\partial x^\varepsilon} g^{\delta\varepsilon}\frac{\partial}{\partial\bar{x}^\nu}\left(\frac{\partial x^\gamma}{\partial\bar{x}^\sigma}\frac{\partial x^\alpha}{\partial\bar{x}^\mu} g_{\gamma\alpha}\right)$$

$$= \frac{\partial\bar{x}^\lambda}{\partial x^\delta}\frac{\partial x^\alpha}{\partial\bar{x}^\mu}\frac{\partial x^\beta}{\partial\bar{x}^\nu} g^{\delta\gamma} g_{\gamma\alpha,\beta} + \frac{\partial\bar{x}^\lambda}{\partial x^\delta}\frac{\partial\bar{x}^\sigma}{\partial x^\varepsilon} g^{\delta\varepsilon} g_{\gamma\alpha}$$

$$\times\left(\frac{\partial^2 x^\gamma}{\partial\bar{x}^\nu\partial\bar{x}^\sigma}\frac{\partial x^\alpha}{\partial\bar{x}^\mu} + \frac{\partial x^\gamma}{\partial\bar{x}^\sigma}\frac{\partial^2 x^\alpha}{\partial\bar{x}^\nu\partial\bar{x}^\mu}\right).$$

When this expression is summed over the various permutations of the indices of $g_{\sigma\mu,\nu}$, four of the terms containing second derivatives cancel, leaving

$$\bar{\Gamma}^\lambda_{\mu\nu} = \frac{1}{2}\frac{\partial\bar{x}^\lambda}{\partial x^\delta}\frac{\partial x^\alpha}{\partial\bar{x}^\mu}\frac{\partial x^\beta}{\partial\bar{x}^\nu} g^{\delta\gamma}\left[g_{\gamma\alpha,\beta} + g_{\gamma\beta,\alpha} - g_{\alpha\beta,\gamma}\right]$$

$$+ \frac{1}{2}\frac{\partial\bar{x}^\lambda}{\partial x^\delta}\frac{\partial\bar{x}^\sigma}{\partial x^\varepsilon} g^{\delta\varepsilon} g_{\gamma\alpha}\left(2\frac{\partial x^\gamma}{\partial\bar{x}^\sigma}\frac{\partial^2 x^\alpha}{\partial\bar{x}^\nu\partial\bar{x}^\mu}\right).$$

Thus

$$\bar{\Gamma}^\lambda_{\mu\nu} = \frac{\partial\bar{x}^\lambda}{\partial x^\delta}\frac{\partial x^\alpha}{\partial\bar{x}^\mu}\frac{\partial x^\beta}{\partial\bar{x}^\nu}\Gamma^\delta_{\alpha\delta} + \frac{\partial\bar{x}^\lambda}{\partial x^\delta}\frac{\partial^2 x^\delta}{\partial\bar{x}^\mu\partial\bar{x}^\nu}. \tag{11.1.13}$$

We can now use this transformation law for the connection coefficients and the transformation laws for A_μ and $A_{\mu,\nu}$ to find how $A_{\mu;\nu}$ transforms. Using the definition (11.1.11) we have

$$\begin{aligned}\bar{A}_{\mu;\nu} &= \bar{A}_{\mu,\nu} - \bar{\Gamma}^{\lambda}_{\mu\nu}\bar{A}_{\lambda}\\ &= \frac{\partial x^{\alpha}}{\partial \bar{x}^{\mu}}\frac{\partial x^{\beta}}{\partial \bar{x}^{\nu}}A_{\alpha,\beta} + \frac{\partial^2 x^{\alpha}}{\partial \bar{x}^{\mu}\partial \bar{x}^{\nu}}A_{\alpha}\\ &\quad - \frac{\partial x^{\alpha}}{\partial \bar{x}^{\mu}}\frac{\partial x^{\beta}}{\partial \bar{x}^{\nu}}\Gamma^{\delta}_{\alpha\beta}A_{\delta} - \frac{\partial^2 x^{\delta}}{\partial \bar{x}^{\mu}\partial \bar{x}^{\nu}}A_{\delta}\\ &= \frac{\partial x^{\alpha}}{\partial \bar{x}^{\mu}}\frac{\partial x^{\beta}}{\partial \bar{x}^{\nu}}A_{\alpha;\beta}.\end{aligned} \tag{11.1.14}$$

This is the transformation law of a tensor field, as claimed.

The connection coefficients can also be used to define covariant differentiation of contravariant vectors:

$$A^{\mu}{}_{;\nu} = A^{\mu}{}_{,\nu} + \Gamma^{\mu}_{\lambda\nu}A^{\lambda}. \tag{11.1.15}$$

Note again that $A^{\mu}{}_{;\nu} = A^{\mu}{}_{,\nu}$ whenever $g_{\mu\nu,\lambda} = 0$. The proof that $A^{\mu}{}_{;\nu}$ is a tensor is left as Problem 2. If we want to differentiate a tensor with more indices, we use more connection coefficients:

$$A^{\mu}{}_{\nu;\sigma} = A^{\mu}{}_{\nu,\sigma} + \Gamma^{\mu}_{\lambda\sigma}A^{\lambda}{}_{\nu} - \Gamma^{\lambda}_{\nu\sigma}A^{\mu}{}_{\lambda}. \tag{11.1.16}$$

Properties of Covariant Differentiation

There are several important properties of ordinary differentiation that carry over to covariant differentiation. These properties can be established most easily by noting that each equation is true at the point x^μ in a "locally inertial" coordinate system, in which $g_{\mu\nu,\lambda} = 0$ at x^μ and thus covariant differentiation reduces to ordinary differentiation. Since the equations are covariant and true in a particular coordinate system, they are true in any coordinate system.

First, covariant differentiation is linear: if a and b are constants, then

$$\left[aA^{\mu\nu} + bB^{\mu\nu}\right]_{;\lambda} = aA^{\mu\nu}{}_{;\lambda} + bB^{\mu\nu}{}_{;\lambda}. \tag{11.1.17}$$

Second, the Leibniz rule for differentiating products still holds:

$$\left[A^{\mu}B^{\nu}\right]_{;\lambda} = A^{\mu}{}_{;\lambda}B^{\nu} + A^{\mu}B^{\nu}{}_{;\lambda}. \tag{11.1.18}$$

Third, the operations of contracting indices can be performed either before or after differentiation:

$$[A^{\mu\alpha}{}_{\alpha}]_{;\lambda}=g^{\alpha}{}_{\beta}[A^{\mu\beta}{}_{\alpha}]_{;\lambda}. \tag{11.1.19}$$

Finally, the covariant derivative of the metric tensor vanishes:

$$g_{\mu\nu;\lambda}=0, \qquad g^{\mu\nu}{}_{;\lambda}=0. \tag{11.1.20}$$

Thus the operation of raising and lowering indices commutes with covariant differentiation; for instance,

$$g^{\mu\nu}A_{\nu;\lambda}=A^{\mu}{}_{;\lambda}. \tag{11.1.21}$$

Conserved Currents and the Graduent of g

Consider a three dimensional surface $\mathbb{S}$ in space-time, described parametricly by $x^{\mu}=x^{\mu}(a,b,c)$. We can perform integrations over this surface by using the covariant surface area differential

$$dS_{\mu}=e_{\mu\alpha\beta\gamma}\frac{\partial x^{\alpha}}{\partial a}\frac{\partial x^{\beta}}{\partial b}\frac{\partial x^{\gamma}}{\partial c}\,da\,db\,dc, \tag{11.1.22}$$

where

$$\begin{aligned} e_{\mu\alpha\beta\gamma} &= g_{\mu\nu}g_{\alpha\delta}g_{\beta\varepsilon}g_{\gamma\kappa}e^{\nu\delta\varepsilon\kappa} \\ &= \sqrt{g}\,\varepsilon_{\mu\alpha\beta\gamma}. \end{aligned} \tag{11.1.23}$$

Apparently dS_{μ} transforms like a covariant vector. Thus if $J^{\mu}(x)$ is a vector field, the integral

$$Q=\int dS_{\mu}J^{\mu}(x) \tag{11.1.24}$$

is invariant under coordinate transformations. If J^{μ} were, say, the electric current and $\mathbb{S}$ were the surface $x^0=0$ in a particular coordinate system, then Q would be the generalization to curved space of the charge at time x^0.

When is Q conserved? That is, when is the integral (11.1.24) independent of the surface $\mathbb{S}$? The question is very similar to the corresponding question posed in Chapter 3; the only difference is the factor $[g(x)]^{1/2}$ in (11.1.23). The arguments of Chapter 3 show that Q is conserved if the integral $\int dS_\mu J^\mu$ over a closed surface always vanishes, and that

$$\int_{\mathbb{S}} dS_\mu J^\mu = \int_{\substack{\text{inside}\\ \mathbb{S}}} d^4x\sqrt{g}\,\frac{1}{\sqrt{g}}\left[\sqrt{g}\,J^\mu\right]_{,\mu}. \tag{11.1.25}$$

Thus conservation of Q is equivalent to the vanishing of $g^{-1/2}[g^{1/2}J^\mu]_{,\mu}$.

Equation (11.1.25) might lead one to guess that

$$\frac{1}{\sqrt{g}}\left[\sqrt{g}\,J^\mu\right]_{,\mu} = J^\mu{}_{;\mu}. \tag{11.1.26}$$

This useful formula is indeed true. We can prove it by comparing

$$\frac{1}{\sqrt{g}}\left[\sqrt{g}\,\right]_{,\mu} = \frac{1}{2}\,\frac{1}{g}\,\frac{\partial g}{\partial g_{\alpha\beta}}\,g_{\alpha\beta,\mu} = \tfrac{1}{2}\,g^{\beta\alpha}g_{\alpha\beta,\mu}$$

with

$$\begin{aligned}\Gamma^\lambda_{\lambda\mu} &= \tfrac{1}{2}\,g^{\lambda\sigma}\left[\,g_{\sigma\lambda,\mu} + g_{\sigma\mu,\lambda} - g_{\lambda\mu,\sigma}\right]\\ &= \tfrac{1}{2}\,g^{\lambda\sigma}g_{\sigma\lambda,\mu}.\end{aligned}$$

We see that

$$\Gamma^\lambda_{\lambda\mu} = \frac{1}{\sqrt{g}}\left[\sqrt{g}\,\right]_{,\mu}. \tag{11.1.27}$$

Thus

$$\begin{aligned}\frac{1}{\sqrt{g}}\left[\sqrt{g}\,J^\mu\right]_{,\mu} &= J^\mu{}_{,\mu} + \frac{1}{\sqrt{g}}\left[\sqrt{g}\,\right]_{,\mu}J^\mu\\ &= J^\mu{}_{,\mu} + \Gamma^\mu_{\mu\sigma}J^\sigma\\ &= J^\mu{}_{;\mu}\end{aligned}$$

as claimed. As a consequence, the condition for conservation of Q is the vanishing of the covariant divergence of J^μ.

This argument applies to conservation of scalar quantities like charge and number of atoms. We will find in Section 11.4 that the interpretation of the local "momentum conservation" equation $\theta^{\mu\nu}{}_{;\nu}=0$ is more complicated.

The identity (11.1.27) is very useful in its own right.

The Curvature Tensor

There is a most important property of ordinary differentiation of fields that does not hold for covariant differentiation: the order in which differentiations are done makes a difference.

Let us consider the second derivative of a vector field, $A^{\lambda}{}_{;\mu\nu} \equiv [A^{\lambda}{}_{;\mu}]_{;\nu}$. The quantity

$$A^{\lambda}{}_{;\mu\nu} - A^{\lambda}{}_{;\nu\mu}$$

is zero in a flat space, in which we can make $g_{\mu\nu} \equiv \eta_{\mu\nu}$ by choosing a suitable coordinate system. However, this quantity need not be zero for a more general choice of $g_{\mu\nu}(x)$. Explicit calculation gives

$$\begin{aligned}
A^{\lambda}{}_{;\mu\nu} &= \left[A^{\lambda}{}_{,\mu} + \Gamma^{\lambda}_{\alpha\mu}A^{\alpha}\right]_{,\nu} \\
&\quad + \Gamma^{\lambda}_{\beta\nu}\left[A^{\beta}{}_{,\mu} + \Gamma^{\beta}_{\alpha\mu}A^{\alpha}\right] \\
&\quad - \Gamma^{\gamma}_{\mu\nu}\left[A^{\lambda}{}_{,\gamma} + \Gamma^{\lambda}_{\alpha\gamma}A^{\alpha}\right] \\
&= A^{\lambda}{}_{,\mu\nu} + \Gamma^{\lambda}_{\alpha\mu,\nu}A^{\alpha} + \Gamma^{\lambda}_{\alpha\mu}A^{\alpha}{}_{,\nu} \\
&\quad + \Gamma^{\lambda}_{\beta\nu}A^{\beta}{}_{,\mu} + \Gamma^{\lambda}_{\beta\nu}\Gamma^{\beta}_{\alpha\mu}A^{\alpha} \\
&\quad - \Gamma^{\gamma}_{\mu\nu}A^{\lambda}{}_{,\gamma} - \Gamma^{\gamma}_{\mu\nu}\Gamma^{\lambda}_{\alpha\gamma}A^{\alpha}.
\end{aligned}$$

When we form $A^{\lambda}{}_{;\mu\nu} - A^{\lambda}{}_{;\nu\mu}$ we find that the first term above does not contribute because of the symmetry of $A^{\lambda}{}_{,\mu\nu}$ in the indices μ, ν; the last two terms do not contribute because of the symmetry of $\Gamma^{\gamma}_{\mu\nu}$; and the contributions from the third and fourth terms cancel each other. This leaves

$$A^{\lambda}{}_{;\mu\nu} - A^{\lambda}{}_{;\nu\mu} = -R^{\lambda}{}_{\alpha\mu\nu}A^{\alpha}, \tag{11.1.28}$$

where

$$R^{\lambda}{}_{\alpha\mu\nu} = \Gamma^{\lambda}_{\alpha\nu,\mu} + \Gamma^{\lambda}_{\beta\mu}\Gamma^{\beta}_{\alpha\nu} - \Gamma^{\lambda}_{\alpha\mu,\nu} - \Gamma^{\lambda}_{\beta\nu}\Gamma^{\beta}_{\alpha\mu}. \tag{11.1.29}$$

We can derive similar formulas involving $R^{\lambda}{}_{\alpha\mu\nu}$ for the derivatives of covariant vectors and of tensors. For instance,

$$A_{\lambda;\mu\nu} - A_{\lambda;\nu\mu} = +R^{\alpha}{}_{\lambda\mu\nu}A_{\alpha},$$

$$A^{\kappa}{}_{\lambda;\mu\nu} - A^{\kappa}{}_{\lambda;\nu\mu} = -R^{\kappa}{}_{\alpha\mu\nu}A^{\alpha}{}_{\lambda} + R^{\alpha}{}_{\lambda\mu\nu}A^{\kappa}{}_{\alpha}.$$

The quantity $R^{\lambda}{}_{\alpha\mu\nu}$ must be a tensor, since it occurs in the covariant equation (11.1.28). To make this argument into a proof, let $\bar{R}^{\lambda}{}_{\alpha\mu\nu}$ be the quantity (11.1.29) calculated in the $\bar{x}$-coordinate system, and let $\tilde{R}^{\lambda}{}_{\alpha\mu\nu}$ be $R^{\lambda}{}_{\alpha\mu\nu}$ transformed to the $\bar{x}$-system according to the tensor transformation law,

$$\tilde{R}^{\lambda}{}_{\alpha\mu\nu} = \frac{\partial\bar{x}^{\lambda}}{\partial x^{\kappa}} \frac{\partial x^{\beta}}{\partial\bar{x}^{\alpha}} \frac{\partial x^{\gamma}}{\partial\bar{x}^{\mu}} \frac{\partial x^{\delta}}{\partial\bar{x}^{\nu}} R^{\kappa}{}_{\beta\gamma\delta}.$$

We must prove that $\bar{R} = \tilde{R}$. On the one hand we have

$$\bar{A}^{\lambda}{}_{;\mu\nu} - \bar{A}^{\lambda}{}_{;\nu\mu} = -\bar{R}^{\lambda}{}_{\alpha\mu\nu}\bar{A}^{\alpha}$$

and on the other hand

$$\bar{A}^{\lambda}{}_{;\mu\nu} - \bar{A}^{\lambda}{}_{;\nu\mu} = -\tilde{R}^{\lambda}{}_{\alpha\mu\nu}\bar{A}^{\alpha}.$$

Thus

$$0 = \left[\bar{R}^{\lambda}{}_{\alpha\mu\nu} - \tilde{R}^{\lambda}{}_{\alpha\mu\nu}\right]\bar{A}^{\alpha}$$

for every vector field $\bar{A}^{\alpha}$. Therefore $\bar{R}^{\lambda}{}_{\alpha\mu\nu} = \tilde{R}^{\lambda}{}_{\alpha\mu\nu}$.

The tensor $R^{\lambda}{}_{\alpha\mu\nu}$ is called the Riemann-Christoffel curvature tensor. Even this brief introduction should have convinced the reader that it plays a fundamental role in tensor analysis. We will use it in the Lagrangian for gravity.

11.2 THE LAGRANGIAN

We can now grapple with the physics of the interaction between matter and gravity. We use Hamilton's principle with an action of the form

$$A = \int d^4x\sqrt{g(x)}\left[\frac{\mathfrak{L}(x)}{\sqrt{g(x)}}\right], \tag{11.2.1}$$

where $\mathfrak{L}$ is a scalar density (that is, $\mathfrak{L}/\sqrt{g}$ is a scalar). Since $d^4x\sqrt{g(x)}$ is a scalar, the action will be invariant under coordinate transformations. Thus the equations of motion are guaranteed to be covariant.

The Lagrangian $\mathfrak{L}$ will be a function of the matter fields $R_a(x)$, the metric field $g_{\mu\nu}(x)$, and their derivatives. Later in this section we include other nongravitational fields, like the electromagnetic potential $A_{\mu}(x)$.

We guess what $\mathcal{L}$ is with the help of the "principle of equivalence" and the assumption that $g_{\mu\nu}(x)$ obeys a second order differential equation of motion. In Section 11.4 we check our guess by making sure that we recover Newton's theory of gravity in the "Newtonian" limit of small velocities and weak gravitational fields. There are some important experimental tests of the full theory, as distinct from its Newtonian limit, but we do not discuss them.

The Matter Part of $\mathcal{L}$

We begin with the fundamental assumption that the equations of motion for the matter fields $R_a(x)$ are the same as in the absence of gravity, except for being made generally covariant. That is, $\eta_{\mu\nu}$ is replaced by $g_{\mu\nu}(x)$ and derivatives are replaced by covariant derivatives. We have seen that for each point X^μ one can choose a "locally inertial" coordinate system such that $g_{\mu\nu}(X)=\eta_{\mu\nu}$ and $g_{\mu\nu}(X)_{,\lambda}=0$. In this coordinate system the equation of motion for $R_a(x)$ at $x^\mu=X^\mu$ will look exactly the same as the equation of motion in the absence of gravity. The ansatz that we should be able to transform away the effects of gravity by making a coordinate transformation is called the "principle of equivalence." It was this principle that originally led Einstein to the general theory of relativity.

According to our assumption, the terms in $\mathcal{L}$ containing $R_a(x)$ should be just the usual Lagrangian for matter, made into a scalar density under general coordinate transformations:

$$\begin{aligned}
\mathcal{L}_M &= -\sqrt{g}\,\rho U(G_{ab}, R_a),\\
G_{ab} &= R_a(x)_{,\mu} R_b(x)_{,\nu}\, g^{\mu\nu}(x), \qquad\qquad (11.2.2)\\
\rho &= n\sqrt{\det G_{ab}}\ .
\end{aligned}$$

The Metric Part of $\mathcal{L}$

We also need a term in $\mathcal{L}$ that does not depend on $R_a(x)$, but only on $g_{\mu\nu}$ and its derivatives,

$$\mathcal{L}_G(g_{\mu\nu}, g_{\mu\nu,\lambda}, g_{\mu\nu,\lambda,\kappa}, \ldots).$$

This term in $\mathcal{L}$ will provide terms in the equation of motion for $g_{\mu\nu}(x)$ that remain nonzero in empty space.

What scalar density $\mathcal{L}_G$ should we pick? We recall that the Newtonian gravitational potential $\phi(t,\mathbf{x})$ obeys a second order differential equation

$\nabla^2\phi = 4\pi G m\rho$. Indeed, almost all of the partial differential equations of physics are second order. Thus we have a strong prejudice that the Lagrangian $\mathcal{L}_G$ should lead to a second order differential equation of motion for $g_{\mu\nu}(x)$. Let us adopt this prejudice as a principle and see what it implies.

The equation of motion corresponding to the Lagrangian $\mathcal{L}_G$ (with, for the moment, $\mathcal{L}_M = 0$) is

$$0 = \int d^4x \left\{ \frac{\partial \mathcal{L}_G}{\partial g_{\mu\nu}} \delta g_{\mu\nu}(x) + \frac{\partial \mathcal{L}_G}{\partial g_{\mu\nu,\lambda}} \delta g_{\mu\nu}(x)_{,\lambda} + \frac{\partial \mathcal{L}_G}{\partial g_{\mu\nu,\lambda\kappa}} \delta g_{\mu\nu}(x)_{,\lambda\kappa} + \cdots \right\}$$

for all $\delta g_{\mu\nu}(x)$. That is,

$$0 = \frac{\partial \mathcal{L}_G}{\partial g_{\mu\nu}} - \left[\frac{\partial \mathcal{L}_G}{\partial g_{\mu\nu,\lambda}} \right]_{,\lambda} + \left[\frac{\partial \mathcal{L}_G}{\partial g_{\mu\nu,\lambda\kappa}} \right]_{,\lambda\kappa} - \cdots. \tag{11.2.3}$$

This would be a second order equation if $\mathcal{L}_G$ were a function of $g_{\mu\nu}$ and $g_{\mu\nu,\lambda}$ only. Unfortunately, it is impossible to make a nonconstant scalar function $\mathcal{L}_G(g_{\mu\nu}, g_{\mu\nu,\lambda})/g^{1/2}$. Whatever function we had in mind would equal the constant $\mathcal{L}_G(\eta_{\mu\nu}, 0)$ because it is possible to make $g_{\mu\nu} = \eta_{\mu\nu}$ and $g_{\mu\nu,\lambda} = 0$ at any point by a coordinate transformation.

The only escape hatch is to allow $\mathcal{L}_g$ to depend on $g_{\mu\nu}$ and its first and second derivatives, but to insist that $\partial \mathcal{L}_G / \partial g_{\mu\nu,\lambda\kappa}$ be a function of $g_{\mu\nu}$ only. According to (11.2.3), the equation of motion for $g_{\mu\nu}$ will then be second order. Thus we choose $\mathcal{L}_G$ of the form

$$\mathcal{L}_G = g_{\mu\nu,\lambda\kappa} B(g_{\alpha\beta})^{\mu\nu\lambda\kappa} + C(g_{\mu\nu}, g_{\mu\nu,\lambda}). \tag{11.2.4}$$

(The notation $B^{\mu\nu\lambda\kappa}$ is not intended to indicate a tensor transformation law under general coordinate transformations.)

We can learn a lot more about $\mathcal{L}_G$ by evaluating it in a locally inertial coordinate system. At a point x^μ which we take to be $x^\mu = 0$, we choose coordinants such that $g_{\mu\nu} = \eta_{\mu\nu}$ and $g_{\mu\nu,\lambda} = 0$. We then have

$$\mathcal{L}_G = g_{\mu\nu,\lambda\kappa} B(\eta_{\alpha\beta})^{\mu\nu\lambda\kappa} + c, \tag{11.2.5}$$

where $c = C(\eta_{\mu\nu}, 0)$. In a new coordinate system related to the x-system by a Lorentz transformation $x^\mu = \Lambda^\mu{}_\nu \bar{x}^\nu$, we still have $\bar{g}_{\mu\nu} = \eta_{\mu\nu} = 0$ and $\bar{g}_{\mu\nu,\lambda} = 0$

at $\bar{x}^{\mu}=0$; but now

$$\bar{g}_{\mu\nu,\lambda\kappa}=\Lambda^{\alpha}{}_{\mu}\Lambda^{\beta}{}_{\nu}\Lambda^{\gamma}{}_{\lambda}\Lambda^{\delta}{}_{\kappa}g_{\alpha\beta,\gamma\delta}.$$

Since $\mathcal{L}_G$ is a scalar under Lorentz transformations,

$$\bar{g}_{\mu\nu,\lambda\kappa}B^{\mu\nu\lambda\kappa}=g_{\mu\nu,\lambda\kappa}B^{\mu\nu\lambda\kappa}.$$

This equation and the symmetry of $g_{\mu\nu,\lambda\kappa}$ under interchange of μ and ν and of λ and κ imply that $\mathcal{L}_G$ has the form

$$\mathcal{L}_G=g_{\mu\nu,\lambda\kappa}\left[a\eta^{\mu\lambda}\eta^{\nu\kappa}-b\eta^{\mu\nu}\eta^{\lambda\kappa}\right]+c, \tag{11.2.6}$$

where a and b are constants.

Now consider $\mathcal{L}_G$ in another coordinate system that is related to the x-system by

$$x^{\mu}=\bar{x}^{\mu}+\tfrac{1}{6}\eta^{\mu\alpha}C_{\alpha\beta\gamma\delta}\bar{x}^{\beta}\bar{x}^{\gamma}\bar{x}^{\delta},$$

where $C_{\alpha\beta\gamma\delta}$ is symmetric in the indices β,γ,δ. Again, $\bar{g}_{\mu\nu}=\eta_{\mu\nu}$ and $\bar{g}_{\mu\nu,\lambda}=0$ at $\bar{x}^{\mu}=0$. But now

$$\bar{g}_{\mu\nu,\lambda\kappa}=g_{\mu\nu,\lambda\kappa}+C_{\mu\nu\lambda\kappa}+C_{\nu\mu\lambda\kappa}$$

at $\bar{x}^{\mu}=0$. Since $\mathcal{L}_G$ must not change under this transformation, we find (using the symmetry of $C_{\mu\nu\lambda\kappa}$) that

$$\begin{aligned}0&=\mathcal{L}_g(\eta_{\mu\nu},0,\bar{g}_{\mu\nu,\lambda\kappa})-\mathcal{L}_g(\eta_{\mu\nu},0,g_{\mu\nu,\lambda\kappa})\\&=(C_{\mu\nu\lambda\kappa}+C_{\nu\mu\lambda\kappa})(a\eta^{\mu\lambda}\eta^{\nu\kappa}-b\eta^{\mu\nu}\eta^{\lambda\kappa})\\&=2C_{\mu\nu\lambda\kappa}\eta^{\mu\nu}\eta^{\lambda\kappa}(a-b).\end{aligned}$$

Thus $a=b$.

We conclude that in a locally inertial coordinate system $\mathcal{L}_G$ must take the form

$$\mathcal{L}_g=ag_{\mu\nu,\lambda\kappa}(\eta^{\mu\lambda}\eta^{\nu\kappa}-\eta^{\mu\nu}\eta^{\lambda\kappa})+c, \tag{11.2.7}$$

where a and c are constants.

What scalar density formed from $g_{\mu\nu}$ and its derivatives reduces to (11.2.7) in a locally inertial coordinate system? Consider the "curvature scalar"*

$$R = R^{\alpha}{}_{\beta\alpha\delta} g^{\beta\delta}. \tag{11.2.8}$$

In a locally inertial coordinate system, R is

$$\begin{aligned} R &= \left[\Gamma^{\alpha}_{\beta\delta,\alpha} - \Gamma^{\alpha}_{\beta\alpha,\delta}\right]\eta^{\beta\delta} \\ &= \tfrac{1}{2}\eta^{\beta\delta}\eta^{\alpha\gamma}\left[\,g_{\beta\gamma,\delta\alpha} + g_{\delta\gamma,\beta\alpha} - g_{\beta\delta,\gamma\alpha}\right] \\ &\quad - \tfrac{1}{2}\eta^{\beta\delta}\eta^{\alpha\gamma}\left[\,g_{\beta\gamma,\alpha\delta} + g_{\alpha\gamma,\beta\delta} - g_{\beta\alpha,\gamma\delta}\right] \\ &= g_{\mu\nu,\lambda\kappa}\left(\eta^{\mu\lambda}\eta^{\nu\kappa} - \eta^{\mu\nu}\eta^{\lambda\kappa}\right). \end{aligned} \tag{11.2.9}$$

Comparing with (11.2.7) we see that

$$\frac{1}{\sqrt{g}}\,\mathfrak{L}_G = aR + c \tag{11.2.10}$$

in a locally inertial coordinate system.† Since R is a scalar and $\mathfrak{L}/\sqrt{g}$, by assumption, is a scalar, this equality holds in any coordinate system.

The constants a and c are conventionally written as

$$a = \frac{1}{16\pi G},$$

$$c = \frac{\lambda}{16\pi G}.$$

The values of λ and G must be determined from experiment. This determination may be made by comparing the Newtonian limit of general relativity with Newton's theory of gravity, as we do in Section 11.5. If $\lambda \neq 0$, space-time would be curved even in the absence of matter and Newtonian mechanics would fail even far away from sources of gravity.

*The curvature tensor $R_{\alpha\beta\gamma\delta}$ is antisymmetric in the indices α,β and the indices γ,δ. Thus all the nonzero scalars that can be formed by contracting $R_{\alpha\beta\gamma\delta}$ with $g^{\mu\nu}$ are equal to R to within a sign.

†An action principle for general relativity with this Lagrangian was first proposed by D. Hilbert, *Kl. Ges. Wiss. Goett., Nachr., Math. Phys. Kl.*, 395–407 (1915); see also D. Hilbert, *Math. Ann.* **92**, 1 (1924). The same Lagrangian, but with $g_{\mu\nu}(x)$ and $\Gamma^{\lambda}_{\mu\nu}(x)$ as independent fields, instead of $\Gamma^{\lambda}_{\mu\nu}(x)$ being defined in terms of $g_{\mu\nu}$, was proposed by A. Palatini, *Rend. Circ. Mat. Palermo* **43**, 203 (1919). (See Problem 4.)

Therefore λ must be very small or zero; we assume that λ is exactly zero.* One then finds that G is the Newtonian gravitational constant, whose experimental value is $G \sim 7\times10^{-8}$ cm^3 g^{-1} sec^{-2}.

To summarize, we have chosen the Lagrangian

$$\mathcal{L} = \mathcal{L}_M + \frac{\sqrt{g}}{16\pi G}R, \tag{11.2.10}$$

where $\mathcal{L}_M$ is the matter Lagrangian made generally covariant.

Other Basic Fields

We have seen in a particular case how to generalize a Lorentz invariant field theory to include the effects of gravity. A similar prescription can be given for extending any Lorentz covariant Lagrangian field theory. Write down the known "flat space" Lagrangian for the basic fields, that is, the fields other than $g_{\mu\nu}(x)$. Then replace $\eta_{\mu\nu}$ by $g_{\mu\nu}(x)$ and ordinary derivatives by covariant derivatives, and multiply by $g^{1/2}$. This gives the "nongravitational" term $\mathcal{L}_{NG}$ in the Lagrangian; to this add $\mathcal{L}_G = [16\pi G]^{-1}g^{1/2}R$ to form the Lagrangian for the full theory including gravity.

This is a "minimal" prescription in the sense that $\mathcal{L}_{NG}$ is to be made into a scalar density in the simplest possible way. One might also add a term proportional to a function of the nongravitational fields times the curvature tensor without changing the theory in flat space.† There are sometimes ambiguities even in the "minimal" prescription, arising from the fact that equivalent flat space Lagrangians can lead to inequivalent final Lagrangians. (See Problem 3 for an example.) Nevertheless, there do not seem to be any difficulties in applying the minimal prescription to the theories covered in this book.

Let us try out the prescription, using the theory of charged matter interacting with the electromagnetic field as an example. The flat space Lagrangian is

$$\mathcal{L}_{NG} = -\rho U(G_{ab}, R_a) - \tfrac{1}{4}F_{\mu\nu}F^{\mu\nu} + \mathcal{J}^{\mu}A_{\mu}, \tag{11.2.11}$$

*The possible effects of letting λ be small but nonzero are sometimes considered in cosmology; λ is called the "cosmological constant."

†For an example of this in quantum field theory, see C. G. Callan, S. Coleman, and R. Jackiw, *Ann. Phys.* **59**, 42 (1970).

with

$$G_{ab}=R_{a,\mu}R_{b,\nu}\eta^{\mu\nu},$$

$$\rho=n\sqrt{\det G_{ab}}\ ,$$

$$F_{\mu\nu}=A_{\nu,\mu}-A_{\mu,\nu},$$

$$\mathcal{J}^{\mu}=q(R_a)n(R_a)\epsilon^{\mu\nu\rho\sigma}R_{1,\nu}R_{2,\rho}R_{3,\sigma}.$$

As we have seen, the curved space generalization of the matter term is obtained by multiplying by $g^{1/2}$ and defining G_{ab} as

$$G_{ab}=R_{a,\mu}R_{b,\nu}g^{\mu\nu}. \tag{11.2.12}$$

There is an important point exemplified by the notation of (11.2.12). When the metric field $g_{\mu\nu}(x)$ is varied while the matter fields $R_a(x)$ are held fixed, the gradient $R_{a,\mu}=\partial R_a/\partial x^{\mu}$ remains fixed while the contravarient version of the gradient, $R_a{}^{,\mu}=g^{\mu\nu}R_{a,\nu}$, changes. We have explicitly displayed the factor $g^{\mu\nu}$ in (11.2.12), rather than writing $R_{a,\mu}R_b{}^{,\mu}$, in order to make it obvious how to differentiate with respect to $g_{\mu\nu}$ when we find the equation of motion for the metric.

The Lagrangian under consideration also involves the vector potential $A_\mu(x)$. We can demand that the action be stationary under independent variations of $A_\mu(x)$ and $g_{\mu\nu}(x)$, or under variations of $A^{\mu}(x)$ and $g_{\mu\nu}(x)$; but we cannot regard A_μ, A^μ, and $g_{\mu\nu}$ as independent. We consider the covariant components $A_\mu(x)$ of the potential to be the quantities to be varied independently of $g_{\mu\nu}(x)$.

Thus, we write the curved space generalization of the second term as

$$\mathcal{L}_E=-\tfrac{1}{4}\sqrt{g}\ F_{\mu\nu}F_{\alpha\beta}g^{\mu\alpha}g^{\nu\beta}, \tag{11.2.13}$$

where now $F_{\mu\nu}=A_{\nu;\mu}-A_{\mu;\nu}$.

A fortunate simplification occurs here: the connection coefficients involved in the definition of $A_{\mu;\nu}$ cancel out of $F_{\mu\nu}$ because they are symmetric in μ,ν:

$$F_{\mu\nu}=A_{\nu,\mu}-A_{\mu,\nu}-\left[\Gamma^{\lambda}_{\nu\mu}-\Gamma^{\lambda}_{\mu\nu}\right]A_\lambda$$
$$=A_{\nu,\mu}-A_{\mu,\nu}. \tag{11.2.14}$$

Finally, the curved space generalization of the interaction term is

$$\mathcal{L}_{E\mathcal{J}}=A_\mu qn\epsilon^{\mu\nu\rho\sigma}R_{1,\nu}R_{2,\rho}R_{3,\sigma}. \tag{11.2.15}$$

Notice that the metric tensor does not occur in $\mathcal{L}_{E\mathcal{J}}$. The factor $1/\sqrt{g}$ in

$e^{\mu\nu\rho\sigma}$ cancels the factor $g^{1/2}$ needed to make $\mathcal{L}_{E\mathcal{J}}$ a scalar density. As we will see shortly, this is the reason that the momentum tensor $\Theta^{\mu\nu}$ for this theory does not contain an interaction term involving the product of matter fields and electromagnetic fields.

11.3 THE EQUATIONS OF MOTION FOR THE BASIC FIELDS

With a Lagrangian in hand, we can easily find the equations of motion. Let us begin with the equations for the fields other than $g_{\mu\nu}(x)$. Consider a tensor field $A_{\alpha\beta\cdots\delta}$ that appears in the nongravitational part of the Lagrangian along with its first derivative:

$$\mathcal{L} = \mathcal{L}_{NG}\left(A_{\alpha\beta\cdots\delta}, A_{\alpha\beta\cdots\delta;\mu}, \ldots\right) + \mathcal{L}_G. \tag{11.3.1}$$

The dots in $\mathcal{L}_{NG}$ indicate its dependence on other fields. If $g_{\mu\nu}(x)=\eta_{\mu\nu}$, the equation of motion for $A_{\alpha\beta\cdots\delta}$ in Minkowski space is

$$0=\frac{\partial\,\mathcal{L}_{NG}}{\partial A_{\alpha\beta\cdots\delta}}-\left[\frac{\partial\,\mathcal{L}_{NG}}{\partial A_{\alpha\beta\cdots\delta;\mu}}\right]_{,\mu}.$$

We show here that the equation of motion in a general space is the same except that the ordinary derivative $\partial/\partial x^\mu$ is replaced by a covariant derivative:

$$0=\frac{1}{\sqrt{g}}\frac{\partial\,\mathcal{L}_{NG}}{\partial A_{\alpha\beta\cdots\delta}}-\left[\frac{1}{\sqrt{g}}\frac{\partial\,\mathcal{L}_{NG}}{\partial A_{\alpha\beta\cdots\delta;\mu}}\right]_{;\mu}. \tag{11.3.2}$$

For instance, when $g_{\mu\nu}=\eta_{\mu\nu}$ the equation of motion for the electromagnetic vector potential is

$$F^{\mu\nu}{}_{,\nu}=\mathcal{J}^\mu.$$

In a curved space this becomes

$$F^{\mu\nu}{}_{;\nu}=\mathcal{J}^\mu.$$

To prove the general formula (11.3.2), we display the arguments of $\mathcal{L}_{NG}$ in more detail:

$$\mathcal{L}_{NG}=\mathcal{L}_{NG}\left(A_{\alpha\beta\cdots\delta},\quad A_{\alpha\beta\cdots\delta,\mu}-\Gamma^\lambda_{\alpha\mu}A_{\lambda\beta\cdots\delta}-\cdots-\Gamma^\lambda_{\delta\mu}A_{\alpha\beta\cdots\lambda},\cdots\right).$$

The equation of motion is

$$\left(\frac{\partial\,\mathcal{L}}{\partial A_{\alpha\beta\cdots\delta}}\right)_{A_{\alpha\cdots\delta,\mu}=\text{const}}=\frac{\partial}{\partial x^\mu}\left(\frac{\partial\,\mathcal{L}}{\partial A_{\alpha\beta\cdots\delta,\mu}}\right)_{A_{\alpha\cdots\delta}=\text{const}}.$$

Since the undifferentiated field occurs in several places in $\mathfrak{L}_{NG}$, this is

$$\frac{\partial \mathfrak{L}}{\partial A_{\alpha\beta\cdots\delta}} = \Gamma^{\alpha}_{\tau\mu} \frac{\partial \mathfrak{L}}{\partial A_{\tau\beta\cdots\delta;\mu}} + \cdots + \Gamma^{\delta}_{\tau\mu} \frac{\partial \mathfrak{L}}{\partial A_{\alpha\beta\cdots\tau;\mu}} + \left(\frac{\partial \mathfrak{L}}{\partial A_{\alpha\beta\cdots\delta;\mu}} \right)_{,\mu}.$$

We divide by $\sqrt{g}$ and, in the last term, use the identity (11.1.27):

$$\frac{1}{\sqrt{g}} (\sqrt{g})_{,\mu} = \Gamma^{\lambda}_{\mu\lambda}.$$

This gives

$$\frac{1}{\sqrt{g}} \frac{\partial \mathfrak{L}}{\partial A_{\alpha\beta\cdots\delta}} = \Gamma^{\alpha}_{\tau\mu} \frac{1}{\sqrt{g}} \frac{\partial \mathfrak{L}}{\partial A_{\tau\beta\cdots\delta;\mu}}$$

$$+ \cdots + \Gamma^{\delta}_{\tau\mu} \frac{1}{\sqrt{g}} \frac{\partial \mathfrak{L}}{\partial A_{\alpha\beta\cdots\tau;\mu}} + \Gamma^{\mu}_{\tau\mu} \frac{1}{\sqrt{g}} \frac{\partial \mathfrak{L}}{\partial A_{\alpha\beta\cdots\delta;\tau}}$$

$$+ \left[\frac{1}{\sqrt{g}} \frac{\partial \mathfrak{L}}{\partial A_{\alpha\beta\cdots\delta;\mu}} \right]_{,\mu}. \tag{11.3.3}$$

Since $g^{-1/2} \partial \mathfrak{L} / \partial A_{\alpha\beta\cdots\delta;\mu} \; \delta A_{\alpha\beta\cdots\delta;\mu} = \delta[g^{-1/2}\mathfrak{L}]$ is a scalar, $g^{-1/2} \, \partial \mathfrak{L} / \partial A_{\alpha\beta\cdots\delta;\mu}$ must be a contravariant tensor. Its covariant derivative

$$\left[\frac{1}{\sqrt{g}} \frac{\partial \mathfrak{L}}{\partial A_{\alpha\beta\cdots\delta;\mu}} \right]_{;\mu}$$

is precisely the right hand side of (11.3.3). This proves (11.3.2).

11.4 EQUATION OF MOTION FOR $g_{\mu\nu}$

The equation of motion for the metric field $g_{\mu\nu}(x)$ is $0 = \delta A / \delta g_{\mu\nu}(x)$, where the "variational derivative" $\delta A / \delta g_{\mu\nu}(x)$ is defined by

$$\delta A = \int d^4x \frac{\delta A}{\delta g_{\mu\nu}(x)} \delta g_{\mu\nu}(x). \tag{11.4.1}$$

That is,

$$\frac{\delta A}{\delta g_{\mu\nu}(x)} = \frac{\partial \mathfrak{L}}{\partial g_{\mu\nu}} - \left[\frac{\partial \mathfrak{L}}{\partial g_{\mu\nu,\lambda}} \right]_{,\lambda}$$

$$+ \left[\frac{\partial \mathfrak{L}}{\partial g_{\mu\nu,\lambda\kappa}} \right]_{,\lambda\kappa}. \tag{11.4.2}$$

Metric Term

We begin with the purely metric term in the action,

$$A_G = \frac{1}{16\pi G}\int d^4x\sqrt{g}\,R. \tag{11.4.3}$$

A straightforward calculation of $\delta A/\delta g_{\mu\nu}(x)$ using the standard Euler-Lagrange formula (11.4.2) would be tedious. A more revealing calculation proceeds by working directly with the variation of A_G due to an infinitesimal variation $\delta g_{\mu\nu}(x)$ in the metric field. For this purpose, it is convenient to write R in the form $R = g^{\mu\nu}R_{\mu\nu}$, where

$$R_{\mu\nu} = R^{\lambda}{}_{\mu\lambda\nu}; \tag{11.4.4}$$

$R_{\mu\nu}$ is called the contracted curvature tensor or the Ricci tensor.

The variation of the integrand in (11.4.3) is

$$\delta\left(\sqrt{g}\;g^{\mu\nu}R_{\mu\nu}\right) = \sqrt{g}\;g^{\mu\nu}\delta R_{\mu\nu} + \sqrt{g}\;(\delta g^{\mu\nu})R_{\mu\nu} + \left(\delta\sqrt{g}\,\right)g^{\mu\nu}R_{\mu\nu}. \tag{11.4.5}$$

Only the first term is complicated enough to cause difficulty. Fortunately, this term is a divergence,

$$\sqrt{g}\;g^{\mu\nu}\delta R_{\mu\nu} = [\cdots]_{,\alpha}, \tag{11.4.6}$$

so it does not contribute to δA_G.

To prove that $g^{1/2}\,g^{\mu\nu}R_{\mu\nu}$ is a divergence, we first show that the variation $\delta\Gamma^{\alpha}_{\beta\gamma}$ is a tensor. Recall from (11.1.13) that the connection coefficients corresponding to the original metric tensor transform according to

$$\bar{\Gamma}^{\alpha}_{\beta\gamma} = \frac{\partial\bar{x}^{\alpha}}{\partial x^{\rho}}\frac{\partial x^{\mu}}{\partial\bar{x}^{\beta}}\frac{\partial x^{\nu}}{\partial\bar{x}^{\gamma}}\Gamma^{\rho}_{\mu\nu} + \frac{\partial\bar{x}^{\alpha}}{\partial x^{\lambda}}\frac{\partial^2 x^{\lambda}}{\partial\bar{x}^{\beta}\partial\bar{x}^{\gamma}}.$$

The varied connection coefficients $\Gamma^{\alpha}_{\beta\gamma} + \delta\Gamma^{\alpha}_{\beta\gamma}$, obey the same transformation law,

$$\bar{\Gamma}^{\alpha}_{\beta\gamma} + \delta\bar{\Gamma}^{\alpha}_{\beta\gamma} = \frac{\partial\bar{x}^{\alpha}}{\partial x^{\rho}}\frac{\partial x^{\mu}}{\partial\bar{x}^{\beta}}\frac{\partial x^{\nu}}{\partial\bar{x}^{\gamma}}\left[\Gamma^{\rho}_{\mu\nu} + \delta\Gamma^{\rho}_{\mu\nu}\right] + \frac{\partial\bar{x}^{\alpha}}{\partial x^{\lambda}}\frac{\partial^2 x^{\lambda}}{\partial\bar{x}^{\beta}\partial\bar{x}^{\gamma}}.$$

When we subtract one equation from the other the inhomogeneous terms cancel and we are left with a tensor transformation law for $\delta\Gamma^{\alpha}_{\beta\gamma}$, as claimed.

Now look at the definition of $R_{\mu\nu}$:

$$R_{\mu\nu}=\Gamma^{\lambda}_{\mu\nu,\lambda}-\Gamma^{\lambda}_{\mu\lambda,\nu}+\Gamma^{\lambda}_{\beta\lambda}\Gamma^{\beta}_{\mu\nu}-\Gamma^{\lambda}_{\beta\nu}\Gamma^{\beta}_{\mu\lambda}.$$

The variation of $R_{\mu\nu}(x)$, calculated in a coordinate system in which $\Gamma^{\alpha}_{\beta\gamma}=0$ at the point x, has the form

$$\delta R_{\mu\nu}=\left[\delta\Gamma^{\lambda}_{\mu\nu}\right]_{,\lambda}-\left[\delta\Gamma^{\lambda}_{\mu\lambda}\right]_{,\nu}.$$

Since the connection coefficients vanish in this coordinate system we can write

$$\delta R_{\mu\nu}=\left[\delta\Gamma^{\lambda}_{\mu\nu}\right]_{;\lambda}-\left[\delta\Gamma^{\lambda}_{\mu\lambda}\right]_{;\nu} \tag{11.4.7}$$

without changing anything. But (11.4.7) is an equality between tensors. Since it is true in one coordinate system, it is true in all coordinate systems.

Finally, we multiply (11.4.7) by $\sqrt{g}\,g^{\mu\nu}$ and use the fact that the covariant derivative of the metric tensor vanishes to obtain

$$\begin{aligned}\sqrt{g}\;g^{\mu\nu}\delta R_{\mu\nu}&=\sqrt{g}\left[g^{\mu\nu}\delta\Gamma^{\lambda}_{\mu\nu}\right]_{;\lambda}-\sqrt{g}\left[g^{\mu\nu}\delta\Gamma^{\lambda}_{\mu\lambda}\right]_{;\nu}\\&=\sqrt{g}\left[g^{\mu\nu}\delta\Gamma^{\alpha}_{\mu\nu}-g^{\mu\alpha}\delta\Gamma^{\lambda}_{\mu\lambda}\right]_{;\alpha}.\end{aligned}$$

Recall that the covariant divergence of a vector field V^{α} can be written in the alternative form (11.1.26)

$$V^{\alpha}{}_{;\alpha}=\frac{1}{\sqrt{g}}\left(\sqrt{g}\,V^{\alpha}\right)_{,\alpha}.$$

Thus

$$\sqrt{g}\;g^{\mu\nu}\delta R_{\mu\nu}=\left[\sqrt{g}\;g^{\mu\nu}\delta\Gamma^{\alpha}_{\mu\nu}-\sqrt{g}\;g^{\mu\alpha}\Gamma^{\lambda}_{\mu\lambda}\right]_{,\alpha}. \tag{11.4.8}$$

Since this term in the integrand of A is a divergence, it does not contribute to δA_G and we can forget about it, as claimed.

The other two terms in the integrand of δA_G are easy to compute:

$$\begin{aligned}\sqrt{g}\,(\delta g^{\mu\nu})R_{\mu\nu}&=-\sqrt{g}\;g^{\mu\alpha}\,\delta g_{\alpha\beta}g^{\beta\nu}R_{\mu\nu}\\&=-\sqrt{g}\;R^{\alpha\beta}\,\delta g_{\alpha\beta},\end{aligned} \tag{11.4.9}$$

and

$$(\delta\sqrt{g}\,)g^{\mu\nu}R_{\mu\nu}=\tfrac{1}{2}\frac{1}{\sqrt{g}}\,\delta[-\det g_{\alpha\beta}]R$$

$$=\tfrac{1}{2}\sqrt{g}\;g^{\alpha\beta}\delta g_{\alpha\beta}R. \qquad (11.4.10)$$

Combining the results (11.4.8), (11.4.9), and (11.4.10), we have

$$\delta A_G=\frac{1}{16\pi G}\int d^4x\left\{[\cdots]_{,\alpha}-\sqrt{g}\,(R^{\alpha\beta}-\tfrac{1}{2}g^{\alpha\beta}R)\delta g_{\alpha\beta}\right\}. \quad (11.4.11)$$

Thus the variational derivative of A_G with respect to $g_{\mu\nu}(x)$ is

$$\frac{\delta A_G}{\delta g_{\mu\nu}(x)}=-\frac{\sqrt{g(x)}}{16\pi G}(R^{\mu\nu}(x)-\tfrac{1}{2}g^{\mu\nu}(x)R(x)). \qquad (11.4.12)$$

The tensor $R^{\mu\nu}-\frac{1}{2}g^{\mu\nu}R$ is often assigned the symbol $G^{\mu\nu}$ and called the Einstein tensor.

Matter Term

We can now turn to the terms in the Lagrangian that involve other fields besides $g_{\mu\nu}(x)$. As a first example, consider the theory of elastic matter interacting with gravity. The action is

$$A=A_G+A_M$$

where

$$A_M=-\int d^4x\sqrt{g}\,\rho U(G_{ab},R_a). \qquad (11.4.13)$$

Since the integrand does not involve derivatives of $g_{\mu\nu}$, the functional derivative $\delta A_M/\delta g_{\mu\nu}(x)$ is simply the ordinary derivative of the integrand with respect to $g_{\mu\nu}$, evaluated at the point x.* This derivative can be easily calculated. First we write

$$\frac{\partial G_{ab}}{\partial g_{\mu\nu}}=\frac{\partial}{\partial g_{\mu\nu}}(R_{a,\alpha}R_{b,\beta}g^{\alpha\beta})$$

$$=-R_{a,\alpha}R_{b,\beta}g^{\alpha\mu}g^{\beta\nu},$$

*Since $\delta g_{\mu\nu}(x)$ is always symmetric its indices we will understand $\delta A/\delta g_{\mu\nu}$ to be the symmetric part of $\partial\mathcal{L}/\partial g_{\mu\nu}$.

so that

$$\frac{\partial U}{\partial g_{\mu\nu}} = -R_a{}^{,\mu} R_b{}^{,\nu} \frac{\partial U}{\partial G_{ab}}.$$

Next we look at the factor $\sqrt{g}\,\rho$. Write ρ in the familiar form

$$\rho = \sqrt{-J^\alpha J^\beta g_{\alpha\beta}},$$

where

$$J^\alpha = n \frac{\epsilon^{\alpha\rho\tau\sigma}}{\sqrt{g}} R_{1,\rho} R_{2,\tau} R_{3,\sigma}.$$

Thus

$$\sqrt{g}\,\rho = \left[-n\epsilon^{\alpha\rho\tau\sigma} R_{1,\rho} R_{2,\tau} R_{3,\sigma} g_{\alpha\beta} n\epsilon^{\beta\kappa\lambda\delta} R_{1,\kappa} R_{2,\lambda} R_{3,\delta} \right]^{1/2}.$$

Notice that $g_{\mu\nu}$ appears in only one place in this expression for $\sqrt{g}\,\rho$. The derivative is

$$\frac{\partial(\sqrt{g}\,\rho)}{\partial g_{\mu\nu}} = -\tfrac{1}{2} \frac{1}{\sqrt{g}\,\rho} \sqrt{g}\,J^\mu \sqrt{g}\,J^\nu$$

$$= -\tfrac{1}{2} \sqrt{g}\,\rho u^\mu u^\nu,$$

where

$$u^\mu = \frac{1}{\rho} J^\mu$$

is the four-velocity of the matter. We conclude that

$$\frac{\delta A_M}{\delta g_{\mu\nu}(x)} = \tfrac{1}{2}\sqrt{g} \left[\rho U u^\mu u^\nu + 2\rho R_a{}^{,\mu} R_b{}^{,\nu} \frac{\partial U}{\partial G_{ab}} \right]. \tag{11.4.14}$$

This derivative $\delta A_M / \delta g_{\mu\nu}(x)$ should look familiar. Except for the factor $g^{1/2}/2$, it is the energy momentum tensor for matter, $\Theta^{\mu\nu}$, that we found in Chapter 5. Here the expression for $\Theta^{\mu\nu}_M$ has merely been made generally covariant by the substitutions $\eta_{\mu\nu} \to g_{\mu\nu}$ and $\epsilon^{\mu\nu\rho\sigma} \to e^{\mu\nu\rho\sigma}$. Thus we have found that

$$\frac{2}{\sqrt{g}} \frac{\delta A_M}{\delta g_{\mu\nu}(x)} = \Theta^{\mu\nu}_M(x). \tag{11.4.15}$$

The equation of motion for the metric field coupled to matter is $0=\delta A_G/\delta g_{\mu\nu}+\delta A_M/\delta g_{\mu\nu}$. Using (11.4.12) and (11.4.15), this is

$$0=-\frac{\sqrt{g}}{16\pi G}\left(R^{\mu\nu}-\tfrac{1}{2}g^{\mu\nu}R\right)+\frac{\sqrt{g}}{2}\Theta_M^{\mu\nu}$$

or

$$R^{\mu\nu}-\tfrac{1}{2}g^{\mu\nu}R=8\pi G\Theta_M^{\mu\nu}. \tag{11.4.16}$$

This equation is Einstein's field equation for gravity in the presence of matter.

Electromagnetic and Other Terms

If we want to consider the effects of more fields interacting with matter and gravity, we simply add the appropriate terms to the Lagrangian. For each new term in the Lagrangian, a term is added to the right hand side of the Einstein field equation.

For instance, we may wish to consider the interaction of charged matter with the electromagnetic field and gravity. The Lagrangian for the system was discussed in Section 11.2. It is

$$\mathcal{L}=\mathcal{L}_G+\mathcal{L}_M+\mathcal{L}_E+\mathcal{L}_{E\mathcal{G}}$$

with

$$\mathcal{L}_G=\frac{\sqrt{g}}{16\pi G}R,$$

$$\mathcal{L}_M=-\sqrt{g}\,\rho U(G_{ab},R_a),$$

$$\mathcal{L}_E=-\tfrac{1}{4}\sqrt{g}\,F_{\mu\nu}F_{\alpha\beta}g^{\mu\alpha}g^{\nu\beta},$$

$$F_{\mu\nu}=A_{\nu,\mu}-A_{\mu,\nu},$$

$$\mathcal{L}_{E\mathcal{G}}=A_\mu qn\epsilon^{\mu\nu\rho\sigma}R_{1,\nu}R_{2,\rho}R_{3,\sigma}. \tag{11.4.17}$$

The contributions of $\mathcal{L}_G$ and $\mathcal{L}_M$ are already included in (11.4.16). The variational derivative of $A_E=\int dx\ \mathcal{L}_E$ with respect to $g_{\mu\nu}$ is

$$\frac{\delta A_E}{\delta g_{\mu\nu}(x)}=\frac{\sqrt{g}}{2}F^{\mu\lambda}F^{\nu}{}_{\lambda}-\frac{\sqrt{g}}{8}g^{\mu\nu}F^{\alpha\beta}F_{\alpha\beta}.$$

We define

$$\Theta_E^{\mu\nu} = \frac{2}{\sqrt{g}}\frac{\delta A_E}{\delta g_{\mu\nu}} = F^{\mu\lambda}F^{\nu}{}_{\lambda} - \tfrac{1}{4}g^{\mu\nu}F^{\alpha\beta}F_{\alpha\beta}. \tag{11.4.18}$$

Then $\Theta_E^{\mu\nu}$ reduces in a flat space to the symmetrized electromagnetic momentum tensor that we found in Chapter 9.

The variational derivative of $A_{E\mathcal{G}} = \int d^4x\, \mathcal{L}_{E\mathcal{G}}$ with respect to $g_{\mu\nu}(x)$ is

$$\frac{2}{\sqrt{g}}\frac{\delta A_{E\mathcal{G}}}{\delta g_{\mu\nu}(x)} = 0, \tag{11.4.19}$$

since $\mathcal{L}_E$ does not depend on $g_{\mu\nu}$. This fits into the previous pattern, since the symmetrized momentum tensor for matter and electromagnetism does not contain a term $\Theta_{E\mathcal{G}}^{\mu\nu}$ (see Table 10.2).

We conclude that the Einstein field equation in this theory is

$$R^{\mu\nu} - \tfrac{1}{2}g^{\mu\nu}R = 8\pi G\Theta^{\mu\nu}, \tag{11.4.20}$$

where

$$\Theta^{\mu\nu}(x) = \frac{2}{\sqrt{g}}\frac{\delta A_{NG}}{\delta g_{\mu\nu}(x)} \tag{11.4.21}$$

is the generally covariant version of the symmetrized momentum tensor that arose in the corresponding flat space theory. We will find in Chapter 12 that $\Theta^{\mu\nu}$ defined by (11.4.21) can always be associated with the energy-momentum carried by the basic (nongravitational) fields, and that it is the covariant version of the symmetrized momentum tensor introduced in Chapter 9.

11.5 THE NEWTONIAN LIMIT

Astronomical observations indicate that Newton's theory of gravitation is very nearly correct when applied to ordinary matter moving slowly in weak gravitational fields. We may therefore demand that general relativity, or any other proposed theory of gravitation, reduce to Newton's theory when applied in this limiting situation. Let us see if general relativity meets the test.

Motion of Matter

Let us define a field

$$h_{\mu\nu}(x)=g_{\mu\nu}(x)-\eta_{\mu\nu} \tag{11.5.1}$$

and suppose that $h_{\mu\nu}(x)$ is small; that is, the gravitional field is weak. How does matter behave in this situation? The equations of motion for the matter fields $R_a(x)$ are derived from the matter action

$$A_M=-\int d^4x\sqrt{g}\,\rho U(G_{ab},R_a).$$

Let us expand A_M in powers of $h_{\mu\nu}(x)$ and throw away the quadratic and higher order terms:

$$A_M\cong[A_M]_{h=0}+\int d^4x\left[\frac{\delta A_M}{\delta g_{\mu\nu}(x)}\right]_{h=0}h_{\mu\nu}(x).$$

The zero order term is the familiar flat space action, with $g_{\mu\nu}$ replaced by $\eta_{\mu\nu}$. By using the expression (11.4.15) for $\delta A_M/\delta g_{\mu\nu}$ in the first order term we obtain

$$A_M\cong[A_M]_{h=0}+\tfrac{1}{2}\int d^4x\Theta_M^{\mu\nu}(x)h_{\mu\nu}(x). \tag{11.5.2}$$

We can further simplify A_M for the application we have in mind: ordinary matter moving at ordinary velocities. The energy density Θ_M^{00} is approximately equal to the rest energy of the atoms of matter:

$$\Theta_M^{00}\cong mJ^0, \tag{11.5.3}$$

where, as usual, m is the mass per atom of matter and J^0 is the number of atoms per unit volume. The other components of $\Theta_M^{\mu\nu}$ are much smaller than mJ^0. The momentum density Θ^{0j} is approximately mJ^0v^j where v^j is the velocity of matter (divided by the velocity of light). The momentum current Θ^{ij} is even smaller: $\Theta^{ij}\sim mJ^0v^iv^j+t^{ij}$, where t^{ij} is the stress in the material. (An ordinary stress $t^{ij}\sim 10^6$ dynes/cm$^2\sim 1$ atm is only $t^{ij}\sim 10^{-15}$ g/cm^3). Thus we put

$$\Theta^{0j},\Theta^{ij}\cong 0. \tag{11.5.4}$$

These approximations give

$$A_M\cong[A_M]_{h=0}+\tfrac{1}{2}\int d^4x\,mJ^0h_{00}(x). \tag{11.5.5}$$

We saw in Chapter 7 on nonrelativistic continuum mechanics that, under the same approximations as made above, the flat space action A_M takes the form

$$[A_M]_{h=0} \to \int d^4x\left[\tfrac{1}{2}mJ^0\mathbf{v}^2 - V(x)\right], \tag{11.5.6}$$

where $V(x)$ is the potential energy density of the matter. Evidently, the effect of gravity in this approximation is to add a new potential energy density

$$V(x)_{\text{grav}} = -mJ^0\tfrac{1}{2}h_{00}. \tag{11.5.7}$$

Thus

$$\phi(x) = -\tfrac{1}{2}h_{00}(x) \tag{11.5.8}$$

is the gravitational potential energy per unit mass—that is, the gravitational potential. With this identification of $\phi(x)$, we recover precisely Newton's law for the motion of falling bodies.

Equation of Motion for the Gravitational Potential

The development of $\phi(x)$ is controlled by the Einstein field equation

$$R_{\mu\nu} - \tfrac{1}{2}g_{\mu\nu}R = 8\pi G\Theta_{M\mu\nu}.$$

It will be convenient to note that the trace of this equation reads

$$R - 2R = 8\pi G\Theta_M{}^\mu{}_\mu.$$

Thus the Einstein equation can be written as

$$R_{\mu\nu} = 8\pi G\left[\Theta_{M\mu\nu} - \tfrac{1}{2}g_{\mu\nu}\Theta_M{}^\alpha{}_\alpha\right]. \tag{11.5.9}$$

A straightforward calculation enables us to rewrite this equation to lowest order in $h_{\mu\nu}$. Since

$$\Gamma^\alpha_{\mu\nu} \cong \tfrac{1}{2}\eta^{\alpha\beta}\left[h_{\beta\mu,\nu} + h_{\beta\nu,\mu} - h_{\mu\nu,\beta}\right], \tag{11.5.10}$$

the Einstein equation becomes

$$\begin{aligned}\tfrac{1}{2}\eta^{\alpha\beta}&\left[h_{\alpha\mu,\beta\nu} + h_{\alpha\nu,\beta\mu} - h_{\alpha\beta,\mu\nu} - h_{\mu\nu,\alpha\beta}\right]\\ &\cong 8\pi G\left[\Theta_{M\mu\nu} - \tfrac{1}{2}\eta_{\mu\nu}\eta^{\alpha\beta}\Theta_{M\alpha\beta}\right].\end{aligned} \tag{11.5.11}$$

It suffices for the moment to examine the (0,0) component of this equation. The (0,0) component of the right hand side is approximately $8\pi G$ times one half the mass density mJ^0. Thus

$$\tfrac{1}{2}\eta^{\alpha\beta}\left[2h_{\alpha 0,\beta 0}-h_{\alpha\beta,00}-h_{00,\alpha\beta}\right]\cong 4\pi GmJ^0.$$

We need one further approximation: since the sources of the gravitational field are presumed to be moving slowly, $h_{\mu\nu}(x)$ will be slowly varying. That is, the time it takes for $h_{\mu\nu}(x)$ to change appreciably is long in comparison to the time it takes for light to travel across the system. Thus we may neglect time derivatives of $h_{\mu\nu}$ in comparison to space derivatives. This gives

$$-\frac{1}{2}\sum_{k=1}^{3} h_{00,kk}\doteq 4\pi GmJ^0.$$

In more familiar notation, this is

$$\nabla^2\phi=4\pi GmJ^0, \tag{11.5.12}$$

which is precisely the standard Newtonian law.

11.6 GRAVITATIONAL WAVES AND THEIR EFFECT ON MATTER

In the preceding section we obtained an approximate linear equation for $h_{\mu\nu}=g_{\mu\nu}-\eta_{\mu\nu}$ in (11.5.11). This equation has wavelike solutions. Although the existence of these gravitational waves has not yet been confirmed experimentally,* they are a subject of substantial experimental and theoretical interest. In this section† we outline the solution of the wave equation (11.5.11) and investigate the effect of a gravitational wave when it passes through matter.

Harmonic Coordinates

Before we begin, we exercise our freedom to choose a convenient coordinate system. Let us look for a coordinate system in which the covariant scalar wave equation $g^{\mu\nu}\phi(x)_{;\mu\nu}=0$ has a simple form. Since

$$g^{\mu\nu}\phi_{;\mu\nu}=g^{\mu\nu}\phi_{,\mu\nu}-g^{\mu\nu}\Gamma^{\lambda}_{\mu\nu}\phi_{,\lambda},$$

*Detection of gravitational waves was reported by J. Weber, *Phys. Rev. Lett.* **22**, 1320 (1969), *Phys. Rev. Lett.* **24**, 276 (1970); J. Weber et al., *Phys. Rev. Lett.* **31**, 779 (1973). However, these results have not been confirmed by other workers: J. L. Levine and R. L. Garwin, *Phys. Rev. Lett.* **31**, 173 (1973); *Phys. Rev. Lett.* **33**, 794 (1974); J. A. Tyson, *Phys. Rev. Lett.* **31**, 326 (1973).

†This section was prepared with the help of Edward Witten.

the wave equation is simplified if

$$g^{\mu\nu}\Gamma^{\lambda}_{\mu\nu}=0. \tag{11.6.1}$$

A coordinate system in which (11.6.1) holds is called a "harmonic" coordinate system. In such a system the functions $\phi(x)=x^{\alpha}$ obey

$$(x^{\alpha})_{;\mu\nu}g^{\mu\nu}=(x^{\alpha})_{,\mu\nu}g^{\mu\nu}-0=0.$$

That is, the harmonic coordinates themselves each obey the covariant scalar wave equation. Thus one can construct a set of harmonic coordinates by solving the wave equation with four different sets of initial conditions on a spacelike surface. Apparently the harmonic condition (11.6.1) does not uniquely specify the coordinate system.

The Equation for $h_{\mu\nu}$

Let us return now to the equation of motion (11.5.11) for $h_{\mu\nu}(x)=g_{\mu\nu}(x)-\eta_{\mu\nu}$. Since we are assuming that $h_{\mu\nu}$ is small and working to lowest order in $h_{\mu\nu}$, we always raise and lower indices with $\eta_{\mu\nu}$. With this convention the equation is

$$\eta^{\alpha\beta}\left[h_{\alpha\mu,\beta\nu}+h_{\alpha\nu,\beta\mu}-h_{\alpha\beta,\mu\nu}-h_{\mu\nu,\alpha\beta}\right]=16\pi G\left[\Theta_{\mu\nu}-\tfrac{1}{2}\eta_{\mu\nu}\Theta^{\alpha}_{\alpha}\right].$$

We choose harmonic coordinates,* so that

$$0=\eta_{\tau\lambda}\eta^{\alpha\beta}\Gamma^{\tau}_{\alpha\beta}=h^{\mu}_{\lambda,\mu}-\tfrac{1}{2}h^{\mu}_{\mu,\lambda} \tag{11.6.2}$$

Then the equation of motion becomes

$$-\eta^{\alpha\beta}h_{\mu\nu,\alpha\beta}=16\pi G\left[\Theta_{\mu\nu}-\tfrac{1}{2}\eta_{\mu\nu}\Theta^{\alpha}_{\alpha}\right]. \tag{11.6.3}$$

Thus each component of $h_{\mu\nu}(x)$ obeys the inhomogeneous wave equation with the corresponding component of $16\pi G[\Theta_{\mu\nu}-\frac{1}{2}\eta_{\mu\nu}\Theta^{\alpha}_{\alpha}]$ as its source.

Plane Waves

The linearized Einstein equation (11.6.3) with no sources present apparently admits plane wave solutions

$$h_{\mu\nu}(x)=\mathrm{Re}\,\epsilon_{\mu\nu}e^{ik_{\alpha}x^{\alpha}} \tag{11.6.4}$$

*If one starts in a nonharmonic coordinate system in which $h_{\mu\nu}$ is small, only a small coordinate change is needed to get to a harmonic coordinate system. Thus $h_{\mu\nu}$ in the new system is still small.

Here $k^\mu k_\mu = 0$ and the "polarization tensor" for the wave, $\epsilon_{\mu\nu}$, is symmetric and satisfies the harmonic coordinate condition

$$\epsilon_{\lambda\mu} k^\mu - \tfrac{1}{2}\epsilon^\mu{}_\mu k_\lambda = 0. \tag{11.6.5}$$

The solution $h_{\mu\nu}(x)$ takes a different form for each choice of $\epsilon_{\mu\nu}$. However, not all of these functional forms for $h_{\mu\nu}(x)$ are physically distinct, since it may be possible to transform one form into the other by means of a small change of coordinates, $x^\mu \to \bar{x}^\mu = x^\mu - \xi^\mu(x)$. The function $h_{\mu\nu}(x)$ is transformed to

$$\bar{h}_{\mu\nu} = \frac{\partial x^\alpha}{\partial \bar{x}^\mu}\frac{\partial x^\beta}{\partial \bar{x}^\nu}(\eta_{\alpha\beta} + h_{\alpha\beta}) - \eta_{\mu\nu}$$

$$\cong h_{\mu\nu} + \xi_{\mu,\nu} + \xi_{\nu,\mu}$$

under this coordinate change. (Notice the similarity between this transformation and the gauge transformation $A_\mu \to A_\mu + \Lambda_{,\mu}$ in electrodynamics.) If we choose

$$\xi_\mu(x) = \mathrm{Re}\left[\frac{1}{i}\xi_\mu e^{ik_\alpha x^\alpha}\right],$$

the transformed wave has the same $\exp(ik_\alpha x^\alpha)$ space-time dependence, but a new polarization tensor:

$$\bar{\epsilon}_{\mu\nu} = \epsilon_{\mu\nu} + \xi_\mu k_\nu + k_\mu \xi_\nu. \tag{11.6.6}$$

This new polarization tensor still satisfies the harmonic coordinate condition (11.6.5), so that the new coordinate system is still harmonic.

Since two polarization tensors related by (11.6.6) really describe the same wave in different coordinate systems, we have the opportunity to choose the coordinate system so as to make the wave look as simple as possible. The reader can easily verify that one can choose ξ_μ so that

$$\bar{\epsilon}_{00} = 0,$$

$$\bar{\epsilon}_{0i} = 0,$$

$$\bar{\epsilon}_{ii} = 0,$$

$$k_i \bar{\epsilon}_{ij} = 0. \tag{11.6.7}$$

The only components of $\epsilon_{\mu\nu}$ that remain nonzero after this adjustment of the coordinates are, for a wave traveling in the z-direction, $\epsilon_{12} = \epsilon_{21}$ and

$\epsilon_{11} = -\epsilon_{22}$. Thus two physical degrees of freedom remain. The conditions (11.6.7) are called the "transverse-traceless" gauge conditions.

A general solution of the linearized Einstein equation can be formed by taking a particular solution $h_{\mu\nu}^{(P)}(x)$ of the inhomogeneous equation, representing the gravitational field of (nearby) sources, and adding any number of plane waves coming from far away:

$$h_{\mu\nu}(x) = h_{\mu\nu}^{(P)}(x) + h_{\mu\nu}^{(W)}(x),$$

$$h_{\mu\nu}^{(W)}(x) = \mathrm{Re}\int d\mathbf{k}\,\epsilon_{\mu\nu}(k)\,e^{i(\mathbf{k}\cdot\mathbf{x} - |\mathbf{k}|t)}. \tag{11.6.8}$$

By an appropriate choice of coordinate system we can, as before, ensure that $\epsilon_{\mu\nu}(\mathbf{k})$ obeys the transverse-traceless gauge conditions (11.6.7) for all wave vectors $\mathbf{k}$. Then the wave part of $h_{\mu\nu}$ will satisfy the conditions

$$h_{00}^{(W)}(x) = 0, \qquad h_{0i}^{(W)}(x) = 0,$$

$$h_{ii}^{(W)}(x) = 0, \qquad h_{ij}^{(W)}(x)_{,i} = 0. \tag{11.6.9}$$

Effect on Matter

What happens when a gravitational wave passes through a piece of matter? The answer is somewhat complicated in general, but is simple in the case of an elastic solid that is displaced only slightly from equilibrium by gravitational and other forces acting on it. It is appropriate in this situation to use the Hooke's law model discussed in Section 6.1 to describe the solid. The internal energy is

$$U(G_{ab}, R_a) = m + \frac{1}{2n} C_{abcd} s_{ab} s_{cd},$$

where the mass per atom m, the number n of atoms per unit $d\mathbf{R}$, and the elastic constants C_{abcd} may in general be functions of R_a. We suppose that the displacement $\phi_a(x) = R_a(x) - x^a$ of the material coordinates from the reference configuration $R_a = x^a$ is small, and we write the equation of motion to first order in ϕ_a.

We assume also, of course, that the gravitational wave represents a small deviation from the background metric $\eta_{\mu\nu}$. Thus $h_{\mu\nu} = g_{\mu\nu} - \eta_{\mu\nu}$ is small, so we neglect terms of order h^2 and $h\phi$ or smaller in the equation of motion.

As the wave passes through the matter it causes the matter to wiggle and send out secondary gravitational waves. Because of the small magnitude of the gravitational coupling constant G, these secondary waves are very weak and can be neglected. Thus we need only consider the matter action, with the metric disturbance $h_{\mu\nu}(x)$ being a given function.

The matter action is

$$A = -\int dx\sqrt{g}\,\rho\left[m + \frac{1}{2n}C_{abcd}s_{ab}s_{cd}\right]. \tag{11.6.10}$$

To obtain the equations of motion for ϕ_a correct to lowest order in ϕ_a and $h_{\mu\nu}$, we must expand A in powers of ϕ_a and $h_{\mu\nu}$ and keep terms of order 1, ϕ, ϕ^2, h, and ϕh. We begin with the potential energy function and write for the strain

$$\begin{aligned} s_{ab} &\equiv \tfrac{1}{2}\left[\delta_{ab} - R_{a,\mu}R_{b,\nu}g^{\mu\nu}\right] \\ &= \tfrac{1}{2}\left[\delta_{ab} - (\delta_{a\mu} + \phi_{a,\mu})(\delta_{b\nu} + \phi_{b,\nu})(\eta^{\mu\nu} - h^{\mu\nu} + \cdots)\right] \\ &= \tfrac{1}{2}\left[\phi_{a,b} + \phi_{b,a} + h_{ab} + \cdots\right]. \end{aligned} \tag{11.6.11}$$

Consider next the density factor $\sqrt{g}\,\rho$. We define

$$\tilde{J}^{\mu} \equiv \frac{\sqrt{g}}{n}J^{\mu} = \epsilon^{\mu\nu\rho\sigma}R_{1,\nu}R_{2,\rho}R_{3,\sigma}$$

and use

$$\begin{aligned} \tilde{J}^{i} &= -\dot{\phi}_i + \cdots, \\ \tilde{J}^{0} &= 1 + \phi_{i,i} + \cdots. \end{aligned}$$

Then

$$\begin{aligned} \sqrt{g}\,\rho &= \left[-n^2\tilde{J}^{\mu}\tilde{J}^{\nu}(\eta_{\mu\nu} + h_{\mu\nu})\right]^{1/2} \\ &= n\left[(\tilde{J}^0)^2 - \tilde{J}^i\tilde{J}^i - \tilde{J}^{\mu}\tilde{J}^{\nu}h_{\mu\nu}\right]^{1/2} \\ &= n\tilde{J}^0 - \tfrac{1}{2}n(\tilde{J}^0)^{-1}\tilde{J}^i\tilde{J}^i - \tfrac{1}{2}n(\tilde{J}^0)^{-1}\tilde{J}^{\mu}\tilde{J}^{\nu}h_{\mu\nu} + \cdots \\ &= n\tilde{J}^0 - \tfrac{1}{2}n\dot{\phi}_i\dot{\phi}_i - \tfrac{1}{2}nh_{00} - \tfrac{1}{2}n\phi_{i,i}h_{00} + n\dot{\phi}_ih_{0i} + \cdots. \end{aligned}$$

By combining the expansions for $\sqrt{g}\,\rho$ and for U we obtain for the Lagrangian

$$\begin{aligned} \mathfrak{L} = &-nm\tilde{J}^0 + \tfrac{1}{2}nm\dot{\phi}_i\dot{\phi}_i + \tfrac{1}{2}nmh_{00} \\ &+ \tfrac{1}{2}nm\phi_{i,i}h_{00} - nm\dot{\phi}_ih_{0i} \\ &- \tfrac{1}{2}C_{ijkl}\left[\phi_{i,j} + \tfrac{1}{2}h_{ij}\right]\left[\phi_{k,l} + \tfrac{1}{2}h_{kl}\right] + \cdots. \end{aligned}$$

(The h^2 part of the last term could be dropped but is kept for aesthetic reasons.) The first term does not contribute to the action because it is a divergence.*

$$nm\tilde{J}^0 = \frac{\partial}{\partial x^\nu}\left[nm\tilde{J}^\nu x^0\right].$$

In each of the other terms except for the third the argument $R_a = x^a + \phi_a$ of $n(\mathbf{R})$, $m(\mathbf{R})$, and $C_{ijkl}(\mathbf{R})$ can be approximated by x^a. In the third term, we must also keep the first correction:

$$\tfrac{1}{2}n(\mathbf{x}+\phi)m(\mathbf{x}+\phi)h_{00}(x) = \tfrac{1}{2}n(\mathbf{x})m(\mathbf{x})h_{00}(x) + \tfrac{1}{2}\phi_i \frac{\partial}{\partial x^i}\left[n(\mathbf{x})m(\mathbf{x})\right]h_{00}(x) + \cdots.$$

The term $\frac{1}{2}n(\mathbf{x})m(\mathbf{x})h_{00}(x)$ can be dropped from the Lagrangian because it does not depend on the fields $\phi_i(x)$. Thus the sum of the third, fourth, and fifth terms is effectively

$$\tfrac{1}{2}\frac{\partial}{\partial x^i}\left[nm\phi_i\right]h_{00} - nm\dot{\phi}_i h_{0i}.$$

The contribution of these terms to the action can be combined by integrating by parts and recognizing the combination

$$h_{0i,0} - \tfrac{1}{2}h_{00,i} = \Gamma^i_{00}.$$

Thus the approximate action† is

$$A = \int dx\left\{\tfrac{1}{2}mn\dot{\phi}_i\dot{\phi}_i + mn\phi_i\Gamma^i_{00} - \tfrac{1}{2}C_{ijkl}\left[\phi_{i,j} + \tfrac{1}{2}h_{ij}\right]\left[\phi_{k,l} + \tfrac{1}{2}h_{kl}\right]\right\}. \tag{11.6.13}$$

The equation of motion follows immediately from the action (11.6.13):

$$mn\ddot{\phi}_i = mn\Gamma^i_{00} + \frac{\partial}{\partial x^j}C_{ijkl}\left[\phi_{k,l} + \tfrac{1}{2}h_{kl}\right]. \tag{11.6.14}$$

This simple formula has been obtained without any assumption about the gauge conditions satisfied by $h_{\mu\nu}$. We now choose coordinates so that the

*Recall that $\tilde{J}^\nu\partial_\nu f(\mathbf{R}) = \tilde{J}^\nu(\partial_\nu R_a)\partial f/\partial R_a = 0$.

†A similar approximate action was obtained by F. J. Dyson, *Astrophys. J.* **156**, 529 (1969). Dyson used (11.5.2) and inserted the nonrelativistic approximation for A_M and $\Theta^{\mu\nu}_M$.

gravitational wave part of $h_{\mu\nu}$ satisfies the transverse-traceless conditions (11.6.9). Then Γ^i_{00} is zero except for the static contribution from the earth's gravitational field, equal to the acceleration of gravity g. Thus a "cloud of dust"—an elastic solid with $C_{ijkl}=0$—would not be affected by the gravitational wave; this is a virtue of the transverse-traceless coordinate system.

We see from (11.6.14) that the only effect of the gravitational wave in the present approximation and coordinate system is to contribute $-\frac{1}{2}h_{kl}$ to the "relativistic strain" $s_{kl}=-\frac{1}{2}(\phi_{k,l}+\phi_{l,k}+h_{kl})$ in the material and thus add

$$T'_{ij}=C_{ijkl}\tfrac{1}{2}h_{kl} \tag{11.6.15}$$

to the stress. The earth's gravitational field also makes a contribution to the stress, but since this contribution is small and time independent it is essentially impossible to observe and is ignored in what follows.

If the material is isotropic the elastic constants are given by (6.3.3):

$$C_{ijkl}=\lambda\delta_{ij}\delta_{kl}+\mu(\delta_{ik}\delta_{jl}+\delta_{il}\delta_{jk}),$$

where λ and μ are the Lame constants of the material. The stress produced by the gravitational wave is

$$T'_{ij}=\mu h_{ij} \tag{11.6.16}$$

The term proportional to λ is zero because of the "traceless" part of the gauge condition: $h_{ii}=0$.* The gravitational force per unit volume inside the material is

$$f^i_{\text{inside}}=-\frac{\partial}{\partial x^j}T'_{ij}=-\mu h_{ij,j}=0 \tag{11.6.17}$$

as long as μ is independent of position, because of the "transverse" part of the gauge condition. (In materials with less symmetry there can be a force inside the material.) The force per unit area on the surface of the material is†

$$f^i_{\text{surface}}=T'_{ij}N_j=\mu h_{ij}N_j, \tag{11.6.18}$$

where N_j is the outward pointing normal to the surface.

*When only small displacements are considered, a fluid is the same as a solid with $\mu=0$. Thus fluids are not affected by gravitational waves in the small displacement approximation.

†For a discussion of boundary conditions, see Section 5.3.

PROBLEMS

1. If $g_{\mu\nu}(x)$ is a covariant tensor field, show that the inverse matrix $g^{\mu\nu}(x)$ is a contravariant tensor field.
2. Prove that $A^{\mu}{}_{;\nu} = A^{\mu}{}_{,\nu} + \Gamma^{\mu}_{\lambda\nu} A^{\lambda}$ is a tensor. (In writing the transformation law for $A^{\mu}{}_{,\nu}$, you may wish to use the identity $\delta(\partial\bar{x}^{\mu}/\partial x^{\alpha}) = -\partial\bar{x}^{\mu}/\partial x^{\beta}\delta(\partial x^{\beta}/\partial\bar{x}^{\lambda})\partial\bar{x}^{\lambda}/\partial x^{\alpha}$.
3. Show that the Lagrangian

$$\mathfrak{L} = B^{\mu}_{,\nu} B^{\nu}_{,\mu}$$

for a vector field $B_{\mu}(x)$ in flat space differs from the Lagrangian

$$\mathfrak{L}' = B^{\mu}_{,\mu} B^{\nu}_{,\nu}$$

by a divergence. Hence $\mathfrak{L}$ and $\mathfrak{L}'$ are equivalent. Then show that the corresponding curved space Lagrangians

$$\mathfrak{L} = \sqrt{g}\, B^{\mu}_{;\nu} B^{\nu}_{;\mu},$$

$$\mathfrak{L}' = \sqrt{g}\, B^{\mu}_{;\mu} B^{\nu}_{;\nu}$$

are not equivalent, but differ by

$$\mathfrak{L} - \mathfrak{L}' = \sqrt{g}\, A^{\nu} A^{\lambda} R^{\mu}_{\lambda\mu\nu}.$$

4. Suppose that the connection coefficients $\Gamma^{\lambda}_{\mu\nu}(x)$ that appear inside of R in the gravitational Lagrangian were treated as independent fields, to be varied independently of $g_{\mu\nu}(x)$. Show that the definition (11.1.12) of $\Gamma^{\lambda}_{\mu\nu}$ in terms of $g_{\mu\nu}$ would emerge as the equation of motion for $\Gamma^{\lambda}_{\mu\nu}$, and that the Einstein equation of motion for $g_{\mu\nu}(x)$ would be unchanged.
5. Solve the wave equation (11.6.3) for the field $h_{\mu\nu}(x)$ produced by a source $\Theta_{\mu\nu} - \frac{1}{2}\eta_{\mu\nu}\Theta^{\alpha}_{\alpha}$, using the retarded Green's function for the scalar wave equation. Show that your solution is consistent with the harmonic coordinate condition (11.6.2) by using the momentum conservation equation, $\Theta^{\mu\nu}_{,\nu} = 0$.

CHAPTER 12

Momentum Conservation in General Relativity

We encountered in the preceding chapter the energy momentum tensor of the basic, nongravitational fields, $\Theta^{\mu\nu}$. In the first section of this chapter we show that this tensor obeys the equation $\Theta^{\mu\nu}{}_{;\nu}=0$, which is the covariant generalization of the conservation equation $\Theta^{\mu\nu}{}_{,\nu}=0$.

The tensor $\Theta^{\mu\nu}$ does not obey $\Theta^{\mu\nu}{}_{,\nu}=0$. Hence it does not define a conserved momentum. However, the action integral is invariant under translations of the coordinates. By Noether's theorem, then, there is a conserved momentum in the theory. One can say more, since the action is actually invariant under *all* coordinate transformations. In Sections 12.2, 12.3, and 12.4 we examine what invariance under general coordinate transformations adds to Noether's theorem.

It is well to mention a caveat. We discuss quantities T_G^{00} which represent the "density of energy carried by the gravitational field." The array $T_G^{\mu\nu}$ to be constructed will not be a tensor under general coordinate transformations, although it will be a tensor under Lorentz transformations.* Thus if we wanted to use conservation of energy to help discuss some isolated physical system—say, a star in the process of formation— we would first choose a sensible coordinate system. We would demand in particular that far away from the star the metric tensor $g_{\mu\nu}(x)$ approach $\eta_{\mu\nu}$. We would then be free to make a Lorentz transformation of the coordinates, but not a more general transformation, while discussing energy transfers within the system.

In the final section of this chapter we verify that the definition of $\Theta^{\mu\nu}$ given in Chapter 11 agrees with the earlier definition of the "symmetrized momentum tensor" given in Chapter 9.

*Indeed, $T_G^{\mu\nu}(x)$ will be a combination of $g_{\mu\nu}$ and $g_{\mu\nu,\lambda}$ which vanishes when $g_{\mu\nu,\lambda}=0$. If $T_G^{\mu\nu}(x)$ were a true tensor, we could evaluate it in a locally inertial coordinate system at the point x and obtain $T_G^{\mu\nu}(x)=0$.

12.1 THE BIANCHI IDENTITIES AND $\Theta^{\mu\nu}{}_{;\nu}=0$

We begin by establishing two identities that follow from the invariance of the action under general coordinate transformations. One of these identities, $\Theta^{\mu\nu}{}_{;\nu}=0$, is the covariant version of the flat space momentum conservation equation, $\Theta^{\mu\nu}{}_{,\nu}=0$.

Consider the gravitational term in the action

$$A_G=\frac{1}{16\pi G}\int d^4\bar{x}\sqrt{g}\,R. \tag{12.1.1}$$

This term is unchanged by a general coordinate transformation $\bar{x}^{\mu}\to x^{\mu}$ with $x^{\mu}=\bar{x}^{\mu}-\epsilon\xi^{\mu}(x)$. However, the metric tensor is changed.

We define the variation $\delta g_{\mu\nu}(x)$ of $g_{\mu\nu}$ to be the first order difference between $g_{\mu\nu}$ evaluated at the point with coordinates x^{γ}, and $\bar{g}_{\mu\nu}$ evaluated at the point with the same coordinate label:

$$\delta g_{\mu\nu}(x)=g_{\mu\nu}(x)-\bar{g}_{\mu\nu}(x). \tag{12.1.2}$$

Since

$$g_{\mu\nu}(x)=\frac{\partial\bar{x}^{\alpha}}{\partial x^{\mu}}\frac{\partial\bar{x}^{\beta}}{\partial x^{\nu}}\bar{g}_{\alpha\beta}(x^{\gamma}+\epsilon\xi^{\gamma}),$$

the variation $\delta g_{\mu\nu}$ is

$$\delta g_{\mu\nu}=\epsilon\xi^{\beta}{}_{,\nu}g_{\mu\beta}+\epsilon\xi^{\alpha}{}_{,\mu}g_{\alpha\nu}+g_{\mu\nu,\gamma}\epsilon\xi^{\gamma}.$$

This can be rewritten in a more compact form by using the factors $g_{\alpha\beta}$ to lower indices:

$$\begin{aligned}\delta g_{\mu\nu}&=\left[\epsilon\xi^{\beta}g_{\mu\beta}\right]_{,\nu}+\left[\epsilon\xi^{\alpha}g_{\alpha\nu}\right]_{,\mu}\\&\quad-\epsilon\xi^{\beta}g_{\mu\beta,\nu}-\epsilon\xi^{\alpha}g_{\alpha\nu,\mu}+\epsilon\xi^{\gamma}g_{\mu\nu,\gamma}\\&=\epsilon\xi_{\mu,\nu}+\epsilon\xi_{\nu,\mu}-2\epsilon\xi_{\delta}\Gamma^{\delta}_{\mu\nu}\end{aligned}$$

or

$$\delta g_{\mu\nu}=\epsilon\xi_{\mu;\nu}+\epsilon\xi_{\nu;\mu}. \tag{12.1.3}$$

We can learn something from the invariance of A_G by using $\delta g_{\mu\nu}$ in this form. The variation of A_G under the coordinate change is

$$0=\delta A_G=\int d^4x\frac{\delta A_G}{\delta g_{\mu\nu}(x)}\delta g_{\mu\nu}(x),$$

where $\delta A_G/\delta g_{\mu\nu}(x)$ is given by (11.4.12):

$$\frac{\delta A_G}{\delta g_{\mu\nu}(x)} = -\frac{\sqrt{g}}{16\pi G}\left(R^{\mu\nu} - \tfrac{1}{2}g^{\mu\nu}R\right).$$

Thus

$$0 = 2\int d^4x \frac{\delta A_G}{\delta g_{\mu\nu}(x)} \epsilon\xi_{\mu;\nu}$$

$$= 2\int d^4x\sqrt{g}\left[\frac{1}{\sqrt{g}}\frac{\delta A_G}{\delta g_{\mu\nu}}\epsilon\xi_\mu\right]_{;\nu}$$

$$-2\int d^4x\sqrt{g}\left[\frac{1}{\sqrt{g}}\frac{\delta A_G}{\delta g_{\mu\nu}}\right]_{;\nu}\epsilon\xi_\mu.$$

The first term vanishes, since it has the form

$$\int d^4x\sqrt{g}\,V^\nu{}_{;\nu} = \int d^4x\left[\sqrt{g}\,V^\nu\right]_{,\nu} = 0,$$

where V^ν is a vector field. This leaves the second term, which must vanish for every choice of $\xi_\mu(x)$. We conclude that the integrand vanishes identically:

$$0 = \left[\frac{1}{\sqrt{g}}\frac{\delta A_G}{\delta g_{\mu\nu}(x)}\right]_{;\nu} \tag{12.1.4}$$

or

$$0 = \left[R^{\mu\nu} - \tfrac{1}{2}g^{\mu\nu}R\right]_{;\nu}. \tag{12.1.5}$$

The identity (12.1.5) is known as the contracted Bianchi identity.* It is a consequence of the structure of $R^{\mu\nu}$ and holds whether or not $g_{\mu\nu}(x)$ satisfies the Einstein field equation.

The same argument can be applied to the nongravitational part of the action. We use one grand field $\phi_J, J=1,2,\ldots,N$, to stand for all of the

*Equation 12.1.5 is a contraction of the full Bianchi identity

$$R_{\mu\nu\alpha\beta;\gamma} + R_{\mu\nu\gamma\alpha;\beta} + R_{\mu\nu\beta\gamma;\alpha} = 0,$$

which we will not have occasion to use in this book.

tensor components of all of the nongravitational fields. With this notation we have

$$0=\delta A_{NG}=\int d^4x\left\{\frac{\delta A_{NG}}{\delta g_{\mu\nu}(x)}\delta g_{\mu\nu}(x)+\frac{\delta A_{NG}}{\delta\phi_J(x)}\delta\phi_J(x)\right\}.$$

When the fields $\phi_J(x)$ satisfy the Euler-Lagrange equations

$$\frac{\delta A_{NG}}{\delta\phi_J(x)}=0,$$

this reduces to

$$0=\int d^4x\frac{\delta A_{NG}}{\delta g_{\mu\nu}(x)}\delta g_{\mu\nu}.$$

As in the case of A_G, we conclude that

$$0=\left[\frac{1}{\sqrt{g}}\frac{\delta A_{NG}}{\delta g_{\mu\nu}}\right]_{;\nu}.$$

Recalling the definition (11.4.21) of $\Theta^{\mu\nu}$, this is

$$0=\Theta^{\mu\nu}{}_{;\nu}. \tag{12.1.6}$$

Equation 12.1.6 holds whether or not $g_{\mu\nu}(x)$ obeys the Einstein field equations, but holds only if the other fields $\phi_J(x)$ obey their field equations. In flat space, it reads $\Theta^{\mu\nu}{}_{,\nu}=0$ and expresses conservation of the momentum

$$P^{\mu}_{NG}=\int d\mathbf{x}\,\Theta^{\mu 0}$$

carried by the nongravitational fields. In a curved space, P^{μ}_{NG} is not conserved. Physically this is because the nongravitational fields can exchange momentum with the gravitational field. In order to obtain a conserved momentum, we must add the momentum of the gravitational field.

12.2 THE GENERAL COVARIANCE IDENTITIES

We now wish to extract the full consequences of the invariance of the action under general coordinate transformations.

The First Order Gravitational Lagrangian

The gravitational part of the action,

$$A_G = \frac{1}{16\pi G}\int d^4x\sqrt{g}\, R,$$

unfortunately contains second derivatives of the metric tensor. For our present purposes it is convenient to eliminate these second derivatives by means of an integration by parts. We write

$$\begin{aligned}\sqrt{g}\,R &= \sqrt{g}\; g^{\alpha\beta}\left[\Gamma^{\gamma}_{\alpha\beta,\gamma} - \Gamma^{\gamma}_{\alpha\gamma,\beta} + \Gamma^{\gamma}_{\alpha\beta}\Gamma^{\delta}_{\gamma\delta} - \Gamma^{\gamma}_{\alpha\delta}\Gamma^{\delta}_{\beta\gamma}\right]\\ &= \left[\sqrt{g}\; g^{\alpha\beta}\Gamma^{\lambda}_{\alpha\beta} - \sqrt{g}\; g^{\alpha\lambda}\Gamma^{\gamma}_{\alpha\gamma}\right]_{,\lambda}\\ &\quad - \left[\sqrt{g}\; g^{\alpha\beta}\right]_{,\gamma}\Gamma^{\gamma}_{\alpha\beta} + \left[\sqrt{g}\; g^{\alpha\beta}\right]_{,\beta}\Gamma^{\gamma}_{\alpha\gamma}\\ &\quad + \sqrt{g}\; g^{\alpha\beta}\left[\Gamma^{\gamma}_{\alpha\beta}\Gamma^{\delta}_{\gamma\delta} - \Gamma^{\gamma}_{\alpha\delta}\Gamma^{\delta}_{\beta\gamma}\right].\end{aligned} \tag{12.2.1}$$

This has the form

$$\frac{1}{16\pi G}\sqrt{g}\; R = V^{\lambda}{}_{,\lambda} + \mathcal{L}'_G. \tag{12.2.2}$$

The integral of $V^{\lambda}{}_{,\lambda}$ does not contribute to A_G,* so that we are left with

$$A_G = \int d^4x\, \mathcal{L}'_G. \tag{12.2.3}$$

The new Lagrangian $\mathcal{L}'_G$ is easier to handle than $\sqrt{g}\; R$ because it contains only $g_{\mu\nu}$ and $g_{\mu\nu,\lambda}$.

By using the identities

$$\sqrt{g}_{\,,\gamma} = \sqrt{g}\;\Gamma^{\lambda}_{\lambda\gamma},$$

$$0 = g^{\alpha\beta}{}_{;\gamma} = g^{\alpha\beta}{}_{,\gamma} + g^{\alpha\lambda}\Gamma^{\beta}_{\lambda\gamma} + g^{\beta\lambda}\Gamma^{\alpha}_{\lambda\gamma},$$

*Recall that in this kind of argument we do not literally mean that $\int dx V^{\lambda}{}_{,\lambda}=0$. Indeed, the integral may not even converge. But in a variational principle we are interested only in the difference $\int dx(V^{\lambda}{}_{,\lambda} - \bar{V}^{\lambda}{}_{,\lambda})$ between the integral calculated with first one and then another choice for $g_{\mu\nu}(x)$ where $g_{\mu\nu}(x) = \bar{g}_{\mu\nu}(x)$ for x outside some compact domain. This difference is zero.

we can rewrite $\mathcal{L}'_G$ in the compact form

$$\mathcal{L}'_G = \frac{\sqrt{g}}{16\pi G} g^{\alpha\beta}\left[\Gamma^{\gamma}_{\alpha\delta}\Gamma^{\delta}_{\beta\gamma} - \Gamma^{\gamma}_{\alpha\beta}\Gamma^{\delta}_{\gamma\delta}\right]. \tag{12.2.4}$$

Transformation Law for $\mathcal{L}'_G$

The Lagrangian $\mathcal{L}'_G$ is not a scalar density. Nevertheless the action $\int d^4x\,\mathcal{L}'_G$ is a scalar, so that we may expect that $\mathcal{L}'_G$ has a simple transformation law under the general coordinate transformation

$$x^\mu = \bar{x}^\mu - \epsilon\xi^\mu(x).$$

Straightforward computation gives, to first order in ϵ,

$$\delta\,\mathcal{L}'_G = \left[\mathcal{L}'_G\,\epsilon\xi^\nu + Q^{\nu\sigma}_\lambda\,(\epsilon\xi^\lambda)_{,\sigma}\right]_{,\nu}, \tag{12.2.5}$$

where

$$Q^{\nu\sigma}_\lambda = \frac{1}{16\pi G}\left[\sqrt{g}\;g^\nu_\lambda g^{\tau\sigma} - \sqrt{g}\;g^\tau_\lambda g^{\nu\sigma}\right]_{,\tau}. \tag{12.2.6}$$

As was to be expected, $\delta A_G = 0$ because $\delta\,\mathcal{L}'_G$ is a divergence.

General Covariance Identities

Consider now a complete theory of some basic fields coupled to each other and to gravity. The fields involved may include the matter fields $R_a(x)$, the electromagnetic potential $A_\mu(x)$, and so forth, in addition to the metric field $g_{\mu\nu}(x)$. We use one grand field $\phi_J(x)$, $J = 1,2,\ldots,N$, to stand for all of the tensor components of all these individual fields (including $g_{\mu\nu}$). We assume that the Lagrangian density $\mathcal{L}$ is a function of the fields ϕ_J and their first derivatives.

Our goal is to derive certain identities that are a consequence of the invariance of the action under general corrdinate transformations.* As a technical device, we let the factor ϵ depend on x, so that the transformation is $\bar{x}^\mu \to x^\mu$,

$$x^\mu = \bar{x}^\mu - \epsilon(x)\xi^\mu(x). \tag{12.2.7}$$

*This program is due to P. G. Bergman and R. Schiller, *Phys. Rev.* **89**, 4 (1953). See also A. Trautman in L. Witten, ed., *Gravitation, An Introduction to Current Research* (Wiley, New York, 1962).

We assume that the nongravitational part of the Lagrangian density transforms like $\sqrt{g}$ times a scalar, so that $\delta \mathfrak{L}_{NG}$ is given by (9.1.10):

$$\delta \mathfrak{L}_{NG} = [\mathfrak{L}_{NG}\epsilon\xi^{\nu}]_{,\nu}.$$

The transformation law for the first order gravitational Lagrangian $\mathfrak{L}'_G$ is given in (12.2.5). Thus the whole Lagrangian transforms like

$$\delta \mathfrak{L} = [\mathfrak{L}\epsilon\xi^{\nu} + Q_{\lambda}^{\nu\sigma}(\epsilon\xi^{\lambda})_{,\sigma}]_{,\nu}. \tag{12.2.8}$$

The variation $\delta\mathfrak{L}$ induced by the coordinate transformation can also be calculated by direct differentiation:

$$\delta \mathfrak{L} = \frac{\partial \mathfrak{L}}{\partial \phi_J}\delta\phi_J + \frac{\partial \mathfrak{L}}{\partial \phi_{J,\nu}}\delta\phi_{J,\nu}.$$

We rewrite this as

$$\delta \mathfrak{L} = \frac{\delta A}{\delta \Phi_J}\delta\phi_J + \left[\frac{\partial \mathfrak{L}}{\partial \phi_{J,\nu}}\delta\phi_J\right]_{,\nu} \tag{12.2.9}$$

in order to be able to use the Euler-Lagrange equations of motion $\delta A/\delta\phi_J = 0$ if we want.

We need an expression giving the form of the variation $\delta\phi_J(x)$. The fields included in $\phi_J(x)$ are assumed to be scalar, vector, or tensor fields. Thus $\phi_J(x)$ obeys a linear transformation law of the form

$$\phi_J(x) = M_J{}^K\,\bar{\phi}_K(x^{\alpha} + \epsilon\xi^{\alpha}),$$

where $M_J{}^K$ is a function of $\partial x^{\alpha}/\partial \bar{x}^{\beta}$. To first order in ϵ, this transformation law takes the form

$$\delta\phi_J(x) = \phi_{J,\lambda}\epsilon\xi^{\lambda} + \phi_K F_{J\lambda}^{K\omega}(\epsilon\xi^{\lambda})_{,\omega}. \tag{12.2.10}$$

The numbers $F_{J\lambda}^{K\omega}$ are certain simple constants. For instance, in the case of a single vector field $\phi_{\alpha}(x)$,

$$\delta\phi_{\alpha} = \phi_{\alpha,\lambda}\epsilon\xi^{\lambda} + \phi_{\gamma}F_{\alpha\lambda}^{\gamma\omega}(\epsilon\xi^{\lambda})_{,\omega},$$

with

$$F_{\alpha\lambda}^{\gamma\omega} = g_{\alpha}^{\omega}g_{\lambda}^{\gamma}. \tag{12.2.11}$$

In the case of a second order tensor field $\phi_{\alpha\beta}(x)$,

$$F^{\gamma\delta\omega}_{\alpha\beta\lambda} = g_\alpha^\gamma g_\beta^\omega g_\lambda^\delta + g_\beta^\delta g_\alpha^\omega g_\lambda^\gamma. \tag{12.2.12}$$

The generalization to an Nth order tensor field $\phi_{\alpha_1 \cdots \alpha_N}$ is

$$F^{\beta_1 \cdots \beta_N \omega}_{\alpha_1 \cdots \alpha_N \lambda} = \sum_{i=1}^{N} g_{\alpha_i}^\omega g_\lambda^{\beta_i} \prod_{\substack{j=1 \\ j \neq i}}^{N} g_{\alpha_j}^{\beta_j}. \tag{12.2.13}$$

The ingredients for our calculation are now assembled. We equate the two forms (12.2.8) and (12.2.9) for $\delta \mathcal{L}$, inserting the expression (12.2.10) for $\delta\phi_J$:

$$\begin{aligned} 0 = &-\left[\mathcal{L}\epsilon\xi^\nu + Q_\lambda^{\nu\sigma}(\epsilon\xi^\lambda)_{,\sigma}\right]_{,\nu} \\ &+ \frac{\delta A}{\delta\phi_J}\left(\phi_{J,\lambda}\epsilon\xi^\lambda + \phi_K F^{K\omega}_{J\lambda}(\epsilon\xi^\lambda)_{,\omega}\right) \\ &+ \left[\frac{\partial \mathcal{L}}{\partial \phi_{J,\nu}}\left(\phi_{J,\lambda}\epsilon\xi^\lambda + \phi_K F^{K\omega}_{J\lambda}(\epsilon\xi^\lambda)_{,\omega}\right)\right]_{,\nu}. \end{aligned} \tag{12.2.14}$$

We write this equation in detail, collecting together the terms proportional to $\epsilon(x)$, $\epsilon(x)_{,\nu}$, and $\epsilon(x)_{,\nu\sigma}$, so that the equation has the form

$$0 = I\epsilon + I^\nu \epsilon_{,\nu} + I^{\nu\sigma}\epsilon_{,\nu\sigma}. \tag{12.2.15}$$

Since $\epsilon(x)$ is arbitrary, each of the terms in (12.2.15) must vanish. This gives three identities:

$$\begin{aligned} &\text{(I)} && I = 0, \\ &\text{(II)} && I^\nu = 0, \\ &\text{(III)} && I^{\nu\sigma} + I^{\sigma\nu} = 0. \end{aligned} \tag{12.2.16}$$

What do these identities tell us? To find out, we compute the coefficients $I, I^\nu, I^{\nu\sigma}$. The first coefficient is

$$I = \frac{\delta A}{\delta \phi_J}\left(\phi_{J,\lambda}\xi^\lambda + \phi_K F^{K\omega}_{J\lambda}\xi^\lambda{}_{,\omega}\right) + J^\nu{}_{,\nu}, \tag{12.2.17}$$

where

$$J^\nu = \frac{\partial \mathcal{L}}{\partial \phi_{J,\nu}}\left(\phi_{J,\lambda}\xi^\lambda + \phi_K F^{K\omega}_{J\lambda}\xi^\lambda{}_{,\omega}\right) - \mathcal{L}\xi^\nu - Q_\lambda^{\nu\sigma}\xi^\lambda{}_{,\sigma}. \tag{12.2.18}$$

We recognize that the identity $I=0$ is just Noether's theorem applied to the transformation $x^\mu=\bar{x}^\mu-\epsilon\xi^\mu(x)$ with $\epsilon=$constant. The theorem asserts that a current $J^\nu(x)$ associated with this transformation is conserved whenever the fields obey their equations of motion, $0=\delta A/\delta\phi_J(x)$. If the action were invariant only under the transformation with a constant ϵ and a fixed $\xi^\lambda(x)$, we would learn nothing more than this Noether's theorem.

The second coefficient, I^ν, is

$$I^\nu=\frac{\delta A}{\delta\phi_J}\phi_K F_{J\lambda}^{K\nu}\xi^\lambda+J^\nu-J^{\nu\sigma}{}_{,\sigma}, \tag{12.2.19}$$

where

$$J^{\nu\sigma}=-\frac{\partial\,\mathcal{L}}{\partial\phi_{J,\sigma}}\phi_K F_{J\lambda}^{K\nu}\xi^\lambda+Q_\lambda^{\sigma\nu}\xi^\lambda. \tag{12.2.20}$$

Thus the second identity, $I^\nu=0$, relates the Noether current J^ν to the divergence of another quantity $J^{\nu\sigma}$, which is often called the superpotential.

The third coefficient is

$$I^{\nu\sigma}=-J^{\nu\sigma}. \tag{12.2.21}$$

Thus the third identity asserts that the superpotential $J^{\nu\sigma}$ is antisymmetric:

$$J^{\nu\sigma}+J^{\sigma\nu}=0. \tag{12.2.22}$$

In a given theory, this fact is readily verified from the computed form of $J^{\nu\sigma}$.

Form of the Superpotential

One can compute the contribution to $J^{\nu\sigma}$ coming from the gravitational part of the action. The definition (12.2.20) says that

$$J_G^{\nu\sigma}=-\frac{\partial\,\mathcal{L}_G'}{\partial g_{\alpha\beta,\sigma}}g_{\gamma\delta}F_{\alpha\beta\lambda}^{\gamma\delta\nu}\xi^\lambda+Q_\lambda^{\sigma\nu}\xi^\lambda.$$

Simple but somewhat laborious computation using the form of $\mathcal{L}_G$, (12.2.4), the appropriate $F_{J\lambda}^{K\nu}$ for a tensor field, (12.2.12), and the form of $Q_\lambda^{\sigma\nu}$, (12.2.6), gives

$$J_G^{\nu\sigma}=-\frac{1}{16\pi G}\frac{1}{\sqrt{g}}\xi_\lambda\left[g\left(g^{\lambda\nu}g^{\omega\sigma}-g^{\lambda\sigma}g^{\omega\nu}\right)\right]_{,\omega}. \tag{12.2.23}$$

We can also give an expression for the nongravitational part of $J^{\nu\sigma}$ that will be useful later. According to (12.2.20), the contribution to $J^{\nu\sigma}$ from $\mathcal{L}_{NG}$ is

$$J^{\nu\sigma}_{NG} = -\sum_{\substack{J,K\\ \text{basic}}} \frac{\partial \mathcal{L}_{NG}}{\partial \phi_{J,\sigma}} \phi_K F^{K\nu}_{J\lambda} \xi^\lambda - \frac{\partial \mathcal{L}_{NG}}{\partial g_{\alpha\beta,\sigma}} g_{\gamma\delta}\left(g^\gamma_\alpha g^\nu_\beta g^\delta_\lambda + g^\delta_\beta g^\nu_\alpha g^\gamma_\lambda \right) \xi^\lambda, \quad (12.2.24)$$

where the sum Σ_{JK} includes only basic, nongravitational fields. The term involving $\partial \mathcal{L}_{NG} / \partial g_{\alpha\beta,\sigma}$ has been written separately. We can rewrite this term if we recall that the nongravitational part of the Lagrangian contains derivatives of $g_{\alpha\beta}$ only where they are needed to form covariant derivatives $\phi_{J;\tau}$. In our present notation, the definition (11.1.16) of the covariant derivatives is

$$\phi_{J;\tau} = \phi_{J,\tau} - \phi_K F^{K\rho}_{J\omega} \Gamma^\omega_{\rho\tau}.$$

Thus

$$-\frac{\partial \mathcal{L}_{NG}}{\partial g_{\alpha\beta,\sigma}} = \frac{\partial \mathcal{L}_{NG}}{\partial \phi_{J;\tau}} \phi_K F^{K\rho}_{J\omega} \frac{\partial \Gamma^\omega_{\rho\tau}}{\partial g_{\alpha\beta,\sigma}}.$$

We can use the explicit form of $\Gamma^\omega_{\rho\tau}$ to evaluate this factor, then insert it into (12.2.24). This computation gives

$$J^{\nu\sigma}_{NG} = \sum_{\substack{J,K\\ \text{basic}}} \xi_\lambda \frac{\partial \mathcal{L}_{NG}}{\partial \phi_{J;\tau}} \phi_K G^{K\lambda\nu\sigma}_{J\tau}, \quad (12.2.25)$$

where

$$\begin{aligned} 2G^{K\lambda\nu\sigma}_{J\tau} = {} & F^{K\lambda}_{J\alpha} g^{\alpha\nu} g^\sigma_\tau - F^{K\nu}_{J\alpha} g^{\alpha\lambda} g^\sigma_\tau \\ & + F^{K\sigma}_{J\alpha} g^{\alpha\lambda} g^\nu_\tau - F^{K\lambda}_{J\alpha} g^{\alpha\sigma} g^\nu_\tau \\ & + F^{K\sigma}_{J\alpha} g^{\alpha\nu} g^\lambda_\tau - F^{K\nu}_{J\alpha} g^{\alpha\sigma} g^\lambda_\tau. \end{aligned} \quad (12.2.26)$$

12.3 GAUSS' LAW FOR CONSERVED QUANTITIES

Let us return to the identity $0 = I^\nu$:

$$J^\nu = J^{\nu\sigma}{}_{,\sigma} - \frac{\delta A}{\delta \phi_J} \phi_K F^{K\nu}_{J\lambda} \xi^\lambda.$$

When the Euler-Lagrange equations, $\delta A/\delta\phi_J = 0$, are satisfied, the Noether current J^ν equals the divergence of the superpotential, $J^{\nu\sigma}$.

We may regard $J^{\nu\sigma}{}_{,\sigma}$ itself as a current for a moment. Because $J^{\nu\sigma}$ is antisymmetric, the current obeys the local conservation equation $(J^{\nu\sigma}{}_{,\sigma})_{,\nu} = 0$, whether or not the fields obey their equations of motion. Such conservation laws, which hold identically, are sometimes called "strong" conservation laws. They are not very useful, since the knowledge that the law holds does not tell us anything about the motion of the system.

In contrast, the conservation law $J^\nu{}_{,\nu} = 0$ for the Noether current is a useful "weak" conservation law that holds when the fields obey the equations of motion.

The fact that $J^\nu = J^{\nu\sigma}{}_{,\sigma}$ when the equations of motion are satisfied does, however, give an alternative way to evaluate the amount of conserved "charge" contained in a region $\mathfrak{R}$ of space bounded by a surface $\mathfrak{S}$:

$$Q_{\mathfrak{R}}(t) = \int_{\mathfrak{R}} d\mathbf{x}\, J^0(t,x)$$

$$= \int_{\mathfrak{R}} d\mathbf{x} \sum_{\sigma=1}^{3} J^{0\sigma}{}_{,\sigma}$$

or

$$Q_{\mathfrak{R}}(t) = \int_{\mathfrak{S}} dS_j J^{0j}(x), \qquad (12.3.1)$$

Thus the total amount of "charge" Q in all space is determined by the integral of J^{0j} over the surface of a sphere of radius R in the limit $R \to \infty$. This total "charge" will be finite and nonzero if $J^{0j}(x)$ falls off like $|\mathbf{x}|^{-2}$ as $|\mathbf{x}| \to \infty$, and will be determined by the coefficient of $|\mathbf{x}|^{-2}$ in an expansion of $J^{0j}(x)$ in powers of $1/|\mathbf{x}|$.

There is one conserved quantity Q for every coordinate transformation $x^\mu = \bar{x}^\mu - \epsilon\xi^\mu(x)$. However, the superpotential $J^{0j}(x)$ appearing in the gravitational Gauss' law (12.3.1) is a function of $\xi^\mu(x)$; thus two transformation functions $\xi^\mu(x)$ that are asymptotically equal as $|\mathbf{x}| \to \infty$ define the same total conserved quantity Q.

A mathematical structure analogous to the one discussed here occurs in electrodynamics. In theories of electrically charged quantum fields used in particle physics, conservation of electric charge is a "weak" conservation law* which results from the invariance of the action under gauge trans-

*In classical electrodynamics as discussed in this book, conservation of charge is a strong conservation law that is due simply to the structure of the current $\mathcal{J}^\mu = qn\epsilon^{\mu\alpha\beta\gamma}R_{1,\alpha}R_{2,\beta}R_{3,\gamma}$.

formations, $A_\mu(x)\to A_\mu(x)-\Lambda(x)_{,\mu}$. Since the gauge function $\Lambda(x)$ is an arbitrary function, one finds that the electric current $\mathcal{J}^\nu(x)$ equals the divergence of a superpotential, $J^{\nu\sigma}$, when the equations of motion are satisfied. The superpotential turns out to be the electromagnetic field $F^{\nu\sigma}$. Thus in the case of electrodynamics, (12.3.1) is just Gauss' law.

Nonuniqueness of the Conserved Quantities

We have seen that Noether's theorem associates a weakly conserved quantity $Q[\xi]$ with every infinitesimal coordinate transformation $\bar{x}^\mu\to x^\mu=\bar{x}^\mu-\epsilon\xi^\mu(x)$. Our derivation gives a specific relation between the transformation ξ and the conserved quantity $Q[\xi]$. However, this relation is not at all unique. What matters is the set of conservation laws, not the relation $\xi\to Q[\xi]$.

To see how one might have derived a different relation $\xi\to Q[\xi]$, consider (12.2.8), which gives the variation of $\mathcal{L}$ under the coordinate transformation:

$$\delta\mathcal{L}=\left[\mathcal{L}\epsilon\xi^\nu+Q_\lambda^{\nu\sigma}(\epsilon\xi^\lambda)_{,\sigma}\right]_{,\nu}.$$

Suppose that $A_\lambda^{\nu\sigma}$ and $B_\lambda^{\nu\sigma\tau}$ are objects constructed from $g_{\alpha\beta}$ and $g_{\alpha\beta,\gamma}$ that are antisymmetric in the indices $\nu\sigma$. Then the modified equation

$$\delta\mathcal{L}=\Big[\mathcal{L}\epsilon\xi^\nu+Q_\lambda^{\nu\sigma}(\epsilon\xi^\lambda)_{,\sigma}-\big(A_\lambda^{\nu\sigma}\epsilon\xi^\lambda+B_\lambda^{\nu\sigma\tau}(\epsilon\xi^\lambda)_{,\tau}\big)_{,\sigma}\Big]_{,\nu} \tag{12.3.2}$$

is also true, since we have merely added zero to the right hand side. If one follows the procedures of Section 12.2 with (12.3.2) as the starting point, he will obtain a different conserved current $J^\nu=J^\nu_{\text{old}}+\Delta J^\nu$ and a new superpotential $J^{\nu\sigma}=J^{\nu\sigma}_{\text{old}}+\Delta J^{\nu\sigma}$, where

$$\Delta J^\nu=(\Delta J^{\nu\sigma})_{,\sigma}$$

identically and

$$\Delta J^{\nu\sigma}=A_\lambda^{\nu\sigma}\xi^\lambda+B_\lambda^{\nu\sigma\tau}\xi^\lambda{}_{,\tau}. \tag{12.3.3}$$

If ΔJ^{0k} does not fall off too fast, the new calculation will lead to a different conserved quantity:

$$Q=Q_{\text{old}}+\int dS_j\,\Delta J^{0j}.$$

There are several reasonable choices for terms $\Delta J^{\nu\sigma}{}_{,\sigma}$ which can be added to the conserved current in this way.* However, our present form is as good as any, so we will stick to it.

12.4 MOMENTUM OF THE GRAVITATIONAL FIELD

The conserved quantity associated with a uniform translation of the coordinates, $\xi^{\lambda}(x) = -\eta^{\lambda\mu}$, is by definition the energy-momentum P^{μ}. The corresponding momentum current contains a contribution from the gravitational Lagrangian

$$T_G^{\mu\nu} = \eta^{\mu\nu}\mathfrak{L}_G' - \eta^{\mu\lambda}\frac{\partial \mathfrak{L}_G'}{\partial g_{\alpha\beta,\nu}}g_{\alpha\beta,\lambda}. \tag{12.4.1}$$

This momentum current was first identified by Einstein in 1916.†

We are, however, free to associate a conserved quantity with any transformation $\xi^{\lambda}(x)$ that equals $-\eta^{\lambda\mu}$ asymptotically as $|\mathbf{x}|\to\infty$. According to the gravitational Gauss' law (12.3.1), the total momentum calculated with $\xi^{\lambda}(x)$ will still be P^{μ}.‡ A convenient choice is $\xi^{\lambda}(x) = -g(x)^{1/2}g(x)^{\mu\lambda}$. As long as the metric tensor $g^{\mu\nu}(x)$ approaches the Minkowski metric $\eta^{\mu\nu}$ as $|\mathbf{x}|\to\infty$, the requirement that $\xi^{\lambda}(x)\to-\eta^{\lambda\mu}$ will be met. With this choice of $\xi^{\lambda}(x)$, the gravitational superpotential is

$$\begin{aligned} J_G^{\nu\sigma} &= -\frac{1}{16\pi G}\frac{1}{\sqrt{g}}\xi_{\lambda}\left[g\left(g^{\lambda\nu}g^{\omega\sigma} - g^{\lambda\sigma}g^{\omega\nu}\right)\right]_{,\omega} \\ &= \frac{1}{16\pi G}\left[g\left(g^{\mu\nu}g^{\sigma\omega} - g^{\mu\sigma}g^{\nu\omega}\right)\right]_{,\omega}. \end{aligned} \tag{12.4.2}$$

The contribution to the momentum current from the gravitational part of the action is, from (12.2.19),

$$\mathcal{T}_G^{\mu\nu} = J_G{}^{\nu} = J_{G,\sigma}^{\nu\sigma} + \frac{\delta A_G}{\delta g_{\alpha\beta}}g_{\gamma\delta}F_{\alpha\beta\lambda}^{\gamma\delta\nu}\sqrt{g}\, g^{\lambda\mu}.$$

*See, for example, P. G. Bergmann, *Phys. Rev.* **112**, 287 (1958), and A. Komar, *Phys. Rev.* **113**, 934 (1958).

†A. Einstein, *Ann. Phys.* **49**, 769 (1916).

‡But the fraction of the total momentum assigned to any particular region in space and the fraction assigned to gravity instead of matter will change.

Using (12.4.2) for $J_G^{\nu\sigma}$, (12.2.12) for $F_{\alpha\beta\lambda}^{\gamma\delta\nu}$, and (11.4.12) for $\delta A_G/\delta g_{\alpha\beta}$ we find

$$\mathfrak{T}_G^{\mu\nu} = \frac{1}{16\pi G}\left[g(g^{\mu\nu}g^{\sigma\omega} - g^{\mu\sigma}g^{\nu\omega})\right]_{,\omega\sigma}$$

$$- \frac{1}{8\pi G} g(R^{\mu\nu} - \tfrac{1}{2} g^{\mu\nu}R). \qquad (12.4.3)$$

This form of gravitational momentum current is due to Landau and Lifshitz.* Its advantage is that it is symmetric. The Landau-Lifschitz momentum current also retains the feature of the Einstein momentum current that it depends only on $g_{\mu\nu}$ and its first derivatives. This feature follows from the definition (12.2.18) of the Noether current J^ν corresponding to *any* transformation $\xi^\lambda(x)$.

There are other possibilities for defining a "momentum current" associated with the gravitational field. The reader is referred to the literature for further examples.†

Momentum of the Basic Fields

The conserved current associated with the transformation $\xi^\lambda(x)$ also contains a contribution from the nongravitational part of the action. This contribution is, from (12.2.19):

$$J_{NG}^{\nu} = -\frac{\delta A_{NG}}{\delta g_{\alpha\beta}} g_{\gamma\delta} F_{\alpha\beta\lambda}^{\gamma\delta\nu}\xi^\lambda + J_{NG,\sigma}^{\nu\sigma} - \sum_{\substack{J,K\\ \text{basic}}} \frac{\delta A_{NG}}{\delta\phi_J}\phi_K F_{J\lambda}^{K\nu}\xi^\lambda. \qquad (12.4.4)$$

The first term is, from (11.4.21) and (12.2.12),

$$-\frac{\delta A_{NG}}{\delta g_{\alpha\beta}} g_{\gamma\delta} F_{\alpha\beta\lambda}^{\gamma\delta\nu}\xi^\lambda = -\sqrt{g}\,\xi_\lambda\Theta^{\lambda\nu}. \qquad (12.4.5)$$

The superpotential appearing in the second term of (12.4.4) is given by (12.2.25),

$$J_{NG}^{\nu\sigma} = \sum_{\substack{J,K\\ \text{basic}}} \xi_\lambda \frac{\partial \mathfrak{L}_{NG}}{\partial\phi_{J,\tau}}\phi_K G_{J\tau}^{K\lambda\nu\sigma},$$

*L. D. Landau and E. M. Lifshitz *The Classical Theory of Fields*, 2nd ed. (Addison-Wesley, Reading, Mass., 1962).

†See A. Trautman, "Conservation Laws in General Relativity," in L. Witten, ed. *Gravitation, an Introduction to Current Research* (Wiley, New York, 1962). See also the textbook of S. Weinberg already cited.

where $G_{J\tau}^{K\lambda\nu\sigma}$ is given in (12.2.26) and is a combination of $F_{J\alpha}^{K\beta}$ and metric tensor factors which is antisymmetric in ν, σ. There is no transformation matrix $F_{J\alpha}^{K\beta}$ for scalar fields. Thus $J_{NG}^{\nu\sigma}=0$ when the basic fields of the theory are all scalar fields like the matter fields $R_a(x)$ and the polarization fields $\mathcal{P}_a(x)$ and $\mathcal{M}_a(x)$. If the electromagnetic potential $A_\mu(x)$ is included in the theory, one finds that

$$J_{NG}^{\nu\sigma} = -\sqrt{g}\, F^{\nu\sigma}A^\lambda\xi_\lambda. \tag{12.4.6}$$

The final term in (12.4.4) also contains contributions from nonscalar fields only. If $A_\mu(x)$ is the only nonscalar field, it is

$$-\sum_{\substack{J,K\\ \text{basic}}} \frac{\delta A_{NG}}{\delta\phi_J}\phi_K F_{J\lambda}^{K\nu}\xi^\lambda = -\frac{\delta A_{NG}}{\delta A_\nu}A_\lambda\xi^\lambda. \tag{12.4.7}$$

When (12.4.5), (12.4.6), and (12.4.7) have been inserted into (12.4.4) we have, for a theory in which $A_\mu(x)$ is the only nonscalar basic field,

$$J_{NG}^{\nu} = -\sqrt{g}\,\xi_\lambda\Theta^{\lambda\nu} - \frac{\delta A_{NG}}{\delta A_\nu}A_\lambda\xi^\lambda - \left(\sqrt{g}\,F^{\nu\sigma}A^\lambda\xi_\lambda\right)_{,\sigma}. \tag{12.4.8}$$

Let us make the Landau-Lifshitz choice for the transformation function:

$$\xi^\lambda(x) = -\sqrt{g}\, g^{\lambda\mu}.$$

The contribution of $\mathcal{L}_{NG}$ to the Landau-Lifshitz momentum current is then

$$\bar{\mathfrak{T}}_{NG}^{\mu\nu} = g\Theta^{\mu\nu} + \sqrt{g}\,\frac{\delta A_{NG}}{\delta A_\nu}A^\mu + (gF^{\nu\sigma}A^\mu)_{,\sigma}. \tag{12.4.9}$$

A bar has been given to $\bar{\mathfrak{T}}_{NG}^{\mu\nu}$ to indicate that it is somewhat unsatisfactory as a momentum tensor, since it is not symmetric and not gauge invariant. Fortunately, the offending terms can be left out of the momentum tensor without doing any damage. The second term contains a factor $\delta A_{NG}/\delta A_\nu(x)$, so that it vanishes when $A_\nu(x)$ obeys its equation of motion. The third term does not vanish, but it does not affect the conservation of momentum since its divergence is zero. Furthermore, the third term (the superpotential term) makes no contribution to the total momentum. Its contribution would be

$$\int dx\,(gF^{0k}A^\mu)_{,k} = \int_{\substack{\text{large}\\ \text{sphere}}} dS_k\, gF^{0k}A^\mu.$$

But if the potential was produced by a localized charge distribution that was stationary before some time t_0 in the past, then $A^\mu \sim |\mathbf{x}|^{-1}$ and $F^{0k} \sim |\mathbf{x}|^{-2}$ so that the surface integral over an infinitely large sphere vanishes. Therefore we can define an improved momentum tensor

$$\mathcal{T}^{\mu\nu}_{NG} = g\Theta^{\mu\nu}, \tag{12.4.10}$$

which gives the same total momentum and has the same divergence as $\bar{\mathcal{T}}^{\mu\nu}_{NG}$.*

The corresponding contribution from $\mathcal{L}_G$ to the Landau-Lifshitz momentum tensor was given in (12.4.3):

$$\mathcal{T}^{\mu\nu}_G = \frac{1}{16\pi G}\left[g\left(g^{\mu\nu}g^{\sigma\omega} - g^{\mu\sigma}g^{\nu\omega}\right)\right]_{,\omega\sigma}$$

$$- \frac{1}{8\pi G} g\left(R^{\mu\nu} - \tfrac{1}{2} g^{\mu\nu}R\right).$$

The theorems of this section state that the total momentum current $\mathcal{T}^{\mu\nu} = \mathcal{T}^{\mu\nu}_{NG} + \mathcal{T}^{\mu\nu}_G$ is conserved when all of the fields obey their equations of motion.

The conservation of $\mathcal{T}^{\mu\nu}$ can be reproved very easily. When the metric tensor obeys its equation of motion $R^{\mu\nu} - \frac{1}{2}g^{\mu\nu}R = 8\pi G\Theta^{\mu\nu}$, the net momentum current is

$$\mathcal{T}^{\mu\nu} = \frac{1}{16\pi G}\left[g\left(g^{\mu\nu}g^{\sigma\omega} - g^{\mu\sigma}g^{\nu\omega}\right)\right]_{,\omega\sigma}.$$

Since the quantity inside the brackets is antisymmetric in the indices ν, σ, we find $\mathcal{T}^{\mu\nu}{}_{,\nu} = 0$. This short proof seems not to use the condition that the nongravitational fields obey their equations of motion, but it should be remembered that the Einstein field equation $R^{\mu\nu} - \frac{1}{2}g^{\mu\nu}R = 8\pi G\Theta^{\mu\nu}$ together with the contracted Bianchi identity $(R^{\mu\nu} - g^{\mu\nu}R)_{;\nu} = 0$ implies that $\Theta^{\mu\nu}_{;\nu} = 0$. Thus some restriction on the motion of the basic fields is implicit in the Einstein equation. We recall from (12.1.6) that $\Theta^{\mu\nu}_{;\nu} = 0$ when the basic fields obey their equations of motion.

12.5 $\Theta^{\mu\nu}$ AND THE "SYMMETRIZED MOMENTUM TENSOR"

In Chapter 9 we discussed the momentum carried by the basic fields in a space with the Minkowski metric. Invariance of the action under translations led to the existence of a conserved momentum tensor $T^{\mu\nu}$, but $T^{\mu\nu}$

*The reader may have noticed that the transition from $\bar{\mathcal{T}}^{\mu\nu}$ to $\mathcal{T}^{\mu\nu}$ is essentially the transition from the canonical momentum tensor to the symmetrized momentum tensor discussed in Chapter 9. We return to this point in the next section.

was not always symmetric. However, the Lorentz invariance of the action implied that $T^{\mu\nu}$ could be modified to form a conserved symmetric momentum tensor, $\Theta_{\mathrm{I}}^{\mu\nu}$.

In this chapter we have found a symmetric momentum tensor $\Theta_{\mathrm{II}}^{\mu\nu}$ which appears on the right hand side of the Einstein field equations:

$$\Theta_{\mathrm{II}}^{\mu\nu} \equiv \frac{2}{\sqrt{g(x)}} \frac{\delta A_{NG}}{\delta g(x)_{\mu\nu}}. \tag{12.5.1}$$

In a space with the Minkowski metric, the "Bianchi" equation $\Theta_{\mathrm{II};\nu}^{\mu\nu}=0$ becomes $\Theta_{\mathrm{II},\nu}^{\mu\nu}=0$, so that $\Theta_{\mathrm{II}}^{\mu\nu}$ is conserved.

We discovered in two examples that the old $\Theta_{\mathrm{I}}^{\mu\nu}$ equals the new $\Theta_{\mathrm{II}}^{\mu\nu}$ with $g_{\mu\nu}(x)$ set equal to $\eta_{\mu\nu}$. We prove here that this equality,

$$\Theta_{\mathrm{I}}^{\mu\nu} = 2\left[\frac{\delta A_{NG}}{\delta g(x)_{\mu\nu}}\right]_{g_{\mu\nu}=\eta_{\mu\nu}}, \tag{12.5.2}$$

holds in general.

The result (12.5.2) is useful even when we are not interested in gravitational interactions. When one has at hand a complicated action like that for polarizable elastic material interacting with the electromagnetic field, he can relatively easily change $\eta_{\mu\nu}$ to $g_{\mu\nu}$ so as to make the action generally covariant, and then differentiate with respect to $g_{\mu\nu}$. The prescription for calculating the old $\Theta_{\mathrm{I}}^{\mu\nu}$ directly is more laborious.

The Old Prescription

Let us first recall what the old prescription was. The canonical momentum tensor $T^{\mu\nu}$ was

$$T_{\mu}^{\ \nu} = g_{\mu}^{\nu} \mathcal{L}_{NG} - \frac{\partial \mathcal{L}_{NG}}{\partial \phi_{J,\nu}} \phi_{J,\mu}. \tag{12.5.3}$$

(In this section implied sums over field components ϕ_J include only the basic fields, not $g_{\mu\nu}$.) To form $\Theta^{\mu\nu}$ we added the divergence of a quantity formed from the spin density $S_{\mu\nu}^{\ \ \sigma}$:

$$S^{\mu\nu\sigma} = \frac{\partial \mathcal{L}_{NG}}{\partial \phi_{J,\sigma}} (\tilde{M}^{\mu\nu})_{JK} \phi_K.$$

The matrix $\tilde{M}^{\mu\nu}$ is the infinitesimal generator of Lorentz transformations on the fields ϕ_J. In our present notation,

$$(\tilde{M}^{\mu\nu})_{JK} = F_{J\alpha}^{K\nu} g^{\alpha\mu} - F_{J\alpha}^{K\mu} g^{\alpha\nu}.$$

The prescription was

$$\Theta_{\mathrm{I}}^{\mu\nu}=T^{\mu\nu}-\tfrac{1}{2}\left[S^{\mu\nu\sigma}+S^{\sigma\mu\nu}+S^{\sigma\nu\mu}\right]_{,\sigma}$$

or

$$\Theta_{\mathrm{I}}^{\mu\nu}=T^{\mu\nu}+\left[\frac{\partial \mathcal{L}_{NG}}{\partial\phi_{J,\tau}}\phi_K G_{J\tau}^{K\mu\nu\sigma}\right]_{,\sigma}, \tag{12.5.4}$$

where

$$\begin{aligned}2G_{J\tau}^{K\mu\nu\sigma}&=F_{J\alpha}^{K\mu}g^{\alpha\nu}g_\tau^\sigma-F_{J\alpha}^{K\nu}g^{\alpha\mu}g_\tau^\sigma\\&+F_{J\alpha}^{K\sigma}g^{\alpha\mu}g_\tau^\nu-F_{J\alpha}^{K\mu}g^{\alpha\sigma}g_\tau^\nu\\&+F_{J\alpha}^{K\sigma}g^{\alpha\nu}g_\tau^\mu-F_{J\alpha}^{K\nu}g^{\alpha\sigma}g_\tau^\mu.\end{aligned} \tag{12.5.5}$$

The New Prescription

The "new" version of $\Theta^{\mu\nu}$, (12.5.1), is related to the Noether momentum current by the general covariance identity (12.2.19):

$$0=-\frac{\delta A_{NG}}{\delta g_{\alpha\beta}}g_{\gamma\delta}F_{\alpha\beta\lambda}^{\gamma\delta\nu}\xi^\lambda-J_{NG}^\nu+J_{NG,\sigma}^{\nu\sigma}-\frac{\delta A_{NG}}{\delta\phi_J}\phi_K F_{J\lambda}^{K\nu}\xi^\lambda. \tag{12.5.6}$$

This identity holds because the action $A=A_{NG}$ is invariant under general coordinate transformations. The metric tensor need not satisfy the Einstein field equations, so we can still use the identity in the particular case of present interest, a flat space with $g_{\mu\nu}=\eta_{\mu\nu}$.

Let $g_{\mu\nu}=\eta_{\mu\nu}$, and choose $\xi^\lambda(x)=-\eta^{\lambda\mu}$. Then the first term in (12.5.6) is $\Theta_{\mathrm{II}}^{\mu\nu}$. The Noether current J_{NG}^ν given by (12.2.18) does not contain the term $-(\partial\mathcal{L}/\partial g_{\alpha\beta,\nu})g_{\alpha\beta,\lambda}\eta^{\lambda\mu}$ since $g_{\alpha\beta,\lambda}=0$; thus J_{NG}^ν is just the canonical momentum tensor $T^{\mu\nu}$. The superpotential $J_{NG}^{\nu\sigma}$ is given by (12.2.25):

$$J_{NG}^{\nu\sigma}=-\frac{\partial\mathcal{L}_{NG}}{\partial\phi_{J,\tau}}\phi_K G_{J\tau}^{K\mu\nu\sigma},$$

with the same array G as in (12.5.4). Thus (12.5.6) reads, in a space with $g_{\mu\nu}=\eta_{\mu\nu}$,

$$0=\Theta_{\mathrm{II}}^{\mu\nu}-T^{\mu\nu}-\left[\frac{\partial\mathcal{L}_{NG}}{\partial\phi_{J,\tau}}\phi_K G_{J\tau}^{K\mu\nu\sigma}\right]_{,\sigma}-\frac{\delta A_{NG}}{\delta\phi_J}\phi_K F_{J\lambda}^{K\nu}\eta^{\mu\lambda}. \tag{12.5.7}$$

When the Euler-Lagrange equations $\delta A_{NG}/\delta\phi_J=0$ are satisfied, this reproduces the old definition (12.5.4).

PROBLEMS

1. In the weak field approximation to general relativity in which one works to lowest order in $h_{\mu\nu}=g_{\mu\nu}-\eta_{\mu\nu}$, it is often convenient to define the "gravitational potentials" $\tilde{h}_{\mu\nu}=h_{\mu\nu}-\frac{1}{2}\eta_{\mu\nu}h^{\alpha}{}_{\alpha}$. Show that the harmonic gauge condition $\eta^{\alpha\beta}\Gamma^{\mu}_{\alpha\beta}=0$ is $\tilde{h}^{\mu}{}_{\nu,\mu}=0$ and that the linearized equation of motion for $\tilde{h}_{\mu\nu}$ in a harmonic coordinate system is $\eta^{\alpha\beta}\tilde{h}_{\mu\nu,\alpha\beta}=-16\pi G\Theta_{\mu\nu}$.
2. Show that to first order in the gravitational potentials $\tilde{h}_{\mu\nu}$ introduced in Problem 1, the superpotential $J^{\nu\sigma}$ corresponding to a coordinate transformation $\bar{x}^{\lambda}\to x^{\lambda}=\bar{x}^{\lambda}-\epsilon\xi^{\lambda}(x)$ is, in a harmonic coordinate system,

$$J^{\nu\sigma}=\frac{1}{16\pi G}\xi_{\lambda}\left[\tilde{h}^{\lambda\nu,\sigma}-\tilde{h}^{\lambda\sigma,\nu}\right].$$

If $\xi^{\lambda}\sim-\eta^{\lambda 0}$ for large $|\mathbf{x}|$, the superpotential gives the energy of the system

$$E=\int dS_i J^{0i}.$$

If $g_{\mu\nu}$ is produced by a weak, time independent source $\Theta^{\mu\nu}$, show that this energy equals the total mass of the source, $\int d\mathbf{x}\Theta^{00}$.
3. Use the results of Problems 1 and 2 to show that, in the weak field approximation with a static source, the conserved quantity corresponding to the transformation function $\xi^{\mu}(x)=\frac{3}{2}x^{k}\delta^{\mu}_{0}$ measures the k-component of the dipole moment of the mass distribution, $\int d\mathbf{x}x^{k}\Theta^{00}$.
4. Use the formula $\Theta^{\mu\nu}=2\delta A_{NG}/\delta g_{\mu\nu}(x)$ to calculate the momentum tensor for an electrically polarized material, using the Lagrangian (10.4.1). Compare with Table 10.2.

CHAPTER 13

Dissipative Processes

Mechanics based on the principle of stationary action is elegant. It is also a powerful tool for the analysis of a wide variety of physical phenomena. However, "friction" cannot always be neglected in the real world. In this chapter we discuss field theories in which "frictional" effects like viscosity, heat flow, and electric current conduction are present.

In the cases we consider, a frictionless, stationary action model serves as a convenient starting point from which a more general model can be constructed. We add dissipative effects to the stationary action model according to certain principles suggested by the classical thermodynamics of systems with a finite number of degrees of freedom.

We investigate the simple viscous relativistic fluid in Sections 13.1 through 13.4, developing insights based on microscopic kinetic theory and following mainly the ideas of Eckart.* In Section 13.5 we refashion this treatment in a more compact form, including electromagnetic effects.

It is not intended that the derivations presented be rigorous deductions from a small number of fundamental laws or that the resulting models fit every material which exists in nature. Indeed, water and air may reasonably be described as viscous fluids in the sense of Section 13.5, but kneaded bread dough may not. The reader is directed to the literature of continuum mechanics for a partial antidote to the deficiencies of this chapter.†

13.1 ENTROPY AND THE SECOND LAW OF THERMODYNAMICS

Consider a small droplet of fluid, containing a fixed number of atoms, as it is carried along in the general motion of a larger sample of the fluid. The internal state of this droplet may be specified by giving its volume per atom, $\mathcal{V} = 1/\rho$, and the entropy per atom, s, as measured in a rest frame of

*C. Eckart, *Phys. Rev.* **58**, 919 (1940).

†See, for example, A. C. Eringen *Mechanics of Continua* (Wiley, New York, 1967).

the droplet. The internal energy per atom is a function $U(s, \mathcal{V})$ of these thermodynamic variables.*

A "perfect fluid," discussed in Chapter 4, is a fluid in which the entropy of each droplet is conserved. Thus in Chapter 4 s was a fixed function of the material coordinates R_a. Now we consider a more realistic model in which entropy increases. Accordingly, s is a field $s(x)$, for which we must give an equation of motion.

This extra equation of motion is the energy conservation equation $\partial_\mu \Theta^{0\mu} = 0$. In a perfect fluid, the three momentum conservation equations $\partial_\mu \Theta^{k\mu}$ determine the motion of the three matter fields $R_a(x)$; conservation of energy then follows as a consequence of conservation of momentum. In a viscous, heat conducting fluid, four energy-momentum conservation equations $\partial_\mu \Theta^{\nu\mu} = 0$ determine the development of four fields, $R_a(x)$ and $s(x)$.

We can identify another important thermodynamic variable, the temperature T, by using the second law of thermodynamics:

$$dU(s, \mathcal{V}) + P\,d\mathcal{V} = T\,ds. \tag{13.1.1}$$

That is,

$$T(s, \mathcal{V}) = \frac{\partial}{\partial s} U(s, \mathcal{V}). \tag{13.1.2}$$

(We have already defined $P = -\partial U / \partial \mathcal{V}$ in Chapter 4.)

The relation (13.1.1) can be rewritten in a useful form involving the conservative momentum tensor $\Theta_{\mathcal{L}}^{\mu\nu}$ derived from the Lagrangian $\mathcal{L} = -\rho U(s, \mathcal{V})$:

$$\Theta_{\mathcal{L}}^{\mu\nu} = \rho U u^\mu u^\nu + P(u^\mu u^\nu + g^{\mu\nu}). \tag{13.1.3}$$

This momentum tensor contains information about entropy conserving forces only, so that it will not be conserved when dissipative forces are added. Let us calculate the energy component of the divergence of $\Theta_{\mathcal{L}}^{\mu\nu}$, as measured in a local rest frame of the fluid, $-u_\mu \partial_\nu \Theta_{\mathcal{L}}^{\mu\nu}$. Since

$$0 = \partial_\nu J^\nu \equiv \partial_\nu (\rho u^\nu)$$

and

$$0 = \tfrac{1}{2} \partial_\nu [u_\mu u^\mu] = u_\mu \partial_\nu u^\mu,$$

*We consider for the moment a homogeneous fluid. Thus U does not depend on the material coordinates R_a in addition to $\mathcal{V}$ and s.

we find

$$-u_\mu \partial_\nu \Theta_{\mathcal{E}}^{\mu\nu} = \rho u^\nu \partial_\nu U + \rho u^\nu \partial_\nu (\mathcal{V} P) - u^\nu \partial_\nu P$$
$$= \rho u^\nu \partial_\nu U + \rho P u^\nu \partial_\nu \mathcal{V}.$$

The definitions $\partial U/\partial s = T$ and $\partial U/\partial \mathcal{V} = -P$ and the chain rule give

$$u^\nu \partial_\nu U = T u^\nu \partial_\nu s - P u^\nu \partial_\nu \mathcal{V}.$$

Thus

$$-u_\mu \partial_\nu \Theta_{\mathcal{E}}^{\mu\nu} = \rho T u^\nu \partial_\nu s. \tag{13.1.4}$$

This identity will prove useful in Section 13.3. In Section 13.5 we prove a more general version of the identity by using a more powerful method.

13.2 DISSIPATIVE FORCES IN A GAS

A "perfect" fluid is subject to conservative forces derived from the momentum tensor $\Theta_{\mathcal{E}}^{\mu\nu}$. In a local rest frame of the fluid, the momentum flow is

$$\Theta^{ij} = \delta_{ij} P.$$

We can understand the origin of this momentum flow, and at the same time see what dissipative corrections need be made, by examining a microscopic model for a gas made of atoms that interact by short range forces.

Consider the plane $x^1 = 0$ in a local rest frame of the gas. Often an atom will move across this plane, carrying its momentum from one side to the other.* The atoms moving from left to right across the plane carry a positive x^1-component of momentum; those moving from right to left carry a negative x^1-component of momentum in the negative x^1-direction. Thus $\Theta^{11} > 0$. If the whole gas is at rest, so that the average velocity $\mathbf{v}$ of the atoms is zero throughout the gas, then the average amount of y- or z-component of momentum crossing the plane is zero. Thus $\Theta^{21} = \Theta^{31} = 0$. Symmetry arguments then give $\Theta^{ij} = (\text{const.})\ \delta^{ij}$. The constant, P, is a function of the average atomic speed, which is related to the temperature.

*Equal numbers of atoms cross the plane going each way. That is what we mean by being in a local rest frame of the gas.

Viscosity

Now suppose that the x^2—component, v_2, of the average velocity of the atoms is a function $v_2(x^1)$ of the coordinate x^1, with $v_2(0)=0$. Then the momentum distribution of the atoms crossing the plane $x^1=0$ is slightly altered (see Figure 13.1). The atoms crossing the plane from left to right will have come, usually, from the region $-\lambda < x^1 < 0$, where λ is the mean free path of atoms in the gas. These atoms will carry an average momentum $mv_2(-\lambda/2) \sim -\frac{1}{2}\lambda m\, \partial v_2/\partial x^1$. The number of these atoms that cross a small area dA of the plane in time dt is roughly $\rho\langle|\mathbf{v}|\rangle dA\, dt$, where $\langle|\mathbf{v}|\rangle$ is the average speed of an atom. Thus the atoms moving from left to right carry momentum

$$dP_2 \sim -\tfrac{1}{2}\lambda m\rho\langle|\mathbf{v}|\rangle \frac{\partial v_2}{\partial x^1}\, dA\, dt.$$

The atoms crossing the surface from right to left carry the opposite momentum in the opposite direction. Thus the total x^2-component of

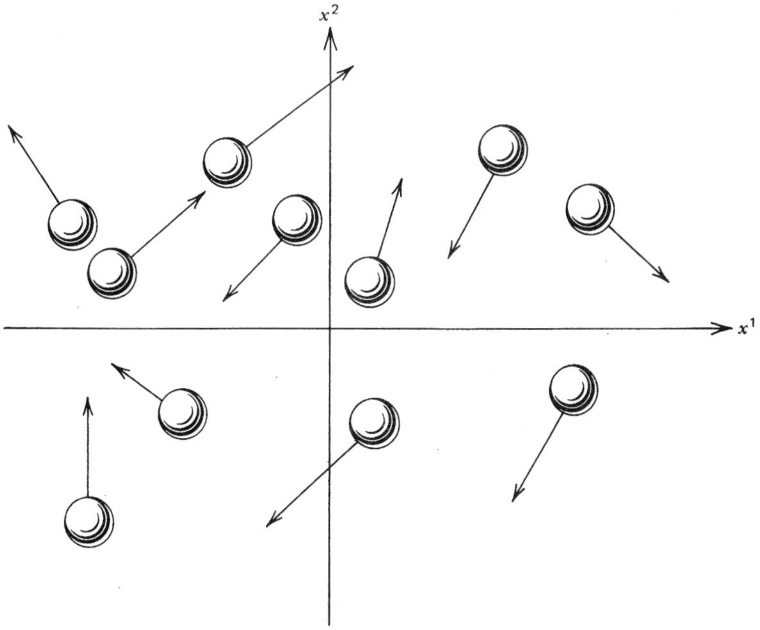

Figure 13.1 Atomic model for viscosity in a gas. The atoms to the left of $x^1=0$ move upward on the average, those to the right move downward. Positive x^2-component of momentum is transferred from left to right.

momentum crossing the surface $x^1=0$, per unit area and unit time, is roughly

$$\Theta^{21}\sim-\lambda m\rho\langle|\mathbf{v}|\rangle\frac{\partial v_2}{\partial x^1}. \qquad (13.2.1)$$

We may conclude from this argument that when the velocity $\mathbf{v}(x)$ of a gas is not uniform, the momentum current Θ^{ij} will contain a term Δ^{ij} in addition to the static pressure term $P\delta_{ij}$:

$$\Theta^{ij}=P\delta_{ij}+\Delta^{ij}. \qquad (13.2.2)$$

Since $\Delta^{ij}=0$ when $\partial_k v_l(x)=0$, a linear relation like (13.2.1) between Δ^{ij} and $\partial_k v_l$ will be a good approximation, provided that $\partial_k v_l$ is small. Here "small" means that $\partial v/\partial x$ should be much less than a typical atomic speed divided by the mean free path of an atom. A typical atomic speed at room temperature is quite large;

$$\langle|\mathbf{v}|\rangle\sim\sqrt{kT/m}\sim10^4\ \text{cm/sec}. \qquad (13.2.3)$$

Atomic mean free paths are generally microscopic. The mean free path is related to the density of atoms and the cross sectional area of an atom σ by

$$\lambda\sim\frac{1}{\rho\sigma}. \qquad (13.2.4)$$

If we take $\rho\sim10^{19}\ \text{cm}^{-3}$ and $\sigma\sim10^{-14}\ \text{cm}^2$ for a typical gas, we obtain

$$\lambda\sim10^{-5}\ \text{cm}.$$

Thus $\partial v/\partial x$ is "small" in a gas when it is much less than $10^9\ \text{sec}^{-1}$. This is certainly the case in laboratory experiments.*

We have derived a rough estimate for the constant of proportionality, η, between Δ^{21} and $\partial v_2/\partial x^1$:

$$\eta\sim\lambda m\rho\langle|\mathbf{v}|\rangle.$$

Using (13.2.3) and (13.2.4), this is

$$\eta\sim\frac{\sqrt{mkT}}{\sigma}. \qquad (13.2.5)$$

We return to this estimate in Section 13.4.

*Note, however, that this argument applies to gases and perhaps some liquids. It certainly cannot be applied to gummy liquids made of big molecules whose size is comparable to a mean free path.

Heat Flow

When the entire gas is at a uniform temperature, as much energy is carried by the atoms moving from left to right as by those moving the other way. The net energy flow across the $x^1=0$ plane is $\Theta^{10}=0$. But if the gas on the left is at a slightly higher temperature than the gas on the right, then the atoms moving from left to right across the plane carry, on the average, slightly more energy than those moving in the other direction. Thus there is an energy current to the right, which we will call Δ^{i0}:

$$\Theta^{i0}=\Delta^{i0}. \tag{13.2.6}$$

This energy current is usually called "heat flow."

For the small temperature gradients $[T^{-1}dT/dx \ll (\text{mean free path})^{-1}]$ normally encountered in the laboratory, a linear approximation $\Delta^{i0}\propto \nabla T$ is adequate. A rough estimate of the coefficient in this linear relation can be obtained by an argument similar to that just given for Δ^{ij}:

$$\Delta^{0i}\sim -\rho\langle|\mathbf{v}|\rangle\lambda\frac{\partial}{\partial x^i}(kT)=-K\frac{\partial T}{\partial x^i},$$

where

$$K\sim\rho\langle|\mathbf{v}|\rangle\lambda k$$

or

$$K\sim\frac{k}{m}\eta. \tag{13.2.7}$$

Momentum Density

In the situation just described, the atoms moving to the right posess more momentum than those moving to the left. Thus the momentum density Θ^{0i} is nonzero. In fact, since the "microscopic" momentum tensor is symmetric, the average, macroscopic momentum tensor $\Theta^{\mu\nu}$ is also symmetric. Thus $\Theta^{0i}=\Theta^{i0}=\Delta^{i0}$.

Conclusion

We abstract from the model the limited conclusion, not necessarily restricted to low density gases, that we should add to the conservative momentum tensor $\Theta^{\mu\nu}_{\mathfrak{L}}$ a dissipative momentum tensor $\Delta^{\mu\nu}$:

$$\Theta^{\mu\nu}=\Theta^{\mu\nu}_{\mathfrak{L}}+\Delta^{\mu\nu}. \tag{13.2.8}$$

The dissipative momentum tensor should be symmetric, and should obey $\Delta^{00}=0$ in a local rest frame of the fluid.*

Later in the development we specify that $\Delta^{\mu\nu}$ be a linear function of velocity and temperature gradients, as suggested by the model.

13.3 ENTROPY FLOW AND THE ENTROPY INEQUALITY

We gain an important insight in the discussion of entropy if we think of entropy as a "fluid" that can flow from place to place in a material at rest. We call the entropy current **S**; that is, $\mathbf{S}\cdot d\mathbf{A}$ is the rate at which entropy crosses a surface element $d\mathbf{A}$. Of course, entropy is not conserved, so we cannot ask for a unique definition of **S** that will make $\nabla\cdot\mathbf{S}=-\partial(\rho s)/\partial t$. Nevertheless, we seek a definition of **S** that comes as close as possible to satisfying this conservation equation.

When an amount of heat dQ flows into a droplet of fluid, the entropy of the droplet increases by $ds=dQ/T$. This suggests that we think of the rate at which entropy crosses a surface as being $1/T$ times the rate at which heat energy crosses the surface;

$$S^j=\frac{1}{T}\Delta^{0j}. \tag{13.3.1}$$

With this definition, the "rate of entropy production" in a stationary fluid is

$$\begin{aligned}\partial_0(\rho s)+\partial_j S^j&=\frac{1}{T}\partial_0(\rho U)+\partial_j S^j\\ &=-\frac{1}{T}\partial_j\Delta^{0j}+\partial_j\left(\frac{1}{T}\Delta^{0j}\right)\\ &=\left(\partial_j\frac{1}{T}\right)\Delta^{0j}.\end{aligned}$$

That is, entropy is produced only where heat flows down a temperature gradient.

This definition of entropy current can be generalized to cover moving fluids. We let the entropy current S^μ be that four-vector which has components $(\rho s, T^{-1}\Delta^{0j})$ in a local rest frame of the fluid:

$$S^\mu=s\rho u^\mu-\frac{1}{T}u_\nu\Delta^{\nu\mu}. \tag{13.3.2}$$

*It would not be obvious nonsense to add a dissipative energy density to $\rho U(s,\mathcal{V})$ in a local rest frame: the dissipative energy might be proportional to $(\nabla T)^2$, for example. We consider this possibility in Section 13.5, but omit it here in order to keep the discussion simple.

Part of the content of the second law of thermodynamics is that the total entropy of an isolated system is a nondecreasing function of the time. This increase of entropy can be ensured in a field theoretic model for a fluid by demanding that the "rate of entropy production" $\partial_\mu S^\mu(x)$ be nonnegative at each point in space-time.*:

$$0 \leqslant \partial_\mu S^\mu(x). \tag{13.3.3}$$

The entropy inequality (13.3.3) restricts the form of the dissipative momentum tensor $\Delta^{\mu\nu}$. Using the definition (13.3.2) of S^μ, we have

$$0 \leqslant \rho u^\mu \partial_\mu s - \frac{1}{T} u_\nu \partial_\mu \Delta^{\nu\mu} - \left(\partial_\mu \frac{u_\nu}{T}\right) \Delta^{\nu\mu}.$$

Energy momentum conservation implies that

$$u_\nu \partial_\mu \Delta^{\nu\mu} = -u_\nu \partial_\mu \Theta_{\mathfrak{L}}^{\nu\mu}.$$

The quantity $-u_\nu \partial_\mu \Theta_{\mathfrak{L}}^{\nu\mu}$ is the rate, measured in a local rest frame of a fluid droplet, at which "heat energy" is being added to the droplet by the dissipative processes. According to the identity (13.1.4), which expresses the relation $dU = T\,ds - P\,d\mathcal{V}$,

$$-u_\nu \partial_\mu \Theta_{\mathfrak{L}}^{\nu\mu} = \rho T u^\mu \partial_\mu s.$$

Combining these results gives

$$0 \leqslant \rho u^\mu \partial_\mu s - \frac{1}{T} \rho T u^\mu \partial_\mu s - \left(\partial_\mu \frac{u_\nu}{T}\right) \Delta^{\nu\mu}$$

or

$$0 \leqslant -\left(\partial_\mu \frac{u_\nu}{T}\right) \Delta^{\nu\mu}. \tag{13.3.4}$$

13.4 THE STRUCTURE OF THE DISSIPATIVE MOMENTUM TENSOR

Analysis of the Entropy Inequality

What does the inequality (13.3.4) imply about the structure of $\Delta^{\mu\nu}$? The analysis of this question can be simplfied by writing the inequality in a

*This is often called the Clausius-Duhem inequality. It was introduced in this form for relativistic fluid mechanics by Eckart, *op. cit.*

local rest frame of the fluid in the form

$$0 \leq -\frac{1}{T^2} T_i \Delta^{0i} - \frac{1}{2T} V_{ij} \Delta^{ij}, \tag{13.4.1}$$

where

$$\begin{aligned} T_i &= \partial_i T + T\partial_0 v_i, \\ V_{ij} &= \partial_i v_j + \partial_j v_i \end{aligned} \tag{13.4.2}$$

and $v_i = u^i/u^0$ is the ordinary three-velocity of the fluid.

Now, to begin the analysis, we write down a list of variables on which Δ^{0i} and Δ^{ij} might depend. First, there are the thermodynamic parameters s and $\mathcal{V}$. If the fluid is inhomogeneous, $\Delta^{\mu\nu}$ may also depend directly on the material coordinates R_a. Then there are the variables T_i and V_{ij} that appear in the entropy inequality. We denote these variables $X_1, X_2, \ldots, X_9$:

$$(X_N) = \left(-\frac{1}{T^2} T_i, -\frac{1}{2T} V_{ij} \right).$$

Finally, there are a number of other variables Y_N on which $\Delta^{\mu\nu}$ might depend. We are already including some first derivatives of T and second derivatives of R_a in the variables X_N. Thus we consider as candidates those combinations of first derivatives of T and first and second derivatives of R_a that are independent of the X_N and are unchanged by the symmetry transformations of a fluid, $R_a \to \overline{R}_a$ with $\partial(R_a)/\partial(\overline{R}_a) = 1$:

$$(Y_N) = (\partial_0 T, \partial_i v_j - \partial_j v_i, \partial_i \mathcal{V}, \partial_0 v_i).$$

The entropy inequality (13.4.1) has the form

$$0 \leq X_N F_N(\mathbf{X}, \mathbf{Y}, s, \mathcal{V}, R_a), \tag{13.4.3}$$

where

$$(F_N) = (\Delta^{0i}, \Delta^{ij}).$$

We require that the inequality hold for all values of the arguments $\mathbf{X}, \mathbf{Y}, s, \mathcal{V}, R_a$. This is certainly sufficient to guarantee that it holds in every physically realizable situation.

The inequality places severe restrictions on the behavior of the functions F_N near $\mathbf{X} = 0$. Letting $\mathbf{X} \to 0$ in (13.4.3), we see* that $F_N = 0$ at $\mathbf{X} = 0$. Thus

*We assume that the functions are continuously differentiable at $X = 0$. If $F_M(0, \mathbf{Y}, s, \mathcal{V}, R_a)$ were, say, bigger than zero for some particular choice of $M, \mathbf{Y}, s, \mathcal{V}$, and R_a, we could choose $\epsilon > 0$ such that $F_M(\mathbf{X}, \mathbf{Y}, s, \mathcal{V}, R_a) > 0$ whenever $|\mathbf{X}| \leq \epsilon$. Then the choice $X_N = -\epsilon\delta_{NM}$ gives $X_N F_N = -\epsilon F_M < 0$, contrary to the inequality (13.4.3).

we can write

$$F_N(\mathbf{X},\mathbf{Y};s,\mathcal{V},R_a)=G_{NM}(\mathbf{X},\mathbf{Y};s,\mathcal{V},R_a)\,X_M, \tag{13.4.4}$$

where the functions G_{NM} are continuous at $\mathbf{X}=0$. Again letting $\mathbf{X}\to 0$ we find

$$0\leqslant X_N G_{NM}(0,\mathbf{Y};s,\mathcal{V},R_a)X_M. \tag{13.4.5}$$

At this point we can introduce an approximation that simplifies the model. We notice that the quantities X_N and Y_N are normally "small" in an appropriate sense involving atomic mean free paths and mean free times, as discussed in Section 13.2. Therefore we assume that a linear approximation in the small variables X_N and Y_N is adequate to describe the functions F_N:

$$F_N=f_N+g_{NM}X_M+h_{NM}Y_M.$$

The general representation (13.4.4) for F_N indicates that the coefficients f_N and h_{NM} are zero. The coefficients g_{NM} are the functions G_{NM} evaluated at $\mathbf{X}=\mathbf{Y}=0$. Therefore we take

$$F_N(\mathbf{X},\mathbf{Y};s,\mathcal{V},R_a)=g_{NM}(s,\mathcal{V},R_a)\,X_M, \tag{13.4.6}$$

where g_{NM} is a positive matrix:

$$0\leqslant X_N g_{NM} X_M \tag{13.3.7}$$

for all choices of the X_N.

Explicit Form of the Expansion

We must now write out the expansion $F_N=g_{MN}X_N$ explicitly. The coefficients g_{MN} depend only on the quantities s, $\mathcal{V}$, and R_a, which are scalars under rotations. Therefore the g_{MN} must be constructed using only the invariant tensors δ_{ij} and ϵ_{ijk} and scalar coefficients.* It is not hard to see that the most general allowed expansion is

$$\begin{aligned}\Delta^{0i}&=-KT_{,i},\\ \Delta^{ij}&=-\tfrac{1}{2}\zeta V_{kk}\delta_{ij}-\eta\left[V_{ij}-\tfrac{1}{3}V_{kk}\delta_{ij}\right],\end{aligned} \tag{13.4.8}$$

where the coefficients K, ζ, and η are functions of s, $\mathcal{V}$, R_a.

*This high degree of symmetry arises because we are considering a fluid. The analysis for a solid would be nearly the same up to here, but the constriction of g_{NM} would be more complicated.

The positivity condition (13.4.7) for the coefficients g_{NM} reads

$$0 \leqslant -\frac{1}{T^2} T_i \Delta^{0i} - \frac{1}{2T} V_{ij} \Delta^{ij}$$

$$= \frac{1}{T^2} K T_i T_i + \frac{1}{4T} \zeta (V_{kk})^2$$

$$+ \frac{1}{2T} \eta \left[V_{ij} - \tfrac{1}{3} V_{ll} \delta_{ij} \right] \left[V_{ij} - \tfrac{1}{3} V_{kk} \delta_{ij} \right].$$

In order for this inequality to hold for all T_i and V_{ij}, it is clearly necessary and sufficient that

$$0 < K, \qquad 0 < \zeta, \qquad 0 < \eta. \tag{13.4.9}$$

Before we go on to interpret the dissipative effects described by the coefficients K, ζ, and η, let us pause to write out (13.4.8) in a form correct in any reference frame:

$$\Delta^{\mu\nu} = A^{\mu\nu\rho\sigma} \left[\left(\partial_\rho \frac{u_\sigma}{T} \right) + \left(\partial_\sigma \frac{u_\rho}{T} \right) \right],$$

$$A^{\mu\nu\rho\sigma} = -T^2 K \left[u^\mu I^{\nu\rho} u^\sigma + u^\nu I^{\mu\rho} u^\sigma \right]$$

$$- \tfrac{1}{2} T \zeta I^{\mu\nu} I^{\rho\sigma}$$

$$- T\eta \left[I^{\mu\rho} I^{\nu\sigma} - \tfrac{1}{3} I^{\mu\nu} I^{\rho\sigma} \right], \tag{13.4.10}$$

$$I^{\mu\nu} = g^{\mu\nu} + u^\mu u^\nu.$$

The Dissipative Momentum Current

The momentum flow we found in (13.4.8) is

$$\Delta^{ij} = -\zeta (\partial_k v_k) \delta_{ij}$$

$$- \eta \left[(\partial_i v_j) + (\partial_j v_i) - \tfrac{2}{3} \delta_{ij} (\partial_k v_k) \right], \tag{13.4.11}$$

where v_k is the ordinary velocity of the fluid. This expression is exact in a local rest frame of the fluid and is an adequate approximation even for moving fluids. (The corrections are of order $\Delta^{ij} v^2/c^2$ and $\Delta^{0i} v/c^2$.)

The coefficient ζ is called the bulk viscosity of the fluid. If $\zeta \neq 0$, then whenever the fluid is moving inward toward a point, so that $\nabla \cdot \mathbf{v} < 0$, an extra pressure $P = \zeta[-\nabla \cdot \mathbf{v}]$ appears that resists the motion. The coefficient η is called the shear viscosity. Imagine a fluid with $\eta > 0$ flowing in the

$\pm x^2$-direction with v_2 a decreasing function of x^1, say $v_j = (0, -\lambda x^1, 0)$. Then

$$\Delta^{ij} = \begin{bmatrix} 0 & +\eta\lambda & 0 \\ +\eta\lambda & 0 & 0 \\ 0 & 0 & 0 \end{bmatrix}.$$

Some two-component of momentum is transferred in the x^1-direction, helping to reduce the velocity gradient. At the same time some one-component of momentum is transferred in the x^2—direction. This momentum transfer tends to start the fluid rotating, so that angular momentum can be conserved.

The dissipative momentum transfers Δ^{ij} cease when the fluid motion consists of a uniform translation plus a rigid rotation:

$$v_i(x) = v(0)_i + \Omega^{ij} x^j$$

with $\Omega^{ij} = -\Omega^{ji}$. (See Problem 1.)

Heat Flow

The energy flow in a local rest frame of the fluid is

$$\Delta^{0i} = -K[\partial_i T + T\partial_0 v_i]. \tag{13.4.12}$$

The first term represents Fourier's law of heat conduction. Heat flows from hot to cold; the magnitude of the heat flow is proportional to the temperature gradient. The proportionality constant K is called the thermal conductivity of the fluid.

The second term, $-KT\partial_0 v_i$, is a relativistic correction to the Fourier law. If we use units in which $c \neq 1$, a factor $1/c^2$ appears in this term. There is no situation known to the author in which the second term is not negligible. Nevertheless it is interesting that the temperature in a rotating fluid in thermal equilibrium is not exactly constant according to (13.4.12). In order to make $\Delta^{0i} = 0$, there must be a temperature gradient $(\partial_i T)$ in the direction opposite to the centripetal acceleration $\partial_0 v_i$. Thus the outside of the spinning fluid will be hotter than the inside.

Values of η, ζ, and K

In Section 13.2 we used kinetic theory to derive rough estimates for η and K in a gas. We found

$$\eta \sim \frac{\sqrt{mkT}}{\sigma}, \qquad K \sim \frac{k}{m}\eta, \tag{13.4.13}$$

where m is the mass of an atom or molecule of the gas, σ is its cross sectional area, k is Boltzman's constant, and T is the temperature of the gas. For a typical gas at room temperature, we might estimate $\sigma \sim 10^{-14}$ cm^2, $m \sim 30m_H \sim 3\times10^{-23}$ g, $k \sim 10^{-16}$ erg K^{-1}, and $T \sim 300$ K; then

$$\eta \sim 10^{-4} \text{g/cm sec},$$

$$K \sim 10^{3} \text{erg/cm sec } K. \tag{13.4.14}$$

The bulk viscosity is harder to estimate without a more detailed analysis. In a low density gas composed of point atoms, which do not have any rotational or vibrational degrees of freedom, kinetic theory gives $\zeta = 0$. There is no reason to expect ζ to vanish in more complicated examples.

The measured values of η, ζ, and K for a few common fluids are given in Table 12.1.

Table 12.1 Viscosities and Thermal Conductivities for Some Familiar Fluids[a]

Substance	Shear Viscosity (g/cm sec)	Bulk Viscosity (g/cm sec)	Thermal Conductivity (erg/cm sec K)
Water	1.0×10^{-2}	3.1×10^{-2}	6.1×10^{4}
Ethyl alcohol	1.2×10^{-2}	5.4×10^{-2}	1.7×10^{4}
Benzene	0.65×10^{-2}	70×10^{-2}	1.4×10^{4}
N_2	1.7×10^{-4}	1.1×10^{-4}	2.6×10^{3}
NH_3	0.98×10^{-4}	1.3×10^{-4}	2.5×10^{3}

[a]Viscosities of the three liquids are from L. N. Lieberman, *Phys. Rev.* **75**, 1415 (1949), and **76**, 440 (1949); viscosities of the two gases are from M. Kohler, *Z. Phys.* **125**, 733 (1949); the thermal conductivities are from Y. S. Touloukian, P. E. Liley, and S. C. Saxena, *Thermal Conductivity, Nonmetallic Liquids and Gases*, Vol. 3 of *Thermophysical Properties of Matter* (IFI/Plenum, New York, 1970.)

13.5 DISSIPATIVE PROCESSES IN A FLUID DIELECTRIC

How can dissipative effects be included in the interaction between matter and the electromagnetic field? In order to keep the number of effects and coefficients that are allowed by the symmetry of the matter from proliferating unduly, we confine our attention to a fluid. The dissipative effects then are viscosity, heat conduction, and conduction of electric current, with six coefficients.

The Conservative Theory

Our starting point is the theory of a perfect fluid dielectric discussed in Chapter 10. The fluid was described by the matter fields $R_a(x)$, the polarization field $M_{\mu\nu}(x)$, and the electromagnetic field $F_{\mu\nu}(x)$. It carried an entropy current $S^{\mu} = s\rho u^{\mu}$ and an electric current $\mathcal{J}^{\mu} = q\rho u^{\mu}$, where the entropy per atom s and the charge per atom q were fixed functions of R. The Lagrangian was

$$
\begin{aligned}
\mathcal{L} = & -\tfrac{1}{4}F_{\mu\nu}F^{\mu\nu} + \tfrac{1}{2}F_{\mu\nu}M^{\mu\nu} + A_{\mu}\mathcal{J}^{\mu} \\
& -\frac{1}{2\kappa}U^{\alpha}M_{\alpha\nu}u^{\beta}M_{\beta}{}^{\nu} \\
& -\frac{1}{4\chi}M_{\mu\alpha}M_{\nu\beta}(g^{\mu\nu}+u^{\mu}u^{\nu})(g^{\alpha\beta}+u^{\alpha}u^{\beta}) \\
& -\rho U. \qquad (13.5.1)
\end{aligned}
$$

In a local rest frame of the fluid, this is

$$
\begin{aligned}
\mathcal{L} = & \tfrac{1}{2}(\mathbf{E}^2 - \mathbf{B}^2) + \mathbf{E}\cdot\mathbf{P} + \mathbf{B}\cdot\mathbf{M} + A_{\mu}\mathcal{J}^{\mu} \\
& -\frac{1}{2\kappa}\mathbf{P}^2 - \frac{1}{2\chi}\mathbf{M}^2 - \rho U.
\end{aligned}
$$

The internal energy of the fluid, U, may be a function of the thermodynamic variables s, $\mathcal{V}$, and q. If the fluid is not homogeneous, U may also depend on the material coordinates R_a. Thus

$$U = U(s, \mathcal{V}, q; R_a). \qquad (13.5.2)$$

We have not previously discussed the dependence of U on q. Imagine adding a small number of electrons to a piece of material. Let the number of electrons added per molecule of material be dN. Then the charge added per molecule is $dq = e\,dN$. The energy density $\frac{1}{2}\mathbf{E}^2$ associated with the macroscopic electric field changes in this process. *In addition*, the internal energy per molecule of the material changes by an amount $(\partial U/\partial q)dq$. Part of this change is due to the rest energy $M_e\,dN$ of the added electrons. The rest of the change reflects the binding energy per electron, W, which binds electrons to the material. Thus

$$\frac{\partial U}{\partial q}e\,dN = (M_e - W)\,dN$$

or

$$\frac{\partial U}{\partial q} = \frac{M_e}{e} - \frac{W}{e}.$$

The electron binding energy W is called the work function of the material. In metals W is usually a few electron volts.

The dielectric coefficients κ and χ may also depend on s, $\mathcal{V}$, q, and R_a. We, ignore their possible dependence on $q(x)$, however, in the spirit of an approximation to lowest order in electromagnetic effects. Thus

$$\begin{aligned} \kappa &= \kappa(s, \mathcal{V}; R_a), \\ \chi &= \chi(s, \mathcal{V}; R_a). \end{aligned} \tag{13.5.3}$$

Changes

Certain changes in the theory are required in order to describe dissipative effects. To allow for the generation and transfer of entropy, the specific entropy must be an independent field $s(x)$ instead of a fixed function of R_a.

The electric current can be decomposed in the form

$$\mathcal{J}^{\mu} = q\rho u^{\mu} + j^{\mu}, \tag{13.5.4}$$

where $u_{\mu} j^{\mu} = 0$. Thus, in a local rest frame of the fluid, $j^0 \equiv 0$ and $j^l = \mathcal{J}^l$ is the electric three-current. We demanded that $j^{\mu} = 0$ in the conservative theory, but now $j^{\mu}(x)$ will be an independent field for which an equation of motion (Ohm's law) will be introduced. The charge per atom, $q(x)$, is also elevated to the status of a field instead of just a fixed function of R_a. Its equation of motion is $\partial_{\mu} \mathcal{J}^{\mu} = 0$. That is,

$$\rho u^{\mu} \partial_{\mu} q = -\partial_{\mu} j^{\mu}. \tag{13.5.5}$$

We retain the equations of motion for $F_{\mu\nu}$ and $M_{\mu\nu}$:

$$\begin{aligned} \partial_{\nu}(F^{\mu\nu} + M^{\mu\nu}) &= \mathcal{J}^{\mu}, \\ \mathbf{P} &= \kappa \mathbf{E}, \\ \mathbf{M} &= \chi \mathbf{B}. \end{aligned} \tag{13.5.6}$$

We allow for dissipative forces and energy transfers within the fluid, however, by adding a new piece to the conservative momentum tensor $\Theta^{\mu\nu}_{\mathfrak{L}}$ derived from the Lagrangian. Thus the matter fields $R_a(x)$ do not obey the equations of motion derived from the Lagrangian (13.5.1).

The Momentum Tensor

We write the complete momentum tensor in the form

$$\Theta^{\mu\nu} = \Theta_{\mathfrak{L}}^{\mu\nu} + \Delta^{\mu\nu} + \rho U_D u^\mu u^\nu, \tag{13.5.7}$$

where $\Delta^{\mu\nu}$ is to be symmetric and obey $u_\mu u_\nu \Delta^{\mu\nu} = 0$. The function U_D represents a possible dissipative energy density in a rest frame of the fluid; we find later that $U_D = 0$. Both $\Delta^{\mu\nu}$ and U_D may be functions of the thermodynamic parameters s, $\mathcal{V}$, and q, the material coordinates R_a, the velocity, specific volume, and temperature gradients $\partial_\mu u_\nu$, $\partial_\mu \mathcal{V}$, and $\partial_\mu T$, and the electromagnetic field $F_{\mu\nu}$. Later, we make a simplifying linearity assumption.

Heat Flow and Entropy Flow

In Section 13.3, we identified the energy current Δ^{0i} in a local rest frame of the fluid as "heat flow", and associated an entropy flow $S^i = (1/T)\Delta^{0i}$ with this heat flow.

In the present case of a conducting fluid, not all of Δ^{0i} represents heat flow. When charge dq moves from one droplet of fluid to another neighboring droplet, the donor droplet loses energy $dU = (\partial U/\partial q)\,dq$ and the recipient droplet gains this energy. Thus an energy current $(\partial U/\partial q)\mathbf{j}$ is associated with the electric current $\mathbf{j}$.

This energy current is not part of $\Theta_{\mathfrak{L}}^{0i}$, that so it must be included in Δ^{0i} in addition to the ordinary heat flow also included in Δ^{0i}. Thus we are led to identify the heat current as

$$\text{heat current} = \Delta^{0i} - \frac{\partial U}{\partial q} j^i. \tag{13.5.8}$$

This reasoning leads us to postulate that the rest frame entropy current has the form

$$S^j = \frac{1}{T}(\Delta^{0i} - \phi j^i). \tag{13.5.9}$$

We expect that $\phi = \partial U/\partial q$, but we leave this choice open for the moment. The temperature T that appears in (13.5.9) also remains to be defined. The form of both T and ϕ is fixed by the entropy inequality, $0 \leqslant \partial_\mu S^\mu$.

The form of the entropy current generalizing (13.5.9) to any reference frame is

$$S^\mu = s\rho u^\mu - \frac{1}{T} u_\nu \Delta^{\nu\mu} - \frac{\phi}{T} j^\mu. \tag{13.5.10}$$

The Entropy Inequality

With the definition (13.5.10) of the entropy current, the entropy inequality reads

$$0 \leqslant \partial_\mu S^\mu = \rho u^\mu \partial_\mu s - \frac{1}{T} u_\nu \partial_\mu \Delta^{\nu\mu} - \left(\partial_\mu \frac{u_\nu}{T}\right)\Delta^{\nu\mu} - \frac{\phi}{T}\partial_\mu j^\mu - j^\mu \partial_\mu\left(\frac{\phi}{T}\right).$$

Momentum conservation enables us to replace the second term with

$$-\frac{1}{T}u_\nu \partial_\mu \Delta^{\nu\mu} = \frac{1}{T}u_\nu \partial_\mu\left[\Theta_{\mathfrak{L}}^{\nu\mu} + \rho u^\nu u^\mu U_D\right]$$

$$= \frac{1}{T}u_\nu \partial_\mu \Theta_{\mathfrak{L}}^{\nu\mu} - \frac{\rho}{T}u^\mu \partial_\mu U_D.$$

Current conservation enables us to replace the fourth term with

$$-\frac{\phi}{T}\partial_\mu j^\mu = \frac{\phi}{T}\partial_\mu(q\rho u^\mu)$$

$$= \frac{\rho\phi}{T}u^\mu \partial_\mu q.$$

Thus the entropy inequality reads

$$0 \leqslant \partial_\mu S^\mu = \rho u^\mu \partial_\mu s + \frac{1}{T}u_\nu \partial_\mu \Theta_{\mathfrak{L}}^{\mu\nu} - \frac{\rho}{T}u^\mu \partial_\mu U_D$$

$$+ \frac{\rho\phi}{T}u^\mu \partial_\mu q - j^\mu \partial_\mu\left(\frac{\phi}{T}\right) - \left(\partial_\mu \frac{u_\nu}{T}\right)\Delta^{\nu\mu}. \qquad (13.5.11)$$

Evaluation of $u_\mu \partial_\nu \Theta_{\mathfrak{L}}^{\nu\mu}$

To proceed, we must calculate $u_\mu \partial_\nu \Theta^{\nu\mu}$. The conservative momentum tensor $\Theta_{\mathfrak{L}}^{\mu\nu}$ for a dielectric fluid was calculated in Chapter 10 [see (10.4.18)]. The momentum tensor is complicated, so that a certain amount of guile is called for in the calculation of $u_\mu \partial_\nu \Theta_{\mathfrak{L}}^{\nu\mu}$.

Recall the general definition of $\Theta_{\mathfrak{L}}^{\mu\nu}$ given in Section 11.4.

$$\Theta^{\mu\nu}(X) = \frac{2}{\sqrt{g(x)}}\frac{\delta A}{\delta g_{\mu\nu}(x)},$$

where A is the generally invariant action formed from the flat space Lagrangian. The covariant divergence of $\Theta_{\mathfrak{L}}^{\mu\nu}$ can be calculated, as in

Section 12.1, by considering the effect of a small change of coordinates $\bar{x}^{\mu} \to x^{\mu} = \bar{x}^{\mu} - \xi^{\mu}(x)$. Define the corresponding variation ϕ_N of the fields by $\delta\phi_N(x) = \phi_N(x) - \bar{\phi}_N(x)$. Since the action is invariant under all coordinate changes, we have

$$0 = \delta A = \int d^4x \sum_N \frac{\delta A}{\delta \phi_N(x)} \delta\phi_N(x). \tag{13.5.12}$$

The Lagrangian (13.5.1) is a function of the fields $g_{\mu\nu}(x)$, $A_\mu(x)$, $M_{\mu\nu}(x)$, $\mathcal{J}^{\mu}(x)$, $R_a(a)$, $s(x)$, and $q(x)$. The specific entropy $s(x)$ appears as an argument of the functions U, κ, and χ. The specific charge $q(x)$ appears as an argument of U. For ease of differentiation, we consider $\mathcal{J}_\mu(x)$ to be independent field here and do not decompose it into $q\rho u^{\mu}$ and j^{μ}. Thus (13.5.12) reads

$$\begin{aligned} 0 = \int d^4x \Bigg\{ & \frac{\delta A}{\delta g_{\mu\nu}(x)} \delta g_{\mu\nu}(x) + \frac{\delta A}{\delta A_\mu(x)} \delta A_\mu(x) \\ & + \frac{\delta A}{\delta M_{\mu\nu}(x)} \delta M_{\mu\nu}(x) + \frac{\delta A}{\delta \mathcal{J}^{\mu}(x)} \delta \mathcal{J}^{\mu}(x) \\ & + \frac{\delta A}{\delta R_a(x)} \delta R_a(x) + \frac{\delta A}{\delta s(x)} \delta s(x) + \frac{\delta A}{\delta q(x)} \delta q(x) \Bigg\}. \end{aligned} \tag{13.5.13}$$

In the first term of (13.5.13) we recognize

$$\frac{\delta A}{\delta g_{\mu\nu}} = +\tfrac{1}{2}\Theta_{\mathfrak{L}}^{\mu\nu},$$

where $\sqrt{g(x)}$ has been set equal to 1 since we are considering a flat Minkowski space. The second and third terms vanish since A_μ and $M_{\mu\nu}$ are assumed to obey their equations of motion (13.5.6).

$$\frac{\delta A}{\delta M_{\mu\nu}} = \frac{\delta A}{\delta A_\mu} = 0.$$

In the fourth term, we can write

$$\frac{\delta A}{\delta \mathcal{J}^{\mu}} = A_\mu$$

as indicated by examination of the Lagrangian (13.5.1). We leave the fifth term undisturbed, but write

$$\frac{\delta A}{\delta s} = \frac{\partial \mathfrak{L}}{\partial s}$$

in the sixth term, since $\mathfrak{L}$ contains only $s(x)$ and no derivatives $\partial_\mu s$. In the final term we write

$$\frac{\delta A}{\delta q} = -\rho \frac{\partial U}{\partial q},$$

since $q(x)$ appears only in U and $\partial_\mu q$ does not appear at all.

Now we must deal with the variations of the fields occurring in (13.5.13). The variation $\delta g_{\mu\nu}(x)$ was calculated in Chapter 12, (12.1.3): $\delta g_{\mu\nu}(x) = \xi_{\mu;\nu} + \xi_{\nu;\mu}$. In a space with the Minkowski metric tensor, this is

$$\delta g_{\mu\nu} = \partial_\nu \xi_\mu + \partial_\mu \xi_\nu.$$

The electromagnetic current $\mathcal{J}^\mu(x)$ is a vector density*:

$$\mathcal{J}^\mu(x) = \frac{\partial(\bar{x})}{\partial(x)} \frac{\partial x^\mu}{\partial \bar{x}^\nu} \bar{\mathcal{J}}^\nu(x+\xi).$$

Thus

$$\delta \mathcal{J}^\mu = (\partial_\alpha \xi^\alpha)\mathcal{J}^\mu - (\partial_\nu \xi^\mu)\mathcal{J}^\nu + (\partial_\nu \mathcal{J}^\mu)\xi^\nu.$$

Finally, the fields $R_a(x)$, $s(x)$, and $q(x)$ are scalar fields. Thus

$$\delta R_a = (\partial_\mu R_a)\xi^\mu,$$

$$\delta s = (\partial_\mu s)\xi^\mu,$$

$$\delta q = (\partial_\mu q)\xi^\mu.$$

We insert this collection of results back into (13.5.13):

$$0 = \int d^4x \Big\{ \Theta_{\mathfrak{L}}^{\mu\nu}(\partial_\nu \xi_\mu) + A_\mu[(\partial_\alpha \xi^\alpha)\mathcal{J}^\mu - (\partial_\nu \xi^\mu)\mathcal{J}^\nu + (\partial_\nu \mathcal{J}^\mu)\xi^\nu]$$

$$+ \frac{\delta A}{\delta R_a}(\partial_\mu R_a)\xi^\mu + \frac{\partial \mathfrak{L}}{\partial s}(\partial_\mu s)\xi^\mu - \rho \frac{\partial U}{\partial q}(\partial_\mu q)\xi^\mu \Big\}.$$

Integration by parts can be used to eliminate derivatives of $\xi(x)$:

$$0 = \int d^4x \Big\{ -\partial_\nu \Theta_{\mathfrak{L}\mu}{}^\nu - (\partial_\mu A_\nu - \partial_\nu A_\mu)\mathcal{J}^\nu$$

$$+ \frac{\delta A}{\delta R_a}(\partial_\mu R_a) + \frac{\partial \mathfrak{L}}{\partial s}(\partial_\mu s) - \rho \frac{\partial U}{\partial q}(\partial_\mu q) \Big\} \xi^\mu(x).$$

*We must define $\mathcal{J}^\mu(x)$ to have exactly the same transformation properties under general coordinate transformations as it had in the conservative theory. Otherwise we get a different expression for $\Theta_{\mathfrak{L}}^{\mu\nu}$. In the conservative theory, $\mathcal{J}^\mu(x) = qn\epsilon^{\mu\nu\rho\sigma}(\partial_\nu R_1)(\partial_\rho R_2)(\partial_\sigma R_3)$, which gives the transformation law stated. [See (11.2.15).]

Since this equation holds for all functions $\xi^\lambda(x)$, the coefficient of $\xi^\lambda(x)$ must vanish:

$$0 = -\partial_\nu \Theta_{\mathfrak{L}}^{\mu\nu} - F^{\mu\nu} \mathcal{J}_\nu + \frac{\delta A}{\delta R_a} \partial^\mu R_a + \frac{\partial \mathfrak{L}}{\partial s} \partial^\mu s - \rho \frac{\partial U}{\partial q} \partial^\mu q. \quad (13.5.14)$$

This result is interesting in itself. But we need only its projection onto the velocity vector u_μ. The third term drops out in this projection since $u^\mu \partial_\mu R_a = 0$; this greatly simplifies the result. In the second term we can write $\mathcal{J}^\nu = q\rho u^\nu + j^\nu$; the part of $\mathcal{J}^\nu$ proportional to u^ν drops out since $u_\mu u_\nu F^{\mu\nu} = 0$. Thus we find

$$u_\mu \partial_\nu \Theta_{\mathfrak{L}}^{\mu\nu} = -u_\mu j_\nu F^{\mu\nu} + \frac{\partial \mathfrak{L}}{\partial s} u^\mu \partial_\mu s - \rho \frac{\partial U}{\partial q} u^\mu \partial_\mu q. \quad (13.5.15)$$

Choice of U_D, ϕ, and T

With the expression for $u_\mu \partial_\nu \Theta_{\mathfrak{L}}^{\mu\nu}$ in hand, we are now prepared to make use of the entropy inequality (13.5.11). We insert (13.5.15) into the entropy inequality and obtain

$$\begin{aligned} 0 \leqslant & \left[T + \mathcal{V} \frac{\partial \mathfrak{L}}{\partial s} \right] \frac{\rho}{T} u^\mu \partial_\mu s \\ & + \left[\phi - \frac{\partial U}{\partial q} \right] \frac{\rho}{T} u^\mu \partial_\mu q - \frac{\rho}{T} u^\mu \partial_\mu U_D \\ & - \left[\frac{1}{T} u^\mu F_{\mu\nu} + \left(\partial_\nu \frac{\phi}{T} \right) \right] j^\nu - \left(\partial_\mu \frac{u_\nu}{T} \right) \Delta^{\nu\mu}. \end{aligned} \quad (13.5.16)$$

We must contrive to make this inequality automatically true. It is easy to make the sum of the last two terms positive by letting j^ν and $\Delta^{\nu\mu}$ be linear functions of the variables

$$\frac{1}{T} u^\mu F_{\mu\nu} + \left(\partial_\nu \frac{\phi}{T} \right), \qquad \left(\partial_\mu \frac{u_\nu}{T} \right)$$

in such a way that the last two terms form a positive quadratic function. We do this presently. No such possibility is available for the third term, so we eliminate it by choosing the "dissipative energy" U_D equal to zero.* The factor $u^\mu \partial_\mu q$ in the second term can be either positive or negative, but

*We could let U_D be a function of s and q alone, absorbing the third term into the first two terms, but such a function U_D might as well be included as part of U in the Lagrangian.

this problem does not affect the inequality if we choose

$$\phi = \frac{\partial U}{\partial q}. \tag{13.5.17}$$

This is the result we expected.

The factor $u^\mu \partial_\mu s$ in the first term can likewise have either sign; hence we define T so as to make the coefficient of $u^\mu \partial_\mu s$ vanish:

$$T = -\mathcal{V}\frac{\partial \mathcal{L}}{\partial s}. \tag{13.5.18}$$

This definition of T is apparently applicable to a wider variety of models than that of a linear dielectric fluid considered here. In the present case the Lagrangian is

$$\begin{aligned}\mathcal{L} &= \tfrac{1}{2}(\mathbf{E}^2 - \mathbf{B}^2) + \mathbf{E}\cdot\mathbf{P} + \mathbf{B}\cdot\mathbf{M} + A_\mu \mathcal{J}^\mu \\ &\quad - \frac{1}{2\kappa}\mathbf{P}^2 - \frac{1}{2\chi}\mathbf{M}^2 - \rho U\end{aligned} \tag{13.5.19}$$

when evaluated in a local rest frame of the material. The variable s appears as an argument of the functions κ, χ, and U. Thus (13.5.18) says that

$$T = -\frac{\mathcal{V}}{2\kappa^2}\mathbf{P}^2\frac{\partial \kappa}{\partial s} - \frac{\mathcal{V}}{2\chi^2}\mathbf{M}^2\frac{\partial \chi}{\partial s} + \frac{\partial U}{\partial s}. \tag{13.5.20}$$

This equation can be written in a more conventional form by using the equations $\mathbf{P} = \kappa\epsilon^{-1}\mathbf{D}, \mathbf{M} = \chi\mathbf{B}, \epsilon = 1 + \kappa, \mu^{-1} = 1 - \chi$:

$$T = +\frac{\mathcal{V}}{2}\mathbf{D}^2\frac{\partial \epsilon^{-1}}{\partial s} + \frac{\mathcal{V}}{2}\mathbf{B}^2\frac{\partial \mu^{-1}}{\partial s} + \frac{\partial U}{\partial s}.$$

The energy per atom in a local rest frame of the material is, according to (10.4.6),

$$\mathcal{V}\Theta^{00} = \frac{\mathcal{V}}{2}\left(\frac{1}{\epsilon}\mathbf{D}^2 + \frac{1}{\mu}\mathbf{B}^2\right) + U.$$

Thus we have found that

$$T = \left[\frac{\partial \mathcal{V}\Theta^{00}}{\partial s}\right]_{\mathcal{V},\mathbf{D},\mathbf{B}=\text{const}} \tag{13.5.21}$$

Form of j^μ and $\Delta^{\nu\mu}$

With the choices $u_D=0$, $\phi=\partial U/\partial q$, and $T=-\mathcal{V}\partial\mathcal{L}/\partial s$, the entropy inequality reduces to

$$0\leqslant -\frac{1}{T}\left[u^\mu F_{\mu\nu}+\left(\partial_\nu\frac{\phi}{T}\right)\right]j^\nu-\left(\partial_\mu\frac{u_\nu}{T}\right)\Delta^{\nu\mu}.$$

In a local rest frame of the fluid this takes the form

$$0\leqslant\left[\frac{1}{T}E_i-\left(\partial_i\frac{\phi}{T}\right)\right]j^i-\frac{1}{T^2}T_i\Delta^{0i}-\frac{1}{2T}V_{ij}\Delta^{ij}, \tag{13.5.22}$$

where

$$T_i=\partial_i T+T\partial_0 v_i,$$

$$V_{ij}=\partial_i v_j+\partial_j v_i.$$

The current j^i and the dissipative momentum tensor $\Delta^{\mu\nu}$ could be functions of the variables: s, $\mathcal{V}$, q, and R_a, the variables X_N that occur in (13.5.22),

$$(X_N)=\left(\frac{1}{T}E_i-\partial_i\frac{\phi}{T},-\frac{T_i}{T^2},-\frac{V_{ij}}{2T}\right),$$

and several other variables Y_N,

$$(Y_N)=(B_i,\partial_i v_j-\partial_j v_i,\partial_i\mathcal{V},\partial_0 v_i,\partial_0 T).$$

We make the same argument we made for a simple viscous fluid in Section 13.4. We assume that the variables X_N and Y_N are small enough so that a linear approximation for the dependence of j^i and Δ^{0i} on $\mathbf{X}$ and $\mathbf{Y}$ is adequate. The functions $(F_N)\equiv(j^i,\Delta^{0i},\Delta^{ij})$ must then have the form $F_N=G_{NM}X_M$, where G_{NM} is a positive matrix which does not depend on $\mathbf{X}$ or $\mathbf{Y}$.

The most general relation $F_N=G_{NM}X_M$ that is consistent with rotational invariance is

$$\left.\begin{aligned} j^i&=A_{11}\left(\frac{1}{T}E_i-\left(\partial_i\frac{\phi}{T}\right)\right)+A_{12}\left(-\frac{T_j}{T^2}\right),\\ \Delta^{0i}&=A_{21}\left(\frac{1}{T}E_i-\left(\partial_i\frac{\phi}{T}\right)\right)+A_{22}\left(-\frac{T_j}{T^2}\right),\end{aligned}\right\} \tag{13.5.23}$$

$$\Delta^{ij}=-\tfrac{1}{2}\zeta V_{kk}\delta_{ij}-\eta\left(V_{ij}-\tfrac{1}{3}V_{kk}\delta_{ij}\right). \tag{13.5.24}$$

The positivity condition on the coefficients A_{NM}, ζ, η is (13.5.22).

$$0 \leqslant A_{11}\left(\frac{1}{T}\mathbf{E}-\nabla\frac{\phi}{T}\right)^2+(A_{12}+A_{21})\left(\frac{1}{T}\mathbf{E}-\nabla\frac{\phi}{T}\right)\cdot\left(-\frac{\mathbf{T}}{T^2}\right)$$
$$+A_{22}\left(\frac{\mathbf{T}}{T^2}\right)^2+\frac{1}{4T}\zeta(V_{kk})^2$$
$$+\frac{1}{2T}\eta\left[V_{ij}-\tfrac{1}{3}V_{kk}\delta_{ij}\right]\left[V_{ij}-\tfrac{1}{3}V_{kk}\delta_{ij}\right].$$

This inequality is satisfied for all values of the variables X_N if and only if (see Problem 3)

$$\left.\begin{aligned} A_{11}\geqslant 0, A_{22}\geqslant 0,\\ A_{11}A_{22}-A_{12}A_{21}-\tfrac{1}{4}(A_{12}-A_{21})^2\geqslant 0, \end{aligned}\right\} \tag{13.5.25}$$

$$\zeta\geqslant 0, \eta\geqslant 0. \tag{13.5.26}$$

Before we proceed to discuss how the coefficients A_{NM} can be measured in simple experiments, and what some of the measured values are, we pause to write the expansion (13.5.23), (13.5.24) in a covariant form applicable to moving media:

$$j^\mu = -A_{11}I^{\mu\nu}\left(\frac{1}{T}u^\alpha F_{\alpha\nu}+\left(\partial_\nu\frac{\phi}{T}\right)\right)$$
$$-A_{12}I^{\mu\nu}u^\alpha\left[\left(\partial_\nu\frac{u_\alpha}{T}\right)+\left(\partial_\alpha\frac{u_\nu}{T}\right)\right], \tag{13.5.27}$$

$$\Delta^{\mu\nu} = -A_{21}(u^\mu I^{\nu\lambda}+u^\nu I^{\mu\lambda})\left(\frac{1}{T}u^\alpha F_{\alpha\lambda}+\left(\partial_\lambda\frac{\phi}{T}\right)\right)$$
$$+\{-A_{22}(u^\mu I^{\nu\rho}u^\sigma+u^\nu I^{\mu\rho}u^\sigma)$$
$$-\tfrac{1}{2}T\zeta I^{\mu\nu}I^{\rho\sigma}-T\eta(I^{\mu\rho}I^{\nu\sigma}-\tfrac{1}{3}I^{\mu\nu}I^{\rho\sigma})\}$$
$$\times\left[\left(\partial_\rho\frac{u_\sigma}{T}\right)+\left(\partial_\sigma\frac{u_\rho}{T}\right)\right],$$

where $I^{\mu\nu}=g^{\mu\nu}+u^\mu u^\nu$.

Thermoelectric effects

The effects represented in (13.5.24) are familiar from Section 13.4. When the fluid flows, it experiences viscous forces given by the viscous stress tensor Δ^{ij}. These forces are not affected by the electrical or thermal properties of the fluid. The coefficients ζ and η are the bulk and shear viscosities as in Section 13.4.

There are several interesting effects embodied in (13.5.23) that involve electric currents and heat currents in a stationary medium. These effects occur in fluids, but are most frequently observed in solids—for example, metal wires. Since the symmetry considerations that led to (13.5.23) apply equally to fluids and to unstrained isotropic solids, we discuss the effects in such solids.

To exhibit the effects most simply, it is helpful to rewrite (13.5.23). We replace T_i by $\partial_i T$, since we are considering a stationary material, and we replace $\partial_i(\phi/T)$ by $T^{-1}\partial_j\phi - \phi T^{-2}\partial_i T$. Then we solve (13.5.23) for the "observed" quantities $E_i - \partial_i\phi$ and $\Delta^{0i} - \phi j^i$ as functions of the "experimentally controlled" quantities j^i and $\partial_i T$. The result is

$$E_i - \partial_i\phi = \frac{1}{\sigma} j^i - S\,\partial_i T, \tag{13.5.28}$$

$$\Delta^{0i} - \phi j^i = -\Pi j^i - K\partial_i T, \tag{13.5.29}$$

where

$$\sigma = \frac{A_{11}}{T}, \qquad S = \frac{1}{T}\left(\phi - \frac{A_{12}}{A_{11}}\right), \tag{13.5.30}$$

$$\Pi = \left(\phi - \frac{A_{21}}{A_{11}}\right), K = \frac{A_{11}A_{22} - A_{12}A_{21}}{A_{11}T^2}.$$

In terms of the phenomenologically observed coefficients σ, S, Π, and K, the positivity conditions (13.5.25) are

$$0 \leqslant \sigma,$$

$$\tfrac{1}{4}\sigma T\left(S - \frac{\Pi}{T}\right)^2 \leqslant K. \tag{13.5.31}$$

As we have already noted in equation (13.5.8), the quantity $\Delta^{0i} - \phi j^i$ in (13.5.29) is the heat current in the material. If a temperature gradient is maintained across an electrically insulated piece of material, heat will flow. In this situation the electric current j^i is zero, so that the heat current will be equal to $-K\partial_i T$. Thus the measured thermal conductivity will be K. According to (13.5.31), K is always nonnegative.

Equation 13.5.29 also predicts that heat will flow in the absence of a temperature gradient if an electric current is present. This is called the Peltier effect, and the corresponding coefficient Π is called the Peltier coefficient. The Peltier effect can be observed by letting a known electric current flow through a junction of two wires made of different metals, say platinum and silver. The wires are kept at a uniform temperature, as shown in Figure 13.2. The electric current I is continuous at the junction. Since the heat current is related to the electric current by $I_Q = -\Pi I$, the heat current will be discontinuous at the junction:

$$I_Q^{(Pt)} - I_Q^{(Ag)} = (\Pi_{Ag} - \Pi_{Pt})I.$$

Thus heat is discharged into the medium surrounding the wires at a rate $(\Pi_{Ag} - \Pi_{Pt})I$. Measurement of this heat flow gives $\Pi_{Ag} - \Pi_{Pt}$.

Consider now (13.5.28). In the absence of temperature gradients, this

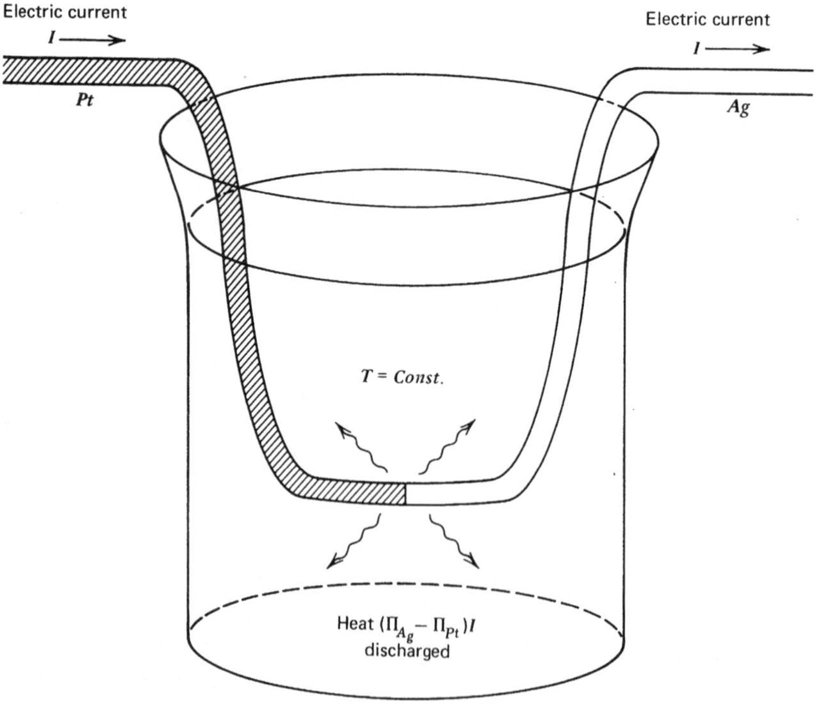

Figure 13.2 The Peltier effect.

equation is Ohm's law:

$$\mathbf{j}=\sigma(\mathbf{E}-\nabla\phi).$$

Notice that the current is not proportional to the true electric field $\mathbf{E}$ but to an "effective electric field" $\mathbf{E}-\nabla\phi$. Thus, for instance, the "effective electrostatic potential", $A^0+\phi$, is continuous at the junction of two wires made from different metals. The true potential has a discontinuity across the junction equal to the difference in ϕ between the metals. This discontinuity is called the contact potential. In principle, the contact potential can be measured by using a test charge to measure the true electric field near a junction.*

Contact potentials do not effect ordinary electric circuits. One simply uses the effective potential $A^0+\phi$ in place of the true potential A^0 in circuit analysis. For instance, potentiometers measure differences in $A^0+\phi$. These statements remain true even if different parts of the circuit are at different temperatures, as long as the temperatures do not vary with time.

The constant σ defined by (13.5.28) is called the electrical conductivity. The values of σ in common materials vary over a wide range. For copper, $\sigma \sim 6\times 10^5$ ohm^{-1} cm^{-1}; for glass, $\sigma \sim 10^{-14}$ ohm^{-1} cm^{-1}.†

Equation 13.5.28 also predicts that in a wire carrying no current, a gradient in the effective potential $A^0+\phi$ will exist along a temperature gradient. This is called the Seebeck effect and is the basis for the thermocouple thermometer. We call the corresponding coefficient S the Seebeck coefficient.

An appropriate device for observing the Seebeck effect is shown in Figure 13.3. A silver and a platinum wire are joined at point c and the junction is held at temperature T_1. The two wires are connected to a

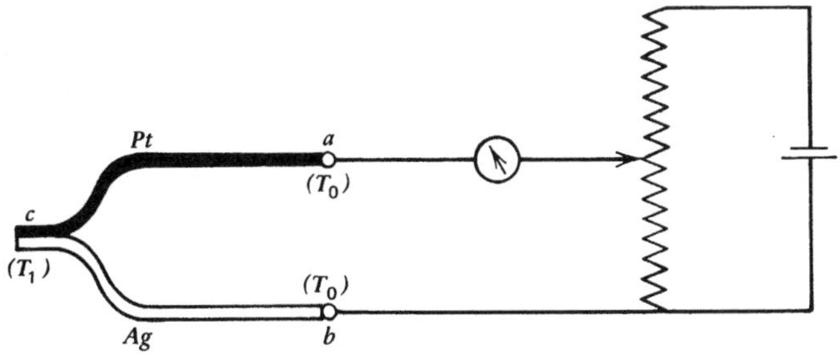

Figure 13.3 The Seebeck effect.

*Actually thermionic emission and photoelectric experiments are used to measure ϕ.

†Values are from the *Handbook of Chemistry and Physics, op. cit.* In the natural units used in this book, 1 ohm is a dimensionless number equal to 1 ohm $=1$ V A^{-1} $[\epsilon_0/\mu_0]^{1/2}\simeq 2.65\times 10^{-3}$.

potentiometer at points a and b, both of which are at temperature T_0. (This is somewhat simplified from a real experiment.) Since no electric current flows through the wires, the effective potential obeys the equation $\partial_i(A^0 + \phi) = S\,\partial_i T$. Thus the potential difference across the potentiometer is

$$\left(A^0(a)+\phi(a)\right)-\left(A^0(b)+\phi(b)\right)$$

$$= \int_c^a S_{Pt} \nabla T \cdot d\mathbf{x} + \int_b^c S_{Ag} \nabla T \cdot d\mathbf{x}$$

$$= \int_{T_1}^{T_0} S_{Pt}(T)\,dT + \int_{T_0}^{T_1} S_{Ag}(T)\,dT$$

$$= \int_{T_1}^{T_0} (S_{Pt}(T) - S_{Ag}(T))\,dT.$$

By measuring this potential difference with a range of temperatures T_0 and T_1, the difference $S_{Pt}(T) - S_{Ag}(T)$ can be determined. When this difference is known, the same device can be used to measure temperatures.

The measured values of the differences in S between two different metals is shown in Table 13.2 for two pairs of metals. Also shown are the differences in Π/T, where Π is the Peltier coefficient. The data suggest the relation

$$S = \frac{\Pi}{T}. \tag{13.5.32}$$

Table 13.2 Difference in the Seebeck Coefficients, and in Peltier Coefficients Divided by *T*, for Two Pairs of Metals[a]

Junction	Seebeck $S_2 - S_1$ (10^{-6} V/K)	Peltier $\frac{\Pi_2}{T} - \frac{\Pi_1}{T}$ (10^{-6} V/K)
Cu-Ni (302 K)	21.7	22.3
Fe-Hg (292 K)	16.7	16.7

[a]Source: M. W. Zemanski, *Heat and Thermodynamics*, 4th ed., (McGraw-Hill, New York, 1957), p. 308.

This is equivalent to the relation $A_{12} = A_{21}$ between the elements of the matrix A_{NM} occurring in (13.5.23). The relation $A_{NM} = A_{MN}$, and other similar relations in other theories, are called the Onsager reciprocal relations. They can be made plausible using an argument in statistical mechanics first proposed by L. Onsager.

PROBLEMS

1. The dissipative momentum transfer Δ^{ij} for a nonrelativistic fluid, (13.4.11), vanishes throughout the fluid if and only if $\partial_i V^j + \partial_j V^i \equiv 0$. Show that the unique solution of this differential equation for $\mathbf{V}(\mathbf{x})$ is a "rigid body" motion, $V^i(x) = V^i(x) = V^i(0) + \Omega^{ij}x^j$, where $V^j(0)$ is a constant vector and Ω^{ij} is a constant antisymmetric matrix.

2. Consider a relativistic fluid that has reached a steady state condition in which the velocity, temperature, and density are independent of time and entropy generation has ceased. That is,

$$\partial_\mu\left(\frac{u_\nu}{T}\right)A^{\mu\nu\rho\sigma}\left[\partial_\rho\left(\frac{u_\sigma}{T}\right)+\partial_\sigma\left(\frac{u_\rho}{T}\right)\right]=0.$$

Show that $\partial_\mu(u_\nu/T)+\partial_\nu(u_\mu/T)=0$. Then show that the equilibrium velocity distribution is that of a "rigid body,"

$$\frac{u^j}{u^0}\equiv V^j(x)=V^j(0)+\Omega^{jk}x^k,$$

where $V^j(0)$ is a constant vector and Ω^{jk} is a constant antisymmetric matrix. Show also that the temperature is

$$T=\frac{T_0}{\sqrt{1-\mathbf{V}(x)^2}}.$$

3. Let $\mathbf{A}$ be a 2×2 real matrix. Show that $0 \leqslant X_i A_{ij} X_i$ for all real vectors X_j if and only if $0 \leqslant A_{11}+A_{22}$ and $0 \leqslant A_{11}A_{22}-\frac{1}{4}(A_{12}+A_{21})^2$. *Hint*: First consider the case of a symmetric matrix.

Index